K. Lenhard Rudolph
Editor

Telomeres and Telomerase in Ageing, Disease, and Cancer

Molecular Mechanisms of Adult Stem Cell Ageing

Springer

Dr. K. Lenhard Rudolph
Institute of Molecular Medicine and Max-Planck-Research-Group
on Stem Cell Aging
University of Ulm
Albert-Einstein-Allee 11, 89081 Ulm
Germany
Lenhard.Rudolph@uni-ulm.de

ISBN 978-3-540-73708-7 e-ISBN 978-3-540-73709-4

Library of Congress Control Number: 2007931191

Cover design: WMX Design GmbH, Heidelberg

Printed on acid-free paper

9 8 7 6 5 4 3 2 1

springer.com

Contents

Contributors

Richard Allsopp
University of Hawaii, Institute for Biogenesis Research, John A Burns School of Medicine, Honolulu, Hawaii, United States, allsopp@hawaii.edu

Steven E. Artandi,
Department of Medicine, Cancer Biology Program, Stanford, California, United States, sartandi@stanford.edu

Monica Bessler
Department of Internal Medicine, Washington University School of Medicine, St. Louis, Missouri, United States

Alan E. Bilsland
Centre for Oncology and Applied Pharmacology, University of Glasgow, Cancer Research UK Beatson Laboratories, Garscube Estate, Switchback Road, Bearsden, Glasgow G61 1BD, United Kingdom, a.bilsland@beatson.gla.ac.uk

María A. Blasco
Telomeres and Telomerase Group, Molecular Oncology Program, Spanish National Cancer Centre (CNIO), 3 Melchor Fernández Almagro, Madrid E-28029, Spain, mblasco@cnio.es

Sandy Chang
Department of Cancer Genetics and Hematopathology, The M.D. Anderson Cancer Center, Houston, Texas, United States, schang@mdanderson.org

Terence Davis
Department of Pathology, School of Medicine, Cardiff University, Heath Park, Cardiff CF14 4XN, United Kingdom

Andrew Dillin
The Salk Institute for Biological Studies, 10010 North Torrey Pines Rd., La Jolla, California, United States, dillin@salk.edu

Hong-Yan Du
Department of Internal medicine, Washington University School of Medicine, 660 S. Euclid Ave., St. Louis, Missouri, United States

Luis Guachalla Gutierrez
Institute of Moleculare Medicine and Max-Planck-Research-Group on Stem Cell Aging, University of Ulm, Albert-Einstein-Allee 11, 89081 Ulm, Germany

Zhenyu Ju
Institute of Moleculare Medicine and Max-Planck-Research-Group on Stem Cell Aging, University of Ulm, Albert-Einstein-Allee 11, 89081 Ulm, Germany, Ju.Zhenyu@mh-hannover.de

Jan Karlseder
The Salk Institute for Biological Studies, 10010 North Torrey Pines Rd., La Jolla, California, United States, karlseder@salk.edu

W. Nicol Keith
Centre for Oncology and Applied Pharmacology, University of Glasgow, Cancer Research UK Beatson Laboratories, Garscube Estate, Switchback Road, Bearsden, Glasgow, United Kingdom, n.keith@beatson.gla.ac.uk

David Kipling
Department of Pathology, School of Medicine, Cardiff University, Heath Park, Cardiff, CF14 4XN, United Kingdom, kipling@cardiff.ac.uk

Ying Liang
Departments of Internal Medicine and Physiology, Markey Cancer Center, University of Kentucky, Lexington, Kentucky, United States

Uwe M. Martens
Medical University Center Freiburg, Department of Hematology/Oncology, D-79106, Freiburg, Germany, uwe.martens@uniklinik-freiburg.de

Philip J. Mason
Department of Internal Medicine, Washington University School of Medicine, 6620 S Euclid Avenue, St. Louis, Missouri, United States

Alison Miller
Departments of Internal Medicine and Physiology, Markey Cancer Center, University of Kentucky, Lexington, Kentucky, United States, amill@uky.edu

Glyn Nelson
Henry Wellcome Laboratory for Biogerontology Research, Institute for Ageing and Health, Center for Integrated Systems Biology of Ageing and Nutrition, University of Newcastle upon Tyne, NE4 6BE, United Kingdom

Erin Oakley
Departments of Internal Medicine and Physiology, Markey Cancer Center, University of Kentucky, Lexington, Kentucky, United States, Erin.James@uky.edu

João F. Passos
Henry Wellcome Laboratory for Biogerontology Research, Institute for Ageing and Health, Center for Integrated Systems Biology of Ageing and Nutrition, University of Newcastle upon Tyne, NE4 6BE, United Kingdom, Joao.Passos@newcastle.ac.uk

K. Lenhard Rudolph
Institute of Moleculare Medicine and Max-Planck-Research-Group on Stem Cell Aging, University of Ulm, Albert-Einstein-Allee 11, 89081 Ulm, Germany, Lenhard.Rudolph@uni-ulm.de

Norman E. Sharpless
Department of Medicine and Genetics, The Lineberger Comprehensive Cancer Center, The University of North Carolina, Chapel Hill, North Carolina, United States, NES@med.unc.edu

Jerry Shay
Department of Cell Biology, University of Texas, Southwestern, Medical Center at Dallas, Texas, United States, Jerry.Shay@UTSouthwestern.edu

Carol Swiderski
Departments of Internal Medicine and Physiology, Markey, Cancer Center, University of Kentucky, Lexington, Kentucky, United States

Gary Van Zant
Departments of Internal Medicine and Physiology, Markey Cancer Center, University of Kentucky, Lexington, Kentucky, United States, gvzant1@uky.edu

Thomas Von Zglinicki
Henry Wellcome Laboratory for Biogerontology Research, Institute for Ageing and Health, Center for Integrated Systems Biology of Ageing and Nutrition, University of Newcastle upon Tyne, NE4 6BE, United Kingdom, t.vonzglinicki@newcastle.ac.uk

Amanda Waterstrat
Departments of Internal Medicine and Physiology, Markey Cancer Center, University of Kentucky, Lexington, Kentucky, United States, AmandaBaker@uky.edu

Stefan Zimmerman
Medical University Center Freiburg, Department of Hematology/Oncology, D-79106, Freiburg, Germany, stefan.zimmermann@uniklinik-freiburg.de

Introduction

Twenty-five years after the disclosure of the telomeric DNA sequence and 22 years after the discovery of telomerase, it has become clear that telomeres and telomerase influence disease of human ageing including cancer. This book summarizes our current knowledge on the role of telomeres and telomerase in ageing, regeneration, and cancer with a special focus on ageing stem cells. Moreover, the book reviews current efforts to target telomeres and telomerase for anticancer treatment or regenerative therapies.

1.1 Ageing

The molecular analysis of ageing is an emerging research field that will have great impact on patients care, lifestyle, and on disease prevention in the human population worldwide. Major molecular mechanisms that influence the ageing process include (1) the accumulation of molecular damage affecting DNA and proteins, (2) alterations in gene expression, including alterations in checkpoint responses, metabolic pathways, and developmental pathways, and (3) the decline in adult stem cell function. All these mechanisms lead to a decrease in organ maintenance and function, thus representing a major factor limiting the quality of life during ageing. Moreover, the accumulation of molecular damage increases genetic alterations and the cancer risk during ageing. Indeed, increased age is the leading cause of cancer. Understanding the molecular basis of ageing will ultimately point to targets for novel therapies aiming to improve the function of cells and organs in the ageing organism, thus allowing increased vitality or what we think of as "healthy ageing."

In this book, Andrew Dillin and Jan Karlseder provide an overview on different model organisms that are currently used to study ageing on the cellular and organismal levels. The authors discuss how different molecular mechanisms (telomere shortening, insulin signaling, caloric restriction, mitochondrial function) may differentially impact ageing of mitotically active versus postmitotic cells. Richard Allsopp summarizes the influence of progressive telomere shortening and checkpoints induced by dysfunctional telomeres on cellular ageing. When telomeres lose capping function, a DNA damage response is induced, and the resultant telomere shortening is one of the factors that leads to an accumulation of unrepaired DNA damage during ageing, which causes cellular growth arrest. Similarly, DNA damage checkpoints have been demonstrated for telomere-based replicative ageing as well. There is now convincing evidence for progressive telomere shortening in various human tissues during ageing, including stem cell compartments. It remains an open debate as to what extent replicative senescence occurs during human ageing in vivo. Sandy Chang summarizes the consequences of telomere dysfunction on ageing in mouse models. He describes evidence that telomere shortening can cooperate with other pathways that are involved in DNA damage accumulation and premature ageing (Werner Syndrome).

Oxidative damage is another factor that is responsible for the accumulation of DNA damage during ageing. João F. Passos and colleagues summarize experimental data indicating that oxidative damage and telomere dysfunction are interconnected and cooperate to induce cellular ageing. It will be interesting to see whether similar cooperation occurs during human ageing, which could then explain the exponential accumulation of DNA damage in late life, when telomeres get short. Hong-Yan Du, Monica Bessler, and Philip Mason conclude the first part of the book on telomeres and ageing by summarizing disease states associated with telomerase mutations and telomere shortening in humans. There is a growing number of human disease syndromes that are associated with organ failure and that are caused by mutations in one allele of genes encoding for telomerase (hTERT or hTR). It is notable that mutation in only one allele of human telomerase leads to impaired organ homeostasis and premature death of these patients, sometimes at an age of 30–50 years. These findings demonstrate that telomere reserves in humans are rather limited and that a reduction in telomerase gene dosage has severe consequences that affect the ability to live a "full" human lifespan. These findings provide strong evidence that telomere shortening can also influence normal ageing in humans without telomerase mutations.

2.1 Stem Cells

Although the molecular mechanisms of impaired organ maintenance during ageing are largely unknown, it has been recognized that impaired organ maintenance correlates with impaired function of adult stem cells during ageing. An understanding of molecular mechanisms that are responsible for adult stem cell ageing is of utmost importance to develop new therapies aiming to improve organ maintenance and function during ageing. The second part of this book focuses on adult stem cell ageing. Garry Van Zant and coworkers provide a summary of our current knowledge of ageing of hematopoietic stem cells. The authors describe age associated changes in stem cell number and function, and describe molecular pathways involved in stem cell ageing. Their chapter shows that both cell intrinsic checkpoints and environmental alterations represent major determinants limiting stem cell function during ageing. Stefan Zimmermann and Uwe Martens also focus on hematopoietic stem cell ageing.

The authors summarize evidence that both telomere-dependent and telomere-independent mechanisms can limit hematopoietic stem cell function in ageing mice and humans. Luis Gutierrez and Zhenyu Ju demonstrate that telomere dysfunction induces cell intrinsic checkpoints and environmental alterations that limit stem cell function in ageing telomerase-deficient mice. Norman E. Sharpless discusses the role of the cell cycle inhibitor p16 on stem cell function during ageing. It is still not entirely clear how the p16 and the Rb- checkpoint are linked to telomere dysfunction. Sharpless summarizes evidence indicating that p16 is upregulated during ageing and impairs the function of ageing stem cells in various compartments. He shows that p16 can be regulated by telomere-dependent and -independent mechanisms. Steven Artandi ends the section on stem cell ageing by reviewing the role of telomerase for stem cell function. He summarizes initial evidence that telomere-independent functions of telomerase can regulate adult stem cell function, and it will be interesting to explore these mechanisms in ageing stem cells.

3.1 Cancer

A potential downside of molecular therapies aiming to improve organ maintenance during ageing could be an increase in the cancer risk. It appears unavoidable that interventions that increase stem cell self-renewal and regeneration would also increase cancer risk if constantly applied. Transient interventions might be one option to avoid such side effects. A detailed understanding of the role of telomeres and telomerase in cancer initiation and progression is a prerequisite to begin to explore telomeres and telomerase as targets for regenerative medicine. Lenhard Rudolph summarizes data from mouse models and human data showing that telomere dysfunction has a dual role in carcinogenesis by increasing the rate of cancer initiation but suppressing cancer progression. The chapter summarizes molecular mechanisms that could explain an increased cancer risk in response to telomere dysfunction, including the induction of chromosomal instability and environmental alterations. In contrast, depending on the checkpoint response, telomere dysfunction may impair cancer progression. Maria Blasco extends the cancer section of the book by summarizing experimental evidence indicating that in addition to telomere length, telomere binding proteins can influence tissue deterioration during ageing and increase the cancer risk specifically in the context of short telomeres.

4.1 Therapies

The growing understanding of the role of telomeres and telomerase in human ageing and cancer biology should ultimately lead to development of novel therapies improving tissue regeneration or impairing cancer growth. The last part of this book summarizes the rationale behind both approaches and shows where the field stands in terms of therapy development. Nicol Keith and Alan E. Bilsland focus on anticancer therapies targeting telomerase or telomere structure. Given that human cancers have very short telomeres and high levels of telomerase compared to non-transformed human tissues, destabilization of telomeres and inhibition of telomerase could both be promising targets in anticancer therapy. It is exciting to see that the first cancer clinical trials have been initiated using a novel telomerase enzyme inhibitor. Somewhat unexpected have been early successes using a telomerase immunotherapy for cancer therapy. In the final chapter Terence Davis and David Kipling introduce therapeutic approaches targeting stress signals downstream of DNA damage and/or telomere dysfunction in Werner syndrome patients; these approaches aim to improve cellular and organismal fitness during ageing. Although they focus on Werner syndrome, it is likely that some of these connections also apply to normal ageing characterized by telomere shortening in various human tissues. In addition, they show that activation of telomerase can prevent the induction of cellular stress responses to some extent, thus representing another therapeutic approach for improving regenerative reserve and organ function in certain premature ageing syndromes and eventually in normal human ageing.

In summary, this book provides a survey of the role of telomeres and telomerase in ageing, diseases, and cancer. The field has made tremendous progress in recent years, and it is becoming ever more clear that telomeres and telomerase have significant influences on human ageing, health, and disease. Telomerase-targeted anticancer therapies are already in human clinical trials, and regenerative therapies are likely to follow soon. It can be anticipated that novel molecular therapies targeting telomeres and telomerase will reach late-stage clinical trials in the near future. This book provides the reader with a fundamental overview of the state of knowledge in basic telomere biology, cellular and organismal ageing, as well as in the emerging field of adult stem cell ageing. Given the progress in translational approaches to target telomeres and telomerase for regenerative and anticancer therapies, this book should be of great interest to students and professionals in both basic science and medicine.

June 2007

Jerry W. Shay
K. Lenhard Rudolph

Part I
Telomere Shortening and Ageing

Chapter 1
Cellular versus Organismal Aging

Andrew Dillin and Jan Karlseder(✉)

Abstract Aging is an extremely complex process, affecting individual cells and organisms as a whole. Here we discuss the impact of telomeres, the natural chromosome ends, on cellular and organismal aging, and how telomere maintenance influences genome stability and tumorigenesis. Since telomeres represent only a fraction of the complexity of the aging process, we discuss how genome integrity and DNA damage response pathways affect postmitotic aging, and which genetic pathways promote survival and elongated life, namely the IGF-1 signaling pathway, dietary restriction, and mitochondrial transport chain. At this point the researchers in the field are intensely studying individual aspects of aging, but it remains a unique challenge to integrate all aspects, with the goal of understanding longevity.

1 Cellular Aging: The Hayflick Limit

The name Leonard Hayflick immediately comes to mind when cellular aging is discussed. In the early 1960s, Hayflick was interested in the division potential of human diploid cells, and performed a number of key experiments at The Wistar Institute in Philadelphia. In two subsequent publications in 1961 and 1965 he described the isolation and growth capacities of primary cells originating from human tissue (Hayflick 1965, Hayflick and Moorhead 1961). He noted that cells are readily recovered from human tissue slices, and can rather easily be subcultivated at a split ratio of 1:2 twice a week. This maintenance schedule can be continued for months, until, after approximately 50 population doublings, the cultures fail to display the numerous mitotic figures characteristic of healthy growth, and cease to divide. Once the cultures arrested, Hayflick found it impossible to recover cell growth, an observation that still holds up today. Based on these results, he suggested

Jan Karlseder
The Salk Institute for Biological Studies, 10010 North Torrey Pines Rd., La Jolla, California, United States.
e-mail: karlseder@salk.edu

K.L. Rudolph (ed.), *Telomeres and Telomerase in Ageing, Disease, and Cancer.*

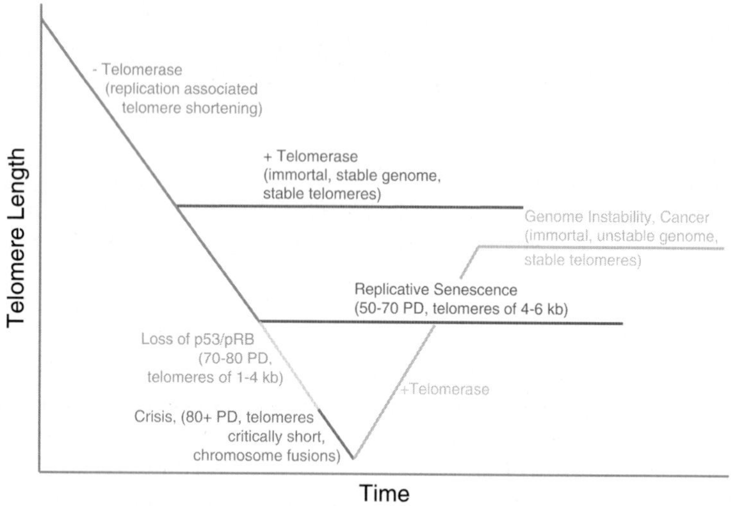

Fig. 1.1 Schematic of telomere dynamics during aging. Telomeres in telomerase-negative cells shorten over time (green line). When telomerase is introduced into such cells, telomere length is stabilized and the cells become immortal without accumulating chromosomal aberrations (purple line). When telomeres become short, cells arrest in senescence (blue line). Suppression of DNA damage pathways allows further telomere shortening (turquoise line), leading to dysfunctional and critically short telomeres, genome fragmentation and crisis (red line). Upregulation of telomerase allows telomere elongation, escape from crisis, and establishment of immortal clones with unstable genomes (orange line) (*See Color Plate*)

that the growth characteristics of such cultures could be divided into three phases. Phase 1, or the early growth phase, is when the culture establishes itself. During Phase 2 the cells grow exponentially, maintain a diploid chromosome set, but eventually stop growing. Phase 3 is termed the degeneration phase, during which tetraploid cells arise occasionally, and nuclei take on various unusual appearances (Hayflick and Moorhead 1961). In more detailed experiments Hayflick demonstrated that primary human cell strains enter Phase 3 after 50 +/−10 passages, no matter when and how long they have been frozen in between passages (Hayflick 1965). Consequently his experiments demonstrate that such cells have a finite lifetime in vitro, and that a cell-intern counting mechanism exists that monitors the accumulative number of population doublings (Fig. 1.1).

2 Cellular Aging: The End Replication Problem

James D. Watson, while investigating the replication pattern of linear molecules of T7 DNA, noted that leading strand replication in the 5′ to 3′ direction should proceed smoothly to the end of its template. However, lagging strand synthesis is unable to copy the parental strand in its entirety (Watson 1972). Lagging strand synthesis is initiated by RNA primers, which are extended to Okazaki fragments that are ligated together after removal of the initiating RNA. This mechanism does not allow for fill-in synthesis of the gap left by the most distal RNA primer, inevitably leaving the daughter strand shorter than the parental strand. The inability to fully replicate the template, leading to terminal sequence loss during each replication cycle, is termed the end replication problem.

Following an independent approach, Alexey Olovnikov at the Russian Academy of Sciences published "A Theory of Marginotomy" (Olovnikov 1973). His hypothesis holds that the limited doubling potential of primary somatic cells can be explained by terminal sequence loss of the daughter DNA strand. To buffer the progressive erosion, Olovnikov proposed that so-called *telogenes* – vital genes without any coding information – are located at opposite ends of the linear chromosome. Once they are lost due to the end replication problem, the coding chromosomal DNA is no longer protected, replication ceases, and the cells enter what Hayflick called Phase 3 (Hayflick and Moorhead 1961).

Now, of course, we know that the ends of chromosomes do not consist of the hypothesized telogenes, but are specialized structures, called telomeres, consisting of G-rich DNA repeats and proteins that bind to these repeats. Olovnikov was partially correct in suggesting that they fulfill a buffer function, and Watson had precisely predicted the reason for terminal sequence loss. However, at the time it was unclear that telomeres and replication-associated telomere loss represent the counting mechanism that monitors and limits the proliferative potential of primary cells as first discovered by Hayflick. At this point in time we can define the Hayflick limit as a function of initial telomere length and rate of terminal sequence loss per cell division, rendering the telomere the genetic clock that measures cellular replicative aging (Fig. 1.1).

3 Telomeres and Telomerase: A Brief Introduction

Telomeres are the specialized structures at the tips of linear chromosomes. They prevent the natural ends from being recognized as damaged DNA, and consequently protect them from enzymatic modification, such as nucleolytic resection and fusion. Telomeres are nucleoprotein complexes consisting of G-rich repeats of DNA, bound by sets of polypeptides that interact with the repeats and each other. The very terminus of telomeres is not blunted, but consists of a single-stranded 3′ protrusion of the G-rich strand, called a G-tail or G-overhang. These overhangs have been observed in yeast, humans, mice, ciliates, plants, and trypanosomes, demonstrating that they are evolutionary conserved and an essential feature of telomeres (Dionne and Wellinger 1996, Hemann and Greider 1999, Jacob et al. 2001, Makarov et al. 1997, Munoz-Jordan et al. 2001, Riha et al. 2000, Wellinger et al. 1996, Wright et al. 1997). Studies in human and mouse cells led to a telomeric loop-model, which suggests that the G-rich telomeric single-stranded overhang can loop back and invade homologous double-stranded telomeric tracts, resulting in a large lasso-like structure, termed telomeric loop, or t-loop (Griffith et al. 1999). This structure provides an attractive mechanism by which chromosome ends can be distinguished from broken DNA ends, and subsequently, in addition to humans and mice, t-loops were observed in trypanosomes, ciliates, plants and in yeast species with artificially long telomeres (de Lange 2004).

Mammalian telomeres are associated with the shelterin complex, an assembly of interdependent telomeric core proteins, consisting of TRF1, TRF2, Tin2, Rap1, TPP1, and POT1 (de Lange 2005). It is becoming increasingly clear that the proteins in this complex display multiple interactions, and disruption of individual members leads to disintegration of the complex and telomere dysfunction. In many model organisms loss of telomere function has multiple consequences, such as loss of the telomeric G-rich overhang, resection of the C-rich strand, increased levels of recombination at chromosome ends, altered gene expression patterns, fusion of chromosomes, genome instability, growth arrest and premature senescence, or cell death. In addition to shelterin, mammalian telomeres interact with a plethora of other factors that can influence chromosome end integrity and dynamics, such as Tankyrase 1 and 2, PARP, the MRN complex, the RecQ helicases WRN and BLM, Ku70, Ku86, DNAPK, ATM, ERCC1, XPF and RAD51D (de Lange 2005), pointing to the repair and recombination machineries as important contributors to telomere function.

To counteract replication-associated telomere shortening, a specialized reverse transcriptase complex evolved, which was named terminal telomere transferase at the time of its discovery in the ciliate *Tetrahymena thermophila* (Greider and Blackburn 1985), and is now known simply as telomerase. Telomerase is capable of adding G-rich telomeric repeats to the very ends of chromosomes using its own RNA, named TERC, as a template (Greider and Blackburn 1987, Greider and Blackburn 1989). The identification of the catalytic telomerase subunits in *Saccharomyces cerevisiae* (Lendvay et al. 1996), the ciliate *Euplotes aediculatus* (Lingner and Cech 1996), and humans (Bodnar et al. 1998, Meyerson et al. 1997) suggests a common reverse transcriptase–based mechanism for telomere length stabilization in most organisms with linear genomes (Lingner et al. 1997). Telomerase is active in the germ line and during early development, ensuring that replicative telomere shortening is suppressed, and telomere length is kept constant. However, in most somatic human cells telomerase is not expressed, consequently subjecting chromosome ends in such cells to shortening every time the cell divides (Harley et al. 1990). Progressive telomere shortening ultimately leads to the generation of telomeres that are too short to fulfill their protective and regulatory properties at chromosome ends. Such dysfunctional telomeres are detected by the cell-internal DNA damage machinery, and the cell responds either with death, or by entering a terminally differentiated state, termed replicative senescence (de Lange 2002; see also Allsopp, this volume). Senescent cells are metabolically active, but most likely irreversibly arrested (Fig. 1.1).

As a result, telomere shortening limits the natural replicative life span of somatic human cells, representing a powerful tumor suppressive mechanism (de Lange and Jacks 1999). The finding that all cancer cells maintain a constant telomere length, mostly via activation of telomerase (Kim et al. 1994), points to telomerase-based telomere length stabilization as a fundamental requirement for immortality (see Rudolph, this volume; Blasco, this volume, Fig. 1.1).

4 Telomere Shortening and Cellular Aging

Telomeres were brought to the forefront of the cellular aging discussion when Olovnikov's hypothesis turned into experimental fact in the model organism *S. cerevisiae*. At the time, it had been suggested that due to the inability of conventional polymerases to fully replicate the 3′ terminus of the G-rich strand, a sequence-specific enzyme would elongate telomeres. Therefore, a mutation screen was designed to identify enzymatic activities responsible for telomere replication via the isolation of mutants defective for the telomere elongation reaction (Lundblad and Szostak 1989). The authors made several predictions, which all turned out to be correct:

- The nature of the screen would allow the identification of enzymatic functions without having to make assumptions about the exact enzymatic activities.
- The identified mutations should not lead to the cells' immediate demise, but rather to lethality or a senescence phenotype over time.
- The *S. cerevisiae* telomeric tracts would become shorter with each generation, and eventually become unable to protect the chromosome ends, resulting in the accumulation of chromosomal abnormalities (Lundblad and Szostak 1989).

The genetic screen, based on the linearization and retention of a circular plasmid carrying telomeric repeats, led to the discovery of a new gene, EST1, which stands for Ever Shorter Telomeres 1. Exactly as predicted by the investigators, a strain mutant for EST1 (the strain estl-1) displays a gradual senescence phenotype, associated with a progressive decrease in telomere length. After 25 generations estl-telomeres are 100 bp shorter than wild type telomeres, and continue to lose terminal sequence for many more generations. Strikingly, the gradual telomere loss is accompanied by a decrease in growth rate and an increase in the frequency of cell death. Finally, after approximately 100 generations the estl-1 strains cease to divide.

The authors had also predicted that mutants in the telomere maintenance pathway should accumulate chromosomal aberrations. Analysis of chromosome loss phenotypes in a strain lacking EST1 demonstrates that mutants do not suffer from chromosome loss in early generations, but as the cells approach senescence, chromosomes are lost at a high frequency, validating the authors' prediction. This landmark publication ends with a statement that finally links Hayflick's observations with telomere biology: " ...a loss of telomeric DNA during each somatic cell division may play a role in determining the number of cell divisions possible in a given lineage."

EST1, a protein that binds single-stranded telomeric DNA, turned out to be a subunit of the telomerase complex and to play an essential role in mediating access of the catalytic subunit of telomerase to telomeres (Evans and Lundblad 1999,

Lundblad 2003). It acts as an adapter between the catalytic core of the enzyme and Cdc13, a single-stranded telomeric DNA binding factor in yeast (Nugent et al. 1996, Pennock et al. 2001).

Discovery of the catalytic subunit of telomerase in model organisms allowed the identification of the human telomerase reverse transcriptase subunit (hTERT) via sequence homology (Kilian et al. 1997, Meyerson et al. 1997, Nakamura et al. 1997). Once identified, it became possible to study the effects of hTERT overexpression on telomere length and the life span of human fibroblasts, which usually do not exhibit telomerase activity, and, as a result, suffer from replication-associated telomere shortening. Since the telomere hypothesis of cellular aging holds that cells become senescent when critically short and dysfunctional telomeres appear, the investigators predicted that expression of active telomerase would extend the replicative life span of such cells (Bodnar et al. 1998). Normal human retinal pigment epithelial cells (RPE) and human BJ foreskin fibroblasts, which both lack telomerase activity, were transfected with a vector expressing the catalytic subunit of human telomerase hTERT, leading to detectable telomerase activity. When telomere length was measured to determine whether in vitro telomerase activity translates into in vivo telomere elongation, the researchers found that telomere length had increased significantly, demonstrating that hTERT reconstituted telomerase activity extends endogenous telomeres in normal human cells. The key experiment was to investigate the effects of hTERT expression on the replicative life span of the hTERT reconstituted cells. hTERT negative RPE or BJ clones showed the expected slowing of growth and consecutive senescence between 50 and 70 population doublings. hTERT-positive RPE and BJ clones, however, exceeded this limit by many population doublings, and kept cycling at the rate of young cells (Bodnar et al. 1998). No senescence markers could be detected in the telomerase-active clones, whereas such markers were highly prevalent in the telomerase-negative populations, demonstrating that the reconstitution of telomerase activity is sufficient to extend the replicative life span of primary human cells. These results corroborate that telomere loss in the absence of telomerase activity is the mitotic clock that controls the number of cell divisions before terminal differentiation in primary human cells.

These life-extending experiments trigger an important question: Would cells that expressed hTERT retain a diploid genome? Immortality in cancer cells is almost always associated with telomerase activity and aneuploidy. In an effort to characterize telomerase-immortalized cells, researchers had a closer look at BJ foreskin fibroblasts that had been maintained in continuous culture for 280 population doublings (Morales et al. 1999). No loss of contact inhibition or acquisition of serum independent growth was noted in the hTERT-expressing clones, suggesting that they had not adopted such features of transformed cells. Another common attribute of cancer cells is the loss of cell cycle checkpoints, hence the investigators scrutinized the pRB and the p53 cascade for functionality in hTERT immortalized primary cells, and found both pathways intact. The final experiment was to monitor chromosomal aberrations, since accumulation of translocations, amplifications, deletions, inversions, and chromosome multiplications is a common feature of cancer cells (Lengauer et al. 1998). However, cells expressing ectopic telomerase activity maintain a normal karyotype. These datasets indicate that introduction of hTERT into normal somatic human cells is not sufficient for transformation, does not bypass cell cycle checkpoint control, and does not lead to genomic instability (Morales et al. 1999).

The discovery of telomerase, and the demonstration that telomere maintenance is required for continued proliferation, was recognized by the 2006 Albert Lasker Award for Basic Medical Research, whereby Elizabeth Blackburn, Carol Greider, and Jack Szostak were honored (Blackburn et al. 2006, de Lange 2006).

5 Organismal Aging and Survival

In the simplest terms, within almost all multicellular organisms two cell types exist: dividing (mitotic) and nondividing (postmitotic). It is abundantly clear that mechanisms to ensure fidelity of the genetic material, DNA damage response, cell cycle checkpoint control, and telomere replication and maintenance are essential for the life span of dividing cells, as defects in these processes lead to increased cellular senescence. However, it is less clear how these processes impinge upon the aging and survival of nondividing cells, such as neurons, muscle cells, and gametes. Therefore, a central question within the aging field has become to determine whether the same processes required for genome integrity of dividing cells are required for the aging of terminally differentiated cells.

5.1 Telomeres and Organismal Aging

Despite the finding that telomere length and telomere shortening rates define the replicative potential of telomerase-negative cells in tissue culture experiments, the in vivo evidence for telomere involvement in human aging is mostly limited to correlations. One informative study suggests that the telomeric restriction fragments isolated from DNA from human sperm cells are significantly longer than such fragments isolated from corresponding replicating cells in vivo (Cooke and Smith 1986). Telomeres in kidney samples from young individuals were found to be significantly longer than telomeres in kidney samples from older individuals (Melk et al. 2000). In a different study, the number of senescent skin fibroblasts in baboons increased exponentially with age, thus linking telomere dysfunction and organismal aging (Herbig et al. 2006). Numerous such correlations have been documented, and they show that telomeres shorten with age, and suggest that telomere shortening is one of the underlying causes of aging in humans.

The targeted deletion of the RNA template of telomerase (TERC) in mice (Blasco et al. 1997) provided an opportunity to study the effects of telomere shortening at the organismal level (Rudolph et al. 1999). An aging population of generation 3 TERC knockout mice displayed an increased incidence of hair graying, hair loss, and skin lesions over time. These observations correlated inversely with telomere length in the animals, suggesting a link between the two. The aged generation 3 TERC-/- animals displayed decreased bodyweight, suffered from an impaired stress response, and a decreased wound healing capacity. Although the mice did not suffer from the full spectrum of age-related pathophysiological symptoms of aging, a critical role of telomeres for the overall fitness and well-being of aging organisms could be demonstrated (Rudolph et al. 1999; see also Chang, this volume).

Recently, it has become increasingly clear that telomere function plays a major role in a variety of human disease syndromes. Classical Dyskeratosis Congenita (DC), a heritable disorder, is characterized by bone marrow failure, frequently resulting in premature mortality. The gene for autosomal-dominant DC localizes to chromosome 3q, the same chromosomal location to which the gene for the RNA subunit of telomerase has been mapped. Mutational analysis of TERC in DC families led to the conclusion that this disease is due to mutations in the telomerase template RNA (Vulliamy et al. 2001; see also Du et al., this volume). The observation that DC patients have very short telomeres (Marciniak and Guarente 2001, Marciniak et al. 2000) suggests that DC arises from diminished telomerase activity. Insufficient telomere elongation accelerates telomere shortening (Mitchell et al. 1999), and cells in highly proliferative tissues are lost, which is consistent with the pathology of the disease. The observation that both the catalytic subunit of telomerase and the RNA template limit telomere elongation (Hao et al. 2005, Liu et al. 2000) suggests that heterozygous mice, missing one allele for either one of these genes, are valid animal models for DC. Indeed, interbreeding of mice heterozygous for TERC with a strain carrying very short telomeres leads to animals suffering from organ failure similar to DC patients (Hao et al. 2005).

Werner Syndrome (WS) is a premature aging disease, in which the patients appear much older than their chronological age, and exhibit many of the clinical signs and symptoms of normal aging at an early stage in life, including an increased incidence of cancer. The WRN gene, mutated in the syndrome, is a member of the RecQ helicase family. Primary cells from WS patients display increased rates of chromosomal aberrations and undergo senescence prematurely, a phenotype reversed by forced telomerase expression (Opresko et al. 2003, Wyllie et al. 2000; see Davis and Kipling, this volume). The involvement of telomerase in the WRN phenotype has been confirmed by the finding that deleting both WRN and telomerase mimics human WRN pathogenesis in mice (Chang et al. 2004). However, accelerated telomere shortening rates in cells lacking WRN cannot explain the slow growth phenotypes of WS cells or the pathological symptoms of WS patients, since no real correlation between the lack of WRN and accelerated telomere erosion has been found (Baird et al. 2004, Schulz et al. 1996, Tahara et al. 1997). However, it has been suggested that individual telomeres are lost at low frequencies from WS cell chromosomes (Crabbe et al. 2004). This telomere loss phenotype is restricted to the telomeric strand synthesized by the lagging-strand replication machinery and depends on the helicase activity of

WRN, leading to a model wherein WRN is required for the efficient replication of the telomeric G strand. Telomerase-based elongation of the critically short telomeres in WS cells is able to suppress the accumulation of genome instability, raising the possibility that telomere loss is linked to the increased incidence of cancer in the syndrome (Crabbe et al. 2007).

While the link between telomeres and replicative aging has now been firmly established (de Lange 1998), little is known about the role played by telomeres during the aging process of differentiated cells and postmitotic organisms. *Caenorhabditis elegans* is an optimal model system to study organismal aging, since this nematode, after passing through its different developmental stages, consists of differentiated cells only. Surprisingly, overexpression of the telomere-binding protein HRP-1 in the worm not only led to elongation of telomeres, but also to an increased life span (Joeng et al. 2004), raising the possibility that telomere length also affects differentiated cells, and in turn, the aging of postmitotic organisms. However, this debate has not been settled yet, since the life span of clonal wild-type nematode strains and of inbred mouse strains is independent of their highly variable telomere length (Hemann and Greider 2000, Raices et al. 2005).

5.2 Genome Integrity/DNA Damage Response in Postmitotic Aging

Several pieces of data from *C. elegans* mutant analysis hint at the idea that genome integrity is indeed essential for the aging of terminally differentiated cells. The soma of *C. elegans* is composed of a fixed number of exclusively postmitotic cells, and somatic cells removed by laser ablation or ectopic cell death are not replaced by cell division, resulting in fewer than the standard 959 cells. The worm also contains a fully formed mitotic germ line; however, laser or genetic ablation of this organ does not alter longevity (Hsin and Kenyon 1999), indicating that worm longevity solely depends upon the maintenance of its 959 postmitotic cells.

Because of the close correlation among cell cycle checkpoint control and cellular senescence of dividing cells, it was interesting to find that mutation or RNAi depletion of genes required for cell cycle checkpoint arrest in response to DNA damage results in extended organismal longevity of the worm, which is composed entirely of postmitotic cells. In these experiments, perturbation of *cid-1* (caffeine-induced cell death), *chk-1* (checkpoint 1), or *cdc-25* (cell division cycle 25) increased longevity, suggesting that an orchestrated sequence of cell cycle checkpoints controls organismal survival (Olsen et al. 2006). Intriguingly, any possible cell cycle involvement of *chk-1* could be unequivocally uncoupled from organismal aging by depleting *chk-1* by RNAi or chemically during the adult stage of the animal's life cycle, long after all developmental cell divisions were completed. The mechanism by which loss of cell cycle checkpoint control in worms extends longevity is still unclear; however, it is appealing to speculate that artificially tricking postmitotic cells into a state where cell cycle checkpoint control is engaged results in upregulation of stress-responsive pathways that lead to increased survival, consistent with the finding that p21 deletion elongates survival of mice with dysfunctional telomeres (Choudhury et al. 2007). The finding that checkpoint mutants upregulate genes that are induced by the ER unfolded protein response, such as the heat shock chaperone HSP-4, supports this hypothesis (Olsen et al. 2006).

Clearly, the cell cycle checkpoint response is involved in organismal survival, and supporting this finding, it has been well documented that adult worms accelerate their aging processes when subjected to DNA damaging agents such as gamma irradiation and UV light (Cypser and Johnson 2002, Johnson et al. 2000). Not surprisingly, large-scale DNA damage reduces survival, but low-level, nonlethal doses of DNA-damaging agents extend the longevity of many postmitotic organisms. Taken together, perhaps ectopic regulation of cell cycle checkpoint pathways by either genetic manipulation or environmental induction may in fact activate other cell stress pathways, such as the JNK, MAPK, or PI3K pathways, to promote survival in the face of oncoming DNA damage.

Because of the close correlation between telomere length regulation and cell cycle checkpoint control, it is intriguing to ask whether the role of telomere length maintenance is also linked to the aging of nondividing cells and possibly organismal aging. In the mouse, it is clear that mitotic cell proliferation supercedes longevity regulation in telomerase knockout mice (Hemann et al. 2001); however, it is not clear how telomere length regulation impacts the aging of nondividing cells. Two reports indicate that there is no impact of somatic aging upon telomere length regulation. In one, knocking out the telomerase catalytic subunit did not alter the life span of worms, although germ line maintenance and survival was greatly altered (Meier et al. 2006). In the second study, wild isolates of *C. elegans* with different telomere lengths had no difference in longevity or stress resistance. Furthermore, clonal isolates of the widely used laboratory strain N2 showed stable variance in telomere length, but life span among these variants was identical (Raices et al. 2005). Taken together, results from both sets of experiments indicate that telomere length regulation is not required for aging of postmitotic organisms. In contrast, one earlier study identified that overexpression of *hrp-1*, a putative telomeric binding protein in *C. elegans*, results in increased telomere length and increased longevity that is dependent upon the insulin/IGF-1 signaling pathway (Joeng et al. 2004). However, clonal isolates from this study were not analyzed and the combination of *hrp-1* overexpression with telomerase knockout mutations was not investigated to definitively determine whether life span extension due to overexpression of *hrp-1* was dependent upon telomere length increase.

5.3 Genetic Pathways of Postmitotic Cell and Organismal Survival

Besides the role of a potential ectopic cell cycle checkpoint pathway helping to modulate the life span of a postmitotic organism such as *C. elegans*, are there other

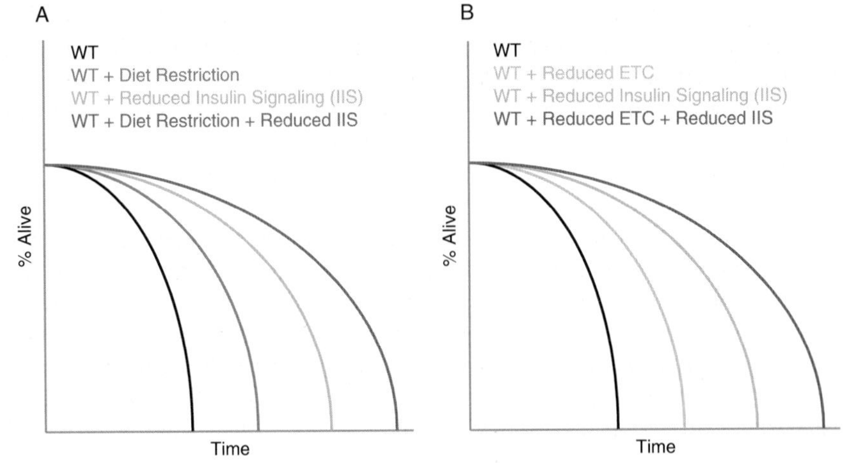

Fig. 1.2 Reduced insulin/IGF-1 signaling (IIS) synergizes with perturbations in other longevity pathways. Representative life span analysis of worms with combinatorial inactivation of IIS, diet restriction, or reduced mitochondrial electron transport chain (ETC) activity is depicted. A) Combination of diet restriction (depicted as red line) with reduced IIS (green line) results in animals that live longer than either single perturbation (blue line). B) Combination of reduced ETC (depicted as yellow line) with reduced IIS (green line) results in animals that live longer than either single perturbation (cayenne line). For more details of separation of these three pathways, please see text (*See Color Plate*)

genetic pathways that determine longevity and survival of postmitotic cells? And, if so, how do they impact genome integrity?

Surprisingly, at least three distinct pathways have been found to promote survival of worms, flies, and mice. These pathways include the insulin/IGF-1 signaling pathway (IIS), dietary restriction pathway (DR), and mitochondrial electron transport chain (ETC) pathway (Fig. 1.2). Below, we have outlined the important discoveries of each pathway and explain in greater detail their independence from each other.

5.3.1 Insulin/IGF-1 Signaling

Well over 15 years ago, single-gene mutations in worms were identified that extended longevity. One such mutation, *age-1*, identified by Michael Klass, turned out to be the vertebrate ortholog of the phosphoinositide 3-kinase essential for transduction of signals from the insulin receptor to downstream kinases, such AKT (Klass 1983, Morris et al. 1996). Since this discovery, mutations in genes for all known members of the insulin/IGF-1 signaling (IIS) pathway have been found to extend longevity of adult worms (reviewed in Wolff and Dillin 2006). Furthermore, all increases in longevity induced by mutations in the IIS depend upon the forkhead-related transcription factor DAF-16, the worm ortholog of the vertebrate Foxo3a. Mutating chico, the fly ortholog of IRS-1, reduces signaling along the IIS pathway and increases longevity in flies. Similarly, reducing expression levels of the IGF-1 receptor (IGF-1R), or knocking out the insulin receptor in adipose tissue, extends longevity of mice. Taken together, reduction of IIS in many distinct organisms extends longevity, suggesting that the IIS pathway plays a highly conserved role in cell survival. Interestingly, knockout of the yeast AKT ortholog, Sch9, extends the replicative life span of yeast cultures (Fabrizio et al. 2001), suggesting that the IIS pathway could potentially upregulate survival cues in both postmitotic and mitotic cells.

How does reduced IIS extend longevity? One possibility is that the IIS pathway feeds back into the DNA damage surveillance pathway and possibly even the regulation of telomere length. DNA microarray analysis of genes regulated by DAF-16 suggests that the IIS pathway (Murphy et al. 2003) intersects with the DNA damage pathway and induces the expression of genes required for resistance to DNA-damaging agents. Consistent with this analysis, IIS mutants are highly resistant to DNA-damaging agents (Larsen 1993). However, a connection of IIS with telomere maintenance is lacking. Analysis of telomere length regulation in animals with reduced IIS, or deletion of DAF-16, found no effect, excluding telomere length regulation as a possible means through which IIS mutant animals extend their life span (Raices et al. 2005). In addition to increased expression of DNA repair and surveillance genes, the IIS pathway also regulates expression of genes required for proper protein folding, innate immunity, and oxidative stress (Murphy et al. 2003). Therefore, in addition to DNA damage surveillance a fine-tuned execution of many different stress-responsive modules is required for a successful increase in survival.

5.3.2 Dietary Restriction

Over 72 years ago it was discovered that reducing food intake results in increased longevity, but the underlying molecular mechanisms have yet to be uncovered. A priori, it would appear that the mechanism by which reduced insulin/IGF-1 signaling mediates a response to increased cell survival should be consistent with the response to reduced food intake. However, many pieces of data suggest otherwise. First, the long life span induced by diet restriction in worms does not depend upon DAF-16 (Houthoofd et al. 2003, Lakowski and Hekimi 1996). Second, long-lived insulin/IGF-1 mutant worms have an even greater life span extension when maintained on a diet-restricted regimen (Lakowski and Hekimi 1996). Third, dietary restriction can be instituted at almost any time in the animal's life cycle, whereas insulin/IGF-1 signaling is required during the early adulthood stages in order to regulate longevity cues (Dillin et al. 2002). Finally, many of the physiological outcomes of animals with reduced IIS compared to animals undergoing DR are similar. Such overt similarities include reduced body size, lower plasma IGF-1 and insulin levels, and increased insulin sensitivity. Furthermore, transcriptional profiling of long-lived dwarf mice with reduced IGF-1 signaling found additional increased expression of multiple liver-specific genes when combined with dietary restriction (Tsuchiya et al. 2004). However, compelling genetic analyses indicate that many key differences among IIS and DR mice exist as well. For example, IGF-1R long-lived heterozygous mice do not show protracted or reduced reproduction as DR animals do (Holzenberger et al. 2003).

To date, the only protein directly implicated in the longevity response to DR is the histone deacetylase, SIR-2. This hypothesis stems from the fact that ectopic expression of SIR-2 extends longevity of yeast, worms, and flies. Furthermore, yeast defective in *sir-2* cannot respond to increased longevity due to growth on low glucose conditions (glucose restriction) (Lamming et al. 2005). However, deletion of *sir-2* does not disrupt a worm's ability to live long under conditions of diet restriction (Wang and Tissenbaum 2006). Furthermore, and most compelling, overexpression of *sir-2* in worms results in increased longevity that is dependent upon *daf-16* (Tissenbaum and Guarente 2001). Taken together, it is not clear where exactly *sir-2* fits into longevity pathways, however, it is interesting to speculate that the role of *sir-2* in telomeric DNA silencing in yeast may be conserved with telomere maintenance in other organisms and thus could account for the longevity increases associated with overexpression of *sir-2* (Moretti et al. 1994). The role of *sir-2* and its paralogs in mice will reveal much about the role of protein deacetylation in longevity determination.

5.3.3 Mitochondrial Electron Transport Chain

Mitochondria have been implicated in the aging process for several decades. Measuring the metabolic rates of several species during the 1920s, Pearl discovered the correlation between metabolic rate and life span: animals with lower metabolic rates lived longer than animals with higher metabolic rates (Pearl 1928). Exceptions have been discovered since Pearl's initial observations, such as the high metabolic rates of some long-lived birds (Holmes et al. 2001); however Pearl's initial observation led to the formulation of the "rate of living theory of aging." The theory suggests that reduced metabolic rates in an animal should result in an increased life span.

Several years later, the "rate of living theory of aging" became synonymous with the "oxygen radical theory of aging" proposed by Denham Harman (Harman 1956; see also Passos et al., this volume). Harman reasoned that enzymatic reactions using molecular oxygen create, on occasion, O_2 radicals. Harman hypothesized that lower levels of oxygen free radicals, by reducing enzymatic activities that utilized molecular oxygen (O_2), would result in increased longevity.

Mitochondria are the primary sites of O_2 consumption within the cell, suggesting that mitochondria are also the primary sites of oxygen-free radical production. Therefore, according to the free radical theory of aging, reduced mitochondrial function should correlate with increased longevity. Indeed, yeast lacking mitochondrial DNA, thereby having to survive on metabolic fermentation, have increased replicative life spans compared to wild-type yeast cells (Jazwinski 1990). Furthermore, long-lived, calorie-restricted mice show reduced expression of several key metabolic genes, including several mitochondrial genes (Lee et al. 1999). Taken together, these studies suggest that reduced mitochondrial function should increase longevity of multicellular organisms.

Research in nematodes identified the mitochondrial electron transport and ATP synthase as regulators of the aging process. Three studies demonstrated that reduction of function of several mitochondrial genes extends the life span of fully developed, adult worms. Siegfried Hekimi's lab showed that a single mutation in an iron sulfur component of Complex III, *isp-1*, increased longevity. This mutation decreases oxygen consumption, suggesting that it lowers the activity of the electron transport chain (Feng 2001). *isp-1* mutant worms display delayed development and reduced rates of other physiological processes, including eating, movement, and defecation (Feng 2001). Two independent RNAi-based screens also showed that components of the mitochondrial electron transport chain increased longevity when they were inactivated using RNAi (Dillin et al. 2002, Lee et al. 2003). It was found that RNAi inactivation of components of Complex I, III, IV, and the ATP synthase increases longevity of wild-type worms (Dillin et al. 2002). These results are consistent with those determined using an RNAi-based screen in Gary Ruvkun's lab (Lee et al. 2003). Similar to the properties of *isp-1* mutant worms, RNAi of the mitochondrial electron transport chain genes reduced ATP levels, O_2 consumption and slowed the rate of development and other physiological processes, including eating, movement and defecation (Dillin et al. 2002, Lee et al. 2003).

Mutations in another worm gene, *clk-1,* lengthens life span (Wong et al. 1995) but does not appear to reduce respiration (Braeckman et al. 1999, Miyadera et al. 2001). *clk-1* mutations prevent the synthesis of ubiquinone (Miyadera et al. 2001), which is required for respiratory chain activity. *clk-1* mutants are viable because they acquire ubiquinone from the bacteria in their culture media (Jonassen et al. 2001).

How does mitochondrial activity affect life span in *C. elegans*? One possibility is that mitochondrial activity inhibits the insulin/IGF-1 signaling pathway (Guarente 2000). Reducing the activity of DAF-2, an insulin/IGF-1 receptor homolog, or downstream signaling components, extends life span approximately two-fold. This life span extension requires activity of the forkhead-family transcription factor DAF-16. However, *isp-1, clk-1* mutations or RNAi of respiratory chain components extend the life span of *daf-16* mutations (Dillin et al. 2002, Feng 2001, Lee et al. 2003, Wong et al. 1995). In addition, the already long life span of *daf-2(e1370)* mutants is further extended by *isp-1, clk-1* mutations, or RNAi of respiratory chain components. Moreover, unlike reduction of respiratory chain activity, reduction of insulin/IGF-1 signaling is known to cause a significant increase in ATP levels (Braeckman et al. 1999, Dillin et al. 2002). Finally, both *daf-2* and *daf-16* act exclusively in adults to regulate life span (Dillin et al. 2002). Together these findings indicate that respiratory-chain RNAi does not increase life span by inhibiting the DAF-2 pathway.

Not all mitochondrial ETC lesions increase longevity. The *mev-1(kn1)* mutation is probably the best example of a mutation that decreases mitochondrial activity, but does not increase longevity. The *mev-1(kn1)* mutation was identified in a genetic screen to identify mutations that resulted in worms that were more sensitive to the drug methyl viologen (Paraquat). Cells treated with Paraquat produce excess oxygen-free radicals. *mev-1(kn1)* mutant animals are hypersensitive to Paraquat, are short lived, and have reduced mitochondrial respiratory rates (Hosokawa et al. 1994, Ishii et al. 1998, Ishii et al. 1990). Additionally, *mev-1(kn1)* mutant animals have higher levels of oxygen-free radicals compared to wild-type animals (Ishii et al. 1998, Senoo-Matsuda et al. 2001). *mev-1* encodes the cytochrome b large subunit of Complex II (Ishii et al. 1998). Similar to *mev-1(kn1)* mutant animals, *gas-1(fc21)* mutant animals are also hypersensitive to Paraquat, short lived, have reduced ETC activity, and have higher levels of oxygen-free radicals compared to wild-type animals (Senoo-Matsuda et al. 2001). *gas-1* encodes the 49 kilo Dalton iron-containing subunit of mitochondrial electron transport chain complex I (Kayser et al. 1999). Besides *mev-1* and *gas-1* mutations, *nuo-1(ua1)* and *atp-2(ua2)* mutant worms also have reduced mitochondrial respiratory function (Tsang et al. 2001). *nuo-1(ua1)* and *atp-2(ua2)* mutant animals arrest during development and do not grow to reproductive adulthood (Tsang et al. 2001). *nuo-1* encodes the NADH- and FMN-binding subunit of complex I and *atp-2* encodes an active site subunit of complex V, the ATP synthase.

Taken together, a clear correlation between metabolic activity and longevity cannot be derived from these studies. For example, *clk-1* mutant animals are long-lived, but have normal respiratory rates, and RNAi of several ETC components results in increased longevity and decreased metabolic rates. Therefore, defining the role that mitochondria play in the aging process will be essential to elucidating the correlation between metabolic rates and longevity, as originally proposed by Pearl in the 1930s.

6 Perspectives

Cellular and organismal aging is influenced by multiple parameters, such as telomere length, telomere shortening rates, the DNA damage response machinery, dietary restriction, mitochondrial metabolism, and the insulin-signaling pathway. Currently it is not well understood how these pathways interact, how the determinants of organismal life span influence cellular aging, and how or if the cellular clock limits organismal longevity. It remains a unique challenge to the aging research community to integrate the various aspects of the field, to realize the connections to age-related diseases, and to exploit these realizations for healthy aging.

Acknowledgements We would like to thank the NIH and NIA for funding, and Marcela Raices and Regina Kirchweger for comments on the manuscript.

References

Baird DM, Davis T, Rowson J, Jones CJ, Kipling D (2004) Normal telomere erosion rates at the single cell level in Werner syndrome fibroblast cells. Hum Mol Genet 13: 1515–24

Blackburn EH, Greider CW, Szostak JW (2006) Telomeres and telomerase: the path from maize, Tetrahymena and yeast to human cancer and aging. Nat Med 12: 1133–38

Blasco MA, Lee HW, Hande MP, Samper E, Lansdorp PM, DePinho RA, Greider CW (1997) Telomere shortening and tumor formation by mouse cells lacking telomerase RNA. Cell 91: 25–34

Bodnar AG, Ouellette M, Frolkis M, Holt SE, Chiu CP, Morin GB, Harley CB, Shay JW, Lichtsteiner S, Wright WE (1998) Extension of life-span by introduction of telomerase into normal human cells. Science 279: 349–52

Braeckman BP, Houthoofd K, De Vreese A, Vanfleteren JR (1999) Apparent uncoupling of energy production and consumption in long-lived Clk mutants of Caenorhabditis elegans. Curr Biol 9: 493–96

Chang S, Multani AS, Cabrera NG, Naylor ML, Laud P, Lombard D, Pathak S, Guarente L, DePinho RA (2004) Essential role of limiting telomeres in the pathogenesis of Werner syndrome. Nat Genet 36: 877–82

Choudhury AR, Ju Z, Djojosubroto MW, Schienke A, Lechel A, Schaetzlein S, Jiang H, Stepczynska A, Wang C, Buer J, Lee HW, von Zglinicki T, Ganser A, Schirmacher P, Nakauchi H, Rudolph KL (2007) Cdkn1a deletion improves stem cell function and lifespan of mice with dysfunctional telomeres without accelerating cancer formation. Nat Genet 39: 99–105

Cooke HJ, Smith BA (1986) Variability at the telomeres of the human X/Y pseudoautosomal region. Cold Spring Harb Symp Quant Biol 51 Pt 1: 213–19

Crabbe L, Jauch A, Naeger CM, Holtgreve-Grez H, Karlseder J (2007) Telomere dysfunction as a cause of genomic instability in Werner syndrome. Proc Natl Acad Sci U S A 104: 2205–10

Crabbe L, Verdun RE, Haggblom CI, Karlseder J (2004) Defective telomere lagging strand synthesis in cells lacking WRN helicase activity. Science 306: 1951–53

Cypser JR, Johnson TE (2002) Multiple stressors in Caenorhabditis elegans induce stress hormesis and extended longevity. J Gerontol A Biol Sci Med Sci 57: B109–14

de Lange T (1998) Telomeres and senescence: ending the debate. Science 279: 334–35

de Lange T (2002) Protection of mammalian telomeres. Oncogene 21: 532–40

de Lange T (2004) T-loops and the origin of telomeres. Nat Rev Mol Cell Biol 5: 323–29

de Lange T (2005) Shelterin: the protein complex that shapes and safeguards human telomeres. Genes Dev 19: 2100–10

de Lange T (2006) Lasker Laurels for telomerase. Cell 126: 1017–20

de Lange T, Jacks T (1999) For better or worse? Telomerase inhibition and cancer. Cell 98: 273–75

Dillin A, Crawford DK, Kenyon C (2002) Timing requirements for insulin/IGF-1 signaling in C. elegans. Science 298: 830–34

Dillin A, Hsu AL, Arantes-Oliveira N, Lehrer-Graiwer J, Hsin H, Fraser AG, Kamath RS, Ahringer J, Kenyon C (2002) Rates of behavior and aging specified by mitochondrial function during development. Science 298: 2398–401

Dionne I, Wellinger RJ (1996) Cell cycle-regulated generation of single-stranded G-rich DNA in the absence of telomerase. Proc Natl Acad Sci U S A 93: 13902–07

Evans SK, Lundblad V (1999) Est1 and Cdc13 as comediators of telomerase access. Science 286: 117–20

Fabrizio P, Pozza F, Pletcher SD, Gendron CM, Longo VD (2001) Regulation of longevity and stress resistance by Sch9 in yeast. Science 292: 288–90

Feng J, Bussiere, F. and S. Hekimi (2001) Mitochondrial electron transport is a key determinant of life span in Caenorhabditis elegans. Developmental Cell 1: 663–644

Greider CW, Blackburn EH (1985) Identification of a specific telomere terminal transferase activity in Tetrahymena extracts. Cell 43: 405–13

Greider CW, Blackburn EH (1987) The telomere terminal transferase of Tetrahymena is a ribonucleoprotein enzyme with two kinds of primer specificity. Cell 51: 887–98

Greider CW, Blackburn EH (1989) A telomeric sequence in the RNA of Tetrahymena telomerase required for telomere repeat synthesis. Nature 337: 331–37

Griffith JD, Comeau L, Rosenfield S, Stansel RM, Bianchi A, Moss H, de Lange T (1999) Mammalian telomeres end in a large duplex loop. Cell 97: 503–14

Guarente L, Kenyon, C (2000) Genetic pathways that regulate ageing in model organisms. Nature 408: 25–32

Hao LY, Armanios M, Strong MA, Karim B, Feldser DM, Huso D, Greider CW (2005) Short telomeres, even in the presence of telomerase, limit tissue renewal capacity. Cell 123: 1121–31

Harley CB, Futcher AB, Greider CW (1990) Telomeres shorten during ageing of human fibroblasts. Nature 345: 458–60
Harman D (1956) Aging: a theory based on free radical and radiation chemistry. J Gerontol 11: 298–300
Hayflick L (1965) The limited in vitro lifetime of human diploid cell strains. Exp Cell Res 37: 614–36
Hayflick L, Moorhead PS (1961) The serial cultivation of human diploid cell strains. Exp Cell Res 25: 585–621
Hemann MT, Greider CW (1999) G-strand overhangs on telomeres in telomerase-deficient mouse cells. Nucleic Acids Res 27: 3964–69
Hemann MT, Greider CW (2000) Wild-derived inbred mouse strains have short telomeres. Nucleic Acids Res 28: 4474–78.
Hemann MT, Strong MA, Hao LY, Greider CW (2001) The shortest telomere, not average telomere length, is critical for cell viability and chromosome stability. Cell 107: 67–77
Herbig U, Ferreira M, Condel L, Carey D, Sedivy JM (2006) Cellular senescence in aging primates. Science 311: 1257
Holmes DJ, Fluckiger R, Austad SN (2001) Comparative biology of aging in birds: an update. Exp Gerontol 36: 869–83
Holzenberger M, Dupont J, Ducos B, Leneuve P, Geloen A, Even PC, Cervera P, Le Bouc Y (2003) IGF-1 receptor regulates lifespan and resistance to oxidative stress in mice. Nature 421: 182–87
Hosokawa H, Ishii N, Ishida H, Ichimori K, Nakazawa H, Suzuki K (1994) Rapid accumulation of fluorescent material with aging in an oxygen- sensitive mutant mev-1 of Caenorhabditis elegans. Mech Ageing Dev 74: 161–70
Houthoofd K, Braeckman BP, Johnson TE, Vanfleteren JR (2003) Life extension via dietary restriction is independent of the Ins/IGF-1 signalling pathway in Caenorhabditis elegans. Exp Gerontol 38: 947–54
Hsin H, Kenyon C (1999) Signals from the reproductive system regulate the life span of C. elegans. Nature 399: 362–66
Ishii N, Fujii M, Hartman PS, Tsuda M, Yasuda K, Senoo-Matsuda N, Yanase S, Ayusawa D, Suzuki K (1998) A mutation in succinate dehydrogenase cytochrome b causes oxidative stress and ageing in nematodes. Nature 394: 694–97
Ishii N, Takahashi K, Tomita S, Keino T, Honda S, Yoshino K, Suzuki K (1990) A methyl viologen-sensitive mutant of the nematode Caenorhabditis elegans. Mutat Res 237: 165–71
Jacob NK, Skopp R, Price CM (2001) G-overhang dynamics at Tetrahymena telomeres. Embo J 20: 4299–308
Jazwinski SM (1990) Aging and senescence of the budding yeast Saccharomyces cerevisiae. Mol Microbiol 4: 337–43
Joeng KS, Song EJ, Lee KJ, Lee J (2004) Long lifespan in worms with long telomeric DNA. Nat Genet 36: 607–11
Johnson TE, Cypser J, de Castro E, de Castro S, Henderson S, Murakami S, Rikke B, Tedesco P, Link C (2000) Gerontogenes mediate health and longevity in nematodes through increasing resistance to environmental toxins and stressors. Exp Gerontol 35: 687–94
Jonassen T, Larsen PL, Clarke CF (2001) A dietary source of coenzyme Q is essential for growth of long-lived Caenorhabditis elegans clk-1 mutants. Proc Natl Acad Sci U S A 98: 421–426
Kayser EB, Morgan PG, Sedensky MM (1999) GAS-1: a mitochondrial protein controls sensitivity to volatile anesthetics in the nematode Caenorhabditis elegans. Anesthesiology 90: 545–54
Kilian A, Bowtell DD, Abud HE, Hime GR, Venter DJ, Keese PK, Duncan EL, Reddel RR, Jefferson RA (1997) Isolation of a candidate human telomerase catalytic subunit gene, which reveals complex splicing patterns in different cell types. Hum Mol Genet 6: 2011–19
Kim NW, Piatyszek MA, Prowse KR, Harley CB, West MD, Ho PL, Coviello GM, Wright WE, Weinrich SL, Shay JW (1994) Specific association of human telomerase activity with immortal cells and cancer. Science 266: 2011–15
Klass MR (1983) A method for the isolation of longevity mutants in the nematode Caenorhabditis elegans and initial results. Mech Ageing Dev 22: 279–86

Lakowski B, Hekimi S (1996) Determination of life-span in Caenorhabditis elegans by four clock genes. Science 272: 1010–13

Lamming DW, Latorre-Esteves M, Medvedik O, Wong SN, Tsang FA, Wang C, Lin SJ, Sinclair DA (2005) HST2 mediates SIR2-independent life-span extension by calorie restriction. Science 309: 1861–64

Larsen PL (1993) Aging and resistance to oxidative damage in Caenorhabditis elegans. Proc Natl Acad Sci U S A 90: 8905–09

Lee SE, Paques F, Sylvan J, Haber JE (1999) Role of yeast SIR genes and mating type in directing DNA double-strand breaks to homologous and non-homologous repair paths. Curr Biol 9: 767–70

Lee SS, Lee RY, Fraser AG, Kamath RS, Ahringer J, Ruvkun G (2003) A systematic RNAi screen identifies a critical role for mitochondria in C. elegans longevity. Nat Genet 33: 40–48

Lendvay TS, Morris DK, Sah J, Balasubramanian B, Lundblad V (1996) Senescence mutants of Saccharomyces cerevisiae with a defect in telomere replication identify three additional EST genes. Genetics 144: 1399–412

Lengauer C, Kinzler KW, Vogelstein B (1998) Genetic instabilities in human cancers. Nature 396: 643–49

Lingner J, Cech TR (1996) Purification of telomerase from Euplotes aediculatus: requirement of a primer 3′ overhang. Proc Natl Acad Sci U S A 93: 10712–17

Lingner J, Hughes TR, Shevchenko A, Mann M, Lundblad V, Cech TR (1997) Reverse transcriptase motifs in the catalytic subunit of telomerase. Science 276: 561–67

Liu Y, Snow BE, Hande MP, Yeung D, Erdmann NJ, Wakeham A, Itie A, Siderovski DP, Lansdorp PM, Robinson MO, Harrington L (2000) The telomerase reverse transcriptase is limiting and necessary for telomerase function in vivo. Curr Biol 10: 1459–62

Lundblad V (2003) Telomere replication: an Est fest. Curr Biol 13: R439–41

Lundblad V, Szostak JW (1989) A mutant with a defect in telomere elongation leads to senescence in yeast. Cell 57: 633–43

Makarov VL, Hirose Y, Langmore JP (1997) Long G tails at both ends of human chromosomes suggest a C strand degradation mechanism for telomere shortening. Cell 88: 657–66

Marciniak R, Guarente L (2001) Human genetics. Testing telomerase. Nature 413: 370–1, 373

Marciniak RA, Johnson FB, Guarente L (2000) Dyskeratosis congenita, telomeres and human ageing. Trends Genet 16: 193–95

Meier B, Clejan I, Liu Y, Lowden M, Gartner A, Hodgkin J, Ahmed S (2006) trt-1 is the Caenorhabditis elegans catalytic subunit of telomerase. PLoS Genet 2: e18

Melk A, Ramassar V, Helms LM, Moore R, Rayner D, Solez K, Halloran PF (2000) Telomere shortening in kidneys with age. J Am Soc Nephrol 11: 444–53

Meyerson M, Counter CM, Eaton EN, Ellisen LW, Steiner P, Caddle SD, Ziaugra L, Beijersbergen RL, Davidoff MJ, Liu Q, Bacchetti S, Haber DA, Weinberg RA (1997) hEST2, the putative human telomerase catalytic subunit gene, is up- regulated in tumor cells and during immortalization. Cell 90: 785–95

Mitchell JR, Wood E, Collins K (1999) A telomerase component is defective in the human disease dyskeratosis congenita. Nature 402: 551–55

Miyadera H, Amino H, Hiraishi A, Taka H, Murayama K, Miyoshi H, Sakamoto K, Ishii N, Hekimi S, Kita K (2001) Altered quinone biosynthesis in the long-lived clk-1 mutants of Caenorhabditis elegans. J Biol Chem 276: 7713–16

Morales CP, Holt SE, Ouellette M, Kaur KJ, Yan Y, Wilson KS, White MA, Wright WE, Shay JW (1999) Absence of cancer-associated changes in human fibroblasts immortalized with telomerase. Nat Genet 21: 115–18

Moretti P, Freeman K, Coodly L, Shore D (1994) Evidence that a complex of SIR proteins interacts with the silencer and telomere-binding protein RAP1. Genes Dev 8: 2257–69

Morris JZ, Tissenbaum HA, Ruvkun G (1996) A phosphatidylinositol-3-OH kinase family member regulating longevity and diapause in Caenorhabditis elegans. Nature 382: 536–39

Munoz-Jordan JL, Cross GA, de Lange T, Griffith JD (2001) t-loops at trypanosome telomeres. Embo J 20: 579–88

Murphy CT, McCarroll SA, Bargmann CI, Fraser A, Kamath RS, Ahringer J, Li H, Kenyon C (2003) Genes that act downstream of DAF-16 to influence the lifespan of Caenorhabditis elegans. Nature 424: 277–83
Nakamura TM, Morin GB, Chapman KB, Weinrich SL, Andrews WH, Lingner J, Harley CB, Cech TR (1997) Telomerase catalytic subunit homologs from fission yeast and human. Science 277: 955–59
Nugent CI, Hughes TR, Lue NF, Lundblad V (1996) Cdc13p: a single-strand telomeric DNA-binding protein with a dual role in yeast telomere maintenance. Science 274: 249–52
Olovnikov AM (1973) A theory of marginotomy. J. Theoretical Biology 41: 181–190
Olsen A, Vantipalli MC, Lithgow GJ (2006) Checkpoint proteins control survival of the postmitotic cells in Caenorhabditis elegans. Science 312: 1381–85
Opresko PL, Cheng WH, von Kobbe C, Harrigan JA, Bohr VA (2003) Werner syndrome and the function of the Werner protein; what they can teach us about the molecular aging process. Carcinogenesis 24: 791–802
Pearl R (1928) The rate of living. Publisher, London
Pennock E, Buckley K, Lundblad V (2001) Cdc13 delivers separate complexes to the telomere for end protection and replication. Cell 104: 387–96
Raices M, Maruyama H, Dillin A, Karlseder J (2005) Uncoupling of Longevity and Telomere Length in C. elegans. PLoS Genet 1: e30
Riha K, McKnight TD, Fajkus J, Vyskot B, Shippen DE (2000) Analysis of the G-overhang structures on plant telomeres: evidence for two distinct telomere architectures. Plant J 23: 633–641
Rudolph KL, Chang S, Lee HW, Blasco M, Gottlieb GJ, Greider C, DePinho RA (1999) Longevity, stress response, and cancer in aging telomerase-deficient mice. Cell 96: 701–12
Schulz VP, Zakian VA, Ogburn CE, McKay J, Jarzebowicz AA, Edland SD, Martin GM (1996) Accelerated loss of telomeric repeats may not explain accelerated replicative decline of Werner syndrome cells. Hum Genet 97: 750–54
Senoo-Matsuda N, Yasuda K, Tsuda M, Ohkubo T, Yoshimura S, Nakazawa H, Hartman PS, Ishii N (2001) A defect in the cytochrome b large subunit in complex II causes both superoxide anion overproduction and abnormal energy metabolism in Caenorhabditis elegans. J Biol Chem 276: 41553–58
Tahara H, Tokutake Y, Maeda S, Kataoka H, Watanabe T, Satoh M, Matsumoto T, Sugawara M, Ide T, Goto M, Furuichi Y, Sugimoto M (1997) Abnormal telomere dynamics of B-lymphoblastoid cell strains from Werner's syndrome patients transformed by Epstein-Barr virus. Oncogene 15: 1911–20
Tissenbaum HA, Guarente L (2001) Increased dosage of a sir-2 gene extends life span in Caenorhabditis elegans. Nature 410: 227–30
Tsang WY, Sayles LC, Grad LI, Pilgrim DB, Lemire BD (2001) Mitochondrial respiratory chain deficiency in Caenorhabditis elegans results in developmental arrest and increased life span. J Biol Chem 276: 32240–6
Tsuchiya T, Dhahbi JM, Cui X, Mote PL, Bartke A, Spindler SR (2004) Additive regulation of hepatic gene expression by dwarfism and caloric restriction. Physiol Genomics 17: 307–15
Vulliamy T, Marrone A, Goldman F, Dearlove A, Bessler M, Mason PJ, Dokal I (2001) The RNA component of telomerase is mutated in autosomal dominant dyskeratosis congenita. Nature 413: 432–35
Wang Y, Tissenbaum HA (2006) Overlapping and distinct functions for a Caenorhabditis elegans SIR2 and DAF-16/FOXO. Mech Ageing Dev 127: 48–56
Watson JD (1972) Origin of concatemeric T7 DNA. Nat New Biol 239: 197–201
Wellinger RJ, Ethier K, Labrecque P, Zakian VA (1996) Evidence for a new step in telomere maintenance. Cell 85: 423–33
Wolff S, Dillin A (2006) The trifecta of aging in Caenorhabditis elegans. Exp Gerontol 41: 894–903
Wong A, Boutis P, Hekimi S (1995) Mutations in the clk-1 gene of Caenorhabditis elegans affect developmental and behavioral timing. Genetics 139: 1247–59

Wright WE, Tesmer VM, Huffman KE, Levene SD, Shay JW (1997) Normal human chromosomes have long G-rich telomeric overhangs at one end. Genes Dev 11: 2801–09
Wyllie FS, Jones CJ, Skinner JW, Haughton MF, Wallis C, Wynford-Thomas D, Faragher RG, Kipling D (2000) Telomerase prevents the accelerated cell ageing of Werner syndrome fibroblasts. Nat Genet 24: 16–17

Chapter 2
Telomere-Induced Senescence of Primary Cells

Richard Allsopp

Abstract Cellular senescence occurs when a cell enters a viable state of irreversible growth arrest. Since the first characterization of cell senescence for human cells, specifically human diploid fibroblasts, in classic experiments by Hayflick and Moorhead in the early 1960s (Hayflick and Moorhead 1961), primary cultures of human cells have been shown to undergo cell senescence in response to a number of different environmental stressors. For example, oxidative stress (von Zglinicki et al. 1995, Finkel and Holbrook 2000), exposure to irradiation (Finkel and Holbrook 2000), and inappropriate stimulation of mitogenic signaling pathways (Ridley et al 1988, Serrano et al. 1997), have all been shown to induce cell senescence in normal human cells grown in vitro. However, even when maintained in a stress-free environment under optimal growth conditions, cultures of normal diploid human cells still ultimately end up in a state of cell senescence after a finite number of population doublings. The cause of this phenomenon, specifically referred to as replicative senescence, or the Hayflick limit, after the discoverer Leonard Hayflick, eluded scientists until the mid-1990s, when researchers at Geron corporation (Bodnar et al. 1998) and the Whitehead Institute (Meyerson et al. 1998) discovered that the gradual erosion of telomeres, essential genetic elements that cap the ends of chromosomes, during proliferation of normal diploid human cells is what ultimately limits the replicative capacity of these cells and induces senescence. In this chapter, I will review the phenomenon of cell senescence, specifically, replicative senescence or telomere-induced senescence, the mechanism of telomere-induced senescence, and the role of replicative senescence in organismal aging.

Richard Allsopp
University of Hawaii, Institute for Biogenesis Research, John A Burns School of Medicine, Honolulu, Hawaii, United States.
e-mail: allsopp@hawaii.edu

K.L. Rudolph (ed.), *Telomeres and Telomerase in Ageing, Disease, and Cancer.*

1 Primary Human Cells and Replicative Senescence

The human diploid fibroblast (HDF) was the first type of human cell used in cell senescence research (Hayflick and Moorhead, 1961), and remains the most well established human cell type for studying replicative senescence in vitro. Prior to the pioneering experiments by Hayflick and Moorhead in the early 1960s, researchers doing tissue culture believed that the limited life span of normal diploid cells in vitro was due to inadequate culture conditions. Work by Hayflick and Moorhead led to a number of key observations that argued that this limited replicative ability in vitro represented true senescence (see Dillin and Karlseder, this volume), not inadequate culture conditions. First, analysis of dozens of different fetal HDF strains revealed that the number of population doublings that these cells could reach before hitting senescence was approximately the same. Second, cryopreserved samples of the same HDF strain that were thawed out at different intervals over a large period of time all had the same replicative life span (in terms of population doublings). Third, when equal numbers of two distinguishable HDF strains, one capable of 20 more population doublings than the other, were mixed and then grown in vitro, only cells from the HDF strain capable of the additional 20 population doublings were detected at the terminal phase of culture (this observation was crucial in ruling out the possibility that infectious agents were responsible for attenuating the proliferative life span). Fourth, when cryopreserved samples of the same HDF strain were sent to be cultured at a number of different labs across the United States, all ceased proliferating at the same population doubling level.

Since the demonstration that the finite life span of HDF was due to replicative senescence, a number of other types of primary human cells have been shown to also have a finite replicative capacity in vitro, including keratinocytes (Rheinwald and Green, 1975), vascular smooth muscle cells (Bierman, 1978), endothelial cells (Mueller et al. 1980), lens epithelial cells (Tassin et al. 1979), muscle satellite cells (Decary et al. 1997), and lymphocytes (Tice et al. 1979). All of these cell types are only capable of ~30–100 population doublings, depending on the cell type and age of the donor, before the onset of replicative senescence. While the growth conditions have yet to be established for the long-term culture of many types of human cells, including various types of stem cells, presently, human embryonic stem cells are the only type of normal human diploid cell that exhibits replicative immortality in vitro (Thomson et al. 1998). Since the cellular and molecular characteristics of the senescent state are best described for HDF, I will focus primarily on these cells for the rest of this section.

The two hallmark features of a senescent cell are that it is viable and has entered a state of irreversible growth arrest. Senescent HDF have been shown to be viable in vitro for periods in excess of one year (Matsumura et al. 1979). The latter hallmark feature of replicative senescence is based on a several studies which have shown that senescent HDF are refractory to a number of mitogenic and regulatory signals (Goldstein, 1990). To date, only the infection of senescent HDF with oncogenic viruses, such as the SV40 virus (Gorman and Cristofalo, 1985), has been shown to allow these cells to overcome the replicative block and resume proliferation. Primary cultures of other types of human cells, such as mammary epithelial cells, have been reported to spontaneously bypass the replicative senescence and acquire replicative immortality, but only with the accompanying loss of karyotypic stability (Romanov et al. 2001).

Senescent HDF primarily exhibit a growth arrest at the G1/S phase of the cell cycle, although a significant fraction, up to 20% for some HDF strains (Sherwood et al. 1988), have a 4N DNA content, indicating the presence of G1 tetraploid cells and/or G2 arrested cells in the senescent cell population. Other features of senescent HDF include increased frequency of karyotypic abnormalities, increased cytoplasmic volume, reduced motility, elevated RNA and protein content, increased granularity, increased frequency of lysosomal and Golgi bodies, and expression of a plethora of genes associated with the senescent state (Goldstein, 1990; Cristofalo et al. 2004). These genes include cell cycle inhibitors, such as p21 and $P16^{INK4a}$ which are members of the CIP-KIP family of kinase inhibitors (Noda et al. 1994; Alcorta et al. 1996); the transcription factor SAG (senescence associated gene) (Wistrom and Villeponteau, 1992); extracellular matrix (ECM)-degrading secreted metalloproteases, including collagenase, stromolysin, and tissue plasminogen activator (t-PA)(Cristofalo, 2004); promyelocytic leukemia-associated protein (PML) (Ferbeyre et al. 2000); senescence-associated beta-galactosidase (SA-β-gal) (Dimri et al. 1995); and, identified in recent studies, plasminogen activator inhibitor-1 (PAI-1), a downstream target of p53, which has been shown to be necessary to maintain cell cycle arrest in senescent HDFs (Kortlever et al. 2006). Of all these genes that are expressed in senescent HDF, SA-β-gal has proven to be the most useful biomarker of replicative senescence because of its ease of detection, both in vitro and in vivo, using a simple histochemical assay and because it also is a marker for replicative senescence of other types of somatic cells (Dimri et al. 1995).

Soon after the landmark experiments by Hayflick and Moorhead, a number of theories were developed to explain the finite replicative capacity of HDF. These theories could be classified into two groups: first, stochastic theories, all of which essentially purported that replicative senescence was due to the random accumulation of damage caused by extrinsic factors during replicative aging; second, genetic theories, which claimed that replicative senescence was a genetically programmed event, intrinsic to the very nature of the cell. In the 1970s, 1980s, and early 1990s, observations from a number of different studies led to increasing support for the genetic programming theories. Some of the first evidence that replicative senescence is a genetically determined process came from cell fusion experiments in the 1970s and 1980s (Norwood et al. 1974, Pereira-Smith and Smith, 1982). These elegant experiments showed that the senescent phenotype was dominant when young HDF were fused with senescent HDF (Norwood et al. 1974), or when immortal human cell lines were fused with senescent HDF (Pereira-Smith and Smith, 1982), to form heterokaryons containing both a senescent and nonsenescent nuclei. This implied that genetic factors inherited from the senescent nuclei were responsible for causing replicative senescence. Other experiments by Wright and Shay in the late 1980s and early 1990s led to a refinement of these genetic theories, when they showed that transformation of HDF with the SV40 large T antigen, allowed these cells to bypass the normal senescence checkpoint and continue to divide for an extended number of doublings (Shay and Wright, 1989). Soon afterward, however, these transformed cells underwent massive cell death (crisis). With continued culture, immortalized clonal cell populations did eventually emerge from crisis, but only at a very low frequency (1 out of every 2–3 $\times 10^7$ transformed cells). This suggested that were at least two genetic barriers or checkpoints in HDF that prevented replicative immortality, which they referred to as M1, the replicative senescence checkpoint, and M2, the checkpoint at crisis.

One of the most intriguing genetic program theories to explain replicative senescence was developed independently by Olovnikov (Olovnikov 1973) and Watson (Watson 1972). This theory, referred to as the end replication problem (Fig. 2.1), provided a detailed mechanistic hypothesis as to why HDF and other primary human cell cultures can only undergo a finite number of population doublings before hitting replicative senescence (see Dillin and Karlseder, this volume). This theory was based on the biochemical properties of DNA polymerase, the key component of the enzymatic complex responsible for replicating

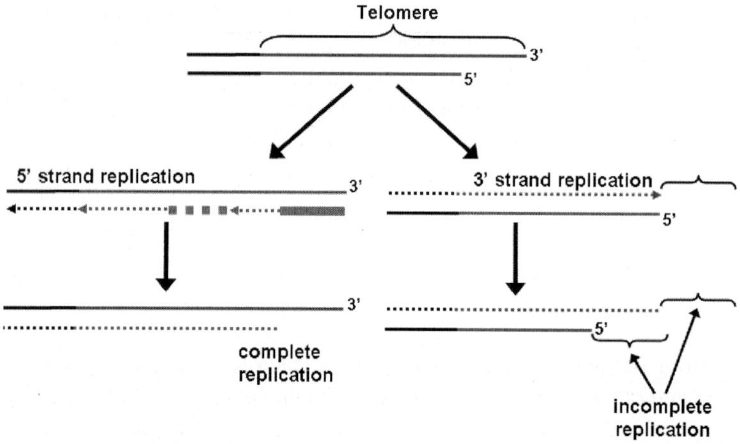

Fig. 2.1 The end replication problem. The replication of the 5′ strand (red) and 3′ strand (blue) of a single telomere are shown, as well as adjacent nontelomeric DNA (black). Newly synthesized DNA (i.e., the daughter strand) is indicated by the dashed lines. Replication of the 5′strand occurs by lagging strand synthesis, where Okazaki fragments are shown in grey. Replication of the 3′ strand occurs by leading strand synthesis. All telomeres end in a 3′overhang, which the normal DNA replication machinery cannot replicate. Degradation of the most 5′ Okazaki fragment presumably allows regeneration of the overhang during replication of the 5′ strand, allowing complete replication of the telomere on this daughter chromosome. However, the overhang cannot be regenerated by *de novo* DNA synthesis during leading strand synthesis, and therefore degradation of the 5′ strand is presumably required, which is hypothesized to account for the telomere shortening observed during cell division in cells that do not express telomerase (*See Color Plate*)

DNA in dividing cells, which precluded it from completing the replication of the ends of a linear DNA molecule (i.e., chromosomes). Both of these hypotheses predicted that the ends of chromosomes would slowly erode in a dividing cell population. Olovnikov went on to further predict that the eventual deletion of essential genes located near the ends of chromosomes would ultimately cause replicative senescence (Olovnikov 1973).

2 The Telomere Hypothesis of Cell Aging and Immortalization

When I started my research on replicative senescence in the early 1990s, the exact nature of the genetic event which accounted for the finite life span of HDFs was still unknown. However, exciting initial data in support of the end replication problem had recently been generated in the lab of my soon-to-be thesis advisor, Calvin Harley (Harley et al. 1990). This early study on the measurement of telomere length in human cells (in this case, HDFs) showed that the size of telomeres gradually shortened during replicative aging in vitro. In fact, prior to this study, there had been some indirect evidence that telomeres shorten during aging in humans. Some of the very earliest evidence in support of the loss of telomeric DNA during aging of primary cells in humans was from a study by Cooke and Smith, who showed the telomeric fragments of chromosomes were longer in the germ line than in the peripheral blood (Cooke and Smith, 1986). Shortly after the sequence of mammalian telomeric DNA was shown to be (TTAGGG)n in the late 1980s, work by de Lange et al. confirmed this earlier finding by directly showing that the terminal (TTAGGG)n tract at the ends of chromosomes is shorter in peripheral blood than the germ line (de Lange et al. 1990). This work by de Lange et al. was soon followed by two other studies which directly demonstrated the reduction in telomere size during replicative aging, one by Harley et al., which demonstrated that the size of the terminal restriction fragment (TRF) gradually shortens during replicative aging of HDF (Harley et al. 1990), and another study which showed that TRF length gradually decreases in size in human colon mucosa and peripheral blood during in vivo aging (Hastie et al. 1990). Another key observation in support of the notion that telomere shortening causes replicative senescence was the initial demonstration of active telomerase, an RNA-protein enzymatic complex that functions to complete the replication of, and even elongate, telomeric DNA, and which was originally discovered in ciliates (Greider and Blackburn, 1985) present in a human immortalized cell line (Morin, 1989).

Together, these initial observations of telomere shortening during replicative aging of primary human cells in vitro and in vivo, along with the end replication problem, led us to propose the telomere hypothesis of cellular aging and immortalization (Fig. 2.2) (Harley, 1991; Harley et al. 1992). This hypothesis made a number of (what were then) predictions: first, due to the end replication problem and possibly other mechanisms, telomeric DNA will shorten during replicative aging of mortal primary cell cultures established from different types of human somatic

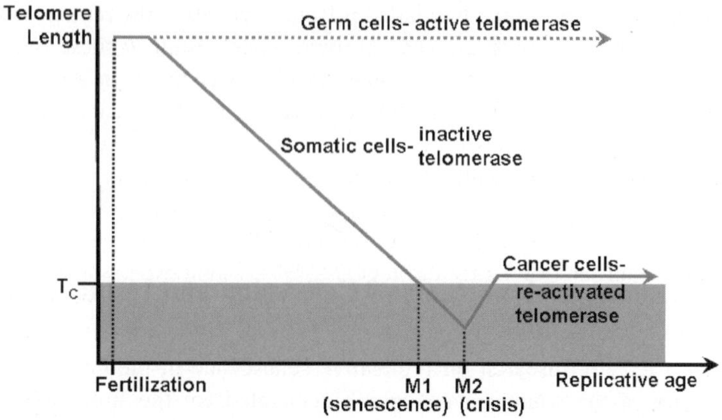

Fig. 2.2 The telomere hypothesis of cell aging and immortalization. The changes in telomere length during replicative aging of cells *in vivo*, beginning at the time of fertilization, is depicted. The change in telomere length during replicative aging for somatic cells and germ cells are indicated by the solid and dashed gray arrows respectively. The green bar shows the telomere size range for which there is insufficient telomeric DNA to allow a stable telomere complex to form. The upper size limit of this region is the critical telomere length (marked as T_C), which is the minimum length of telomeric DNA required to allow formation of a stable telomere complex at the end of a chromosome. The status of telomerase activity is also indicated. See text for further details (*See Color Plate*)

cells; second, telomere length will be maintained in germ line cells from generation to generation by telomerase; third, once a critically short size is reached on one or more telomeres, a signal will be sent to initiate replicative senescence, or the M1 checkpoint (the "Hayflick limit"); fourth, telomeres will continue to shorten in primary cells transformed by DNA tumor viruses, or transforming genes from these viruses (e.g., SV40 large T antigen) until nearly all cells have acquired multiple chromosomal ends with critically short telomeres. At this point, the M2 checkpoint, or crisis, will be signaled; fifth, in the rare cells which survive crisis and acquire replicative immortality, telomere length will be maintained by telomerase.

3 Validation of the Hypothesis

Since the initial proposal of the telomere hypothesis for cell aging and immortalization, all of the predictions of this hypothesis have been shown, by many independent studies, to be correct.

The first prediction, that telomeres will shorten in human somatic cells during replicative aging, has not only been demonstrated in HDFs (Harley et al. 1990, Allsopp et al. 1992), but also in other types of primary human cell cultures, including, but not limited to, keratinocytes (Ramirez et al. 2001), melanocytes

(Bandyopadhyay et al. 2001), endothelial cells (Chang and Harley 1995), vascular smooth muscle cells (Minamino et al. 2001), muscle satellite cells (Decary et al. 1997), mammary epithelial cells (Stampfer et al. 1997), retinal pigment epithelial cells (RPE) (Bodnar et al. 1998), lymphocytes (Hastie et al. 1990; Weng et al. 1995), and hematopoietic stem cells (HSC) (Vaziri et al. 1994). The rate of telomere shortening is variable (~30–250bp/PD), not only for different cell types, but among strains of the same cell type established from different donors. The only types of cultured human cells in which telomeres do not shorten are, as mentioned above, human embryonic stem cells (Thomson et al. 1998), which express telomerase and are immortal, and cells which are maintained in a quiescent (nonproliferating) state for extended periods of time, such as neuronal cells, or quiescent HDF (Allsopp, et al. 1995). The lack of telomere shortening over time in quiescent cells is exactly what is predicted by the end replication problem—if DNA does not replicate, then the amount of telomeric DNA should remain constant.

The second prediction, that telomere length is maintained in the germ line by telomerase, has also been confirmed. Studies have shown that telomerase is readily detectable in human testes and ovaries (Kim et al. 1994, Wright et al. 1995). Furthermore, telomeres in mature spermatozoa are relatively long, and do not shorten during aging in vivo (Allsopp et al. 1992). However, whether telomere length shortens during aging in mature oocytes has yet to be confirmed.

The third prediction, that shortening of telomeres below a critical size signals the senescence checkpoint, has also been shown to be correct in a number of elegant studies. The principal demonstration that telomere shortening causes replicative senescence and limits the proliferative life span of human somatic cells was demonstrated once the catalytic component of telomerase, telomerase reverse transcriptase (Tert), was cloned in the mid-1990s (Nakamura et al. 1997; Meyerson et al. 1998). The human Tert gene was first cloned independently by two groups, who searched a then newly established human EST data base for sequences homologous to the previously cloned Tert gene in *Saccharomyces cerevisiae* (Nakamura et al. 1997, Meyerson et al. 1998). Shortly after the human Tert gene was identified and cloned, experiments were performed to assess the affects of ectopic expression of Tert in a number of different primary human cell cultures, including HDF, endothelial cells, and RPE (Bodnar et al. 1998). In all of these different types of primary human cells, which were also shown to normally lack telomerase activity and detectable expression of endogenous Tert, ectopic expression of Tert was sufficient not only to restore active telomerase and prevent telomere shortening during replicative aging, but also to endow these cells with replicative immortality. Since these initial studies, many other types and strains of primary human cells have been immortalized by ectopic expression of Tert (Harley 2001).

Studies have also been performed to assess whether the shortening of just one or a few telomeres below length that is critical for telomere function, or the reduction in the average size of all telomeres in a cell below a certain length, is required to signal replicative senescence. Together, the data suggest that the former event is correct. First, analysis of karyotype stability during replicative aging of HDF has shown that the frequency of dicentric chromosomes involving end-to-end fusion events, which are presumed to arise due to the critical shortening and loss of the protective cap structure of a telomere, are elevated in HDF near senescence (Benn 1976). Second, a number of studies have observed, based on the analysis of signal intensity at the ends of metaphase chromosomes stained with a fluorescent telomeric probe using fluorescent in situ hybridization (FISH), interchromosomal variability in telomere length within primary human cells (Henderson et al. 1996. Lansdorp et al. 1996). This finding demonstrates that some telomeres are shorter than others within a cell, and therefore, if all telomeres shorten at the same rate, the shortest telomere in a cell will become critically short before the rest. Third, experiments examining karyotype stability in human transformed cell lines have demonstrated the presence of dicentric chromosomes at a low intracellular frequency (~0–2 per cell) in transformed human cell lines that are approaching crisis (Counter et al. 1992). Fourth, and perhaps most convincingly, a detailed analysis of individual telomeres in primary mouse cells using a combination of spectral karyotyping (SKY) and quantitative telo-FISH (q-FISH) not only confirmed intracellular variability in the size of telomeres, but also showed that cells from two different groups of mice, both of which had critically short telomeres but differed considerably in average telomere length, had attenuated replicative life spans (Hemann et al. 2001). This experiment was done using a mouse strain deficient in the telomerase RNA component (mTR), which is essential for activity (see Chang, this volume). Specifically, when late generation (G6) mTR knockout (–/–) mice, which have an accelerated aging phenotype and in which the average telomere length is short and where most cells contain one or more chromosomal ends that lack a detectable telomere, were bred with mTR heterozygous (+/–) mice, in which the average telomere length is considerably longer, the F1 mTR–/– progeny had a similar accelerated aging phenotype as the G6 mTR–/– mice despite a marked difference in average telomere length between these two groups of mice. Furthermore, a normal healthy phenotype was restored to the F1 mTR+/– progeny from this cross, even though the average telomere length for these mice was essentially identical to that of the F1 mTR–/– mice. This latter observation is very important since it suggests that telomerase functions in a cell *in vivo* to selectively maintain or heal telomeres that are, or are becoming, critically short.

The fourth prediction, that telomeres will continue to shorten as transformed primary human cells continue to proliferate beyond the Hayflick limit, or M1 checkpoint, until most telomeres become critically short, has also been demonstrated. In the initial study by Counter et al., which confirmed this prediction, primary cultures of human embryonic kidney cells were transformed with SV40 large T antigen, and telomere length and chromosome stability were assessed by Southern analysis of terminal restriction length (TRF) and karyotype analysis respectively (Counter et al. 1992). It was found that these transformed cells were able to proliferate an additional finite number of population doublings beyond the M1 checkpoint before hitting crisis (the M2 checkpoint), during which time telomeres continued to shorten. More importantly, end-to-end chromosomal fusions, as well as signal free ends, both indicators of the loss of telomeric DNA from the ends of chromosomes, could be detected throughout the extended life span phase of these cells, and increased in frequency as these cells approached crisis.

Finally, the fifth prediction, that telomeres will cease shortening and telomerase will be activated in the rare immortalized clonal lines that emerge from crisis has also been validated. This prediction was also initially demonstrated by Counter et al., who found that 293 cells (a human embryonic kidney cell line immortalized with SV40 large T antigen) maintained telomere length over a large number of population doublings (> 200), and also contained high levels of telomerase activity (Counter et al. 1992). A more thorough analysis of telomerase activity in many different types of human tumor cell lines found that >90% of human tumors expressed readily detectable levels of telomerase (Kim et al. 1994). Interestingly, while most human tumor cell lines maintain telomere length at a relatively short length (close to the average telomere size for cells when they hit crisis, ~1–2 Kb), the average telomere size in some cell lines is quite long (Takada et al. 1992, Klingelhutz, et al. 1994). The reason why only some tumor cell lines extend and maintain telomeres at a relatively large size is still poorly understood.

4 Exceptions to the Hypothesis

There are a number of important exceptions to the telomere hypothesis of cell aging and immortalization that should be mentioned. First, there is at least one type of normal somatic cell that the author is aware of, urogenital epithelial cells, which has a finite replicative life span and yet does not exhibit attrition of telomeres during replicative aging (Belair et al. 1997). However, in this study, analysis of telomere length was only assessed for these cells over a relatively small number of population doublings, and calculation of the average telomere length was not performed, leaving open the possibility that the rate of telomere shortening is perhaps just slower in these cells as compared to what is typically observed for other types of primary human cells. Furthermore, the primary cell cultures analyzed in this study did express low levels of telomerase, which could also account for the lack of detectable telomere shortening. It is possible that these primary cells senesced due to inadequate culture conditions, as has been demonstrated for certain other types of primary human cell cultures (Ramirez et al. 2001).

Second, while most primary human cell strains as well as biopsies from different tissues from adults lack telomerase activity (Nam et al. 1994), there is one notable exception–the hematopoietic system (Broccoli et al. 1995, Chiu et al. 1996). Telomerase activity has been detected in most types of hematopoietic cells, including mature cells (Broccoli et al. 1995), progenitors (Chiu et al. 1996), and HSC (Chiu et al. 1996), although the data indicate that there is slightly less activity in HSC than progenitor cells (Chiu et al. 1996). Why telomeres still shorten during replicative aging of HSC and other types of hematopoietic cells is still a mystery; however, studies which have compared the rate of telomere shortening in HSC from telomerase-deficient and wild-type mice (Samper et al. 2002; Allsopp et al. 2003), as well as studies examining the size of telomeres in the dominant form of the syndrome Dyskeratosis Congenita (Vulliamy, 2001), in which the essential RNA component of telomerase is mutated, together suggest that the key function of telomerase in the hematopoietic system, which is a relatively rapidly proliferating tissue, is to prevent accelerated telomere loss and premature senescence. The fact that hematopoietic cells do contain active telomerase and yet lose telomeric DNA during replicative aging suggests that there are one or more additional levels of regulation of telomerase function present in these cells. In support of this is a recent study in which human HSC and progenitors from cord blood were transfected ex vivo with a retroviral Tert expression construct and then grown in vitro (Wang et al. 2005). Analysis of telomeres at different population doublings of the transfected and nontransfected cells showed that overexpression of Tert is not sufficient to maintain telomere length despite elevated levels of telomerase activity in the former. Furthermore, other studies have shown that telomerase is regulated not just by the expression of Tert in hematopoietic cells, but also by the intracellular localization of telomerase. Specifically, these studies have shown that Tert is localized primarily in the cytoplasm of resting hematopoietic cells, in this case, T lymphocytes, and is then shuttled into the nucleus following mitogenic stimulation (Liu et al. 2001, Allsopp et al. 2002). However, this finding in itself is not sufficient to explain why telomeres shorten in replicating hematopoietic cells. Experiments to determine exactly why telomerase does not complete telomere replication in hematopoietic cells will be an exciting area of future research.

Third, while telomere shortening does indeed appear to be the ultimate cause of replicative senescence in nearly all types of primary human cells, other mechanisms do exist that also induce cellular senescence. These include elevated levels of oxidative stress (von Zglinicki et al. 1995, Finkel and Holbrook 2000), exposure to UV or gamma irradiation (Finkel and Holbrook, 2000), and inappropriate mitogenic signaling, such as the constitutive activation of Ras (Ridley et al. 1988, Serrano et al. 1997). There is an interesting caveat to the first mechanism mentioned above, namely, oxidative stress is not only capable of inducing cell senescence, but in some instances can also cause a considerable increase in the rate of telomere shortening (von Zglinicki et al. 1995). This topic will be discussed in much greater detail in Chapter 3 in this book by von Zglinicki and colleagues. While there are a number of similarities among cells that have entered replicative senescence as a result of these different mechanisms, significant differences has also been revealed in studies that have compared the gene expression profile of these cells (Shelton et al. 1999).

Fourth, a small number of human tumor cell lines, both derived from tumor cell explants and also created by transforming primary cells, have been discovered; both are immortal and also lack telomerase activity (Bryan et al. 1995, Muntoni and Reddel 2005). Southern analysis of TRF length in these cells shows that the distribution of telomere length is much more heterogeneous than that observed in telomerase expressing tumor cell lines, but is nevertheless stable. These studies clearly demonstrate the existence of an alternative mechanism, referred to as ALT, for maintenance of telomere length in some human tumors. This ALT pathway operates via a DNA recombination type of mechanism to replicate telomeres, although the exact details of this mechanism have yet to be fully elucidated. While the existence of the ALT pathway has important implications for future telomerase-based therapies to treat cancer, ALT has not yet been detected in normal human somatic cells.

5 Mechanism of Telomere-Induced Senescence

Despite these exceptions to the telomere hypothesis of cell aging and immortalization, the fact that numerous studies have now shown, in many different types of human primary cell cultures, that ectopic expression of Tert is sufficient to restore telomerase activity, prevent telomere shortening, and endow the cells with replicative immortality (Harley 2001) clearly shows that telomere shortening is the primary mechanism accounting for the finite replicative life span of normal human cells, at least in vitro. Furthermore, as described above, there is now considerable evidence suggesting that the shortening of just one or perhaps a few telomeres below a critical length, as opposed to the erosion of all telomeres in a cell below an average length, is what triggers cell senescence. The question remains, how does telomere shortening below a critical length signal the induction of replicative senescence ?

To understand how this signal is initiated, it is important to consider why telomere shortening ultimately causes replicative senescence. It has been known for some time, from cytogenetic studies by Barbra McClintock in maize, that chromosomal ends in eukaryotic cells are capped by essential genetic elements (i.e., the telomere) (McClintock 1941). We now know that the telomere is a nucleoprotein complex composed of telomeric DNA, the sequence of which is (TTAGGG)n in mammals (Moyzis et al. 1988) and several different telomeric binding proteins (de Lange 2005; see Blasco, this volume). The telomeric binding proteins associate with the telomeric DNA to form a higher-ordered structure that caps the ends of chromosomes and protects the ends from erosion and aberrant end-to-end recombination (de Lange 2005; see Blasco, this volume). At the very end of the telomere there is a 3′ overhang of the G-rich (TTAGGG) strand (Chai et al. 2006), which allows the telomere end to loop back on itself and form a structure referred to as a t-loop (Griffith et al. 1999). Thus it is tempting to speculate that telomere shortening signals replicative senescence once there is no longer a sufficient amount of telomeric DNA to form a stable t-loop; however, this has yet to be directly demonstrated. Therefore, other possible signaling mechanisms, including the loss of the 3′ overhang, or inability of the telomeric binding proteins to form a stable nucleoprotein complex with a limited amount of telomeric DNA, cannot be ruled out. Nevertheless, there is considerable evidence suggesting that normal somatic cells recognize a critically shortened telomere as damaged DNA, akin to the end of a broken chromosome, for reasons described below.

We have known for over 20 years that the tumor suppressors p53 and the retinoblastoma gene product, Rb, are important for arresting the cell cycle in senescent cells. Early studies showed that DNA tumor viruses that prevent activation of p53 or Rb allow primary human cells to proliferate past the normal Hayflick limit, or M1 checkpoint (Ide et al. 1983, Shay et al. 1991). Furthermore, studies have shown that both p53 and Rb exist in senescent cells in their activated form, phosphorylated (Webley et al. 2000) and nonphosphorylated (Stein et al. 1990), respectively, whereas in mitotically active early passage cells, in S phase, p53 and Rb are present, but in the inactive state. Subsequent studies have shown that in some cells, inhibition of p53 with dominant negative p53 mutants (Bond et al. 1994) or with antisense oligonucleotides that target p53 mRNA (Hara et al. 1991) is sufficient to allow cells to bypass the M1 checkpoint, whereas in other types of cells, inhibition of both p53 and Rb is required (Beausejour et al. 2003). This latter study shows that the genetic pathway for replicative senescence is not exactly the same for all types of human somatic cells.

How does activation of p53 and Rb cause cell cycle arrest? Most senescent primary human cells are blocked at the G1/S phase of the cell cycle, because the cyclin-dependent kinases (CDK) that are essential for entry into S phase are inactive (Fukami et al. 1995) or expressed at low levels (Afshari et al. 1993) (Fig. 2.3). A number of different CDK inhibitors (CKI) have now been identified, including p21, p16, p27, and p57. It is now well established that activation of p53, in different types of damaged or stressed cells, including senescent cells, allows the enhanced

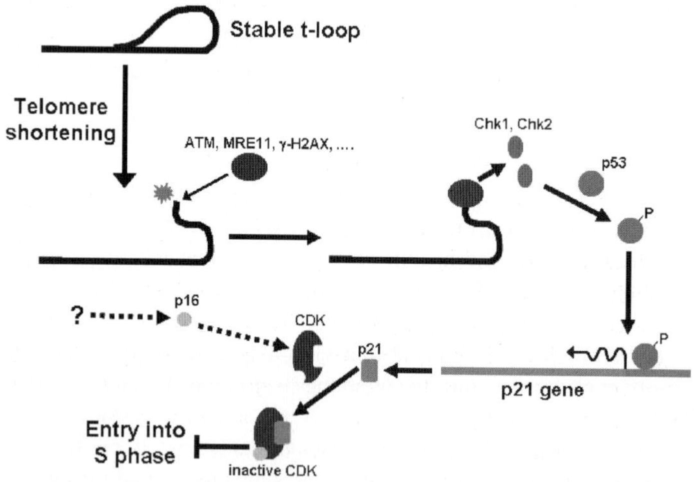

Fig. 2.3 The mechanism of telomere-induced replicative senescence. A stable telomere is shown in a t-loop confirmation, in which the end of the chromosome is protected. After a finite number of replication events, the telomere can no longer maintain a stable structure, represented as unfolding of the loop, and consequently the end of the telomere is recognized by DNA damage response sensors. See text for further details (*See Color Plate*)

transcription of the p21 gene (Herbig and Sedivy 2006) (Fig. 2.3). It has also been shown that in the absence of functional p53, p21 is not induced during culture of HDF (Dulic et al. 2000). Furthermore, an elegant study by Brown et al. has shown that knocking out the p21 gene in HDFs is sufficient to enable these cells to bypass the M1 checkpoint, and allow them to continue to proliferate to crisis (M2 checkpoint) (Brown et al. 1997). Regarding Rb, in most proliferating cells, it is a downstream target of CDKs, which act to phosphorylate and inactivate Rb prior to entry into S phase. The main function of Rb is to sequester the transcription factor E2F, which is responsible for transactivation of a number of genes that are essential for entry into S phase (Fig. 2.3). In senescent cells, Rb is hypophosphorylated (Stein et al. 1990), and therefore E2F activity remains inhibited (Dimri et al. 1994), thereby blocking entry into S phase. Thus in many, but not all, primary human cells, Rb acts downstream of p53 to block entry into S phase in senescent cells.

In addition to p21, another CKI that has been shown to play an important role in maintaining cell cycle arrest in at least some types of senescent cells is $p16^{ink4a}$ (Alcorta et al. 1996). Levels of $p16^{ink4a}$ have been shown to be elevated in senescent human cells (Alcorta et al. 1996). However, in contrast to p21, the kinetics of the appearance of $p16^{ink4a}$ following telomere-induced senescence (rapid; 0–48 hours) is different than that observed when cells are exposed to DNA damage inducing agents (slow; 1–2 weeks) (Sedivy 2006). Therefore, the present view regarding the role of $p16^{ink4a}$ in replicative senescence is that it is activated indirectly, via a DNA damage-independent mechanism, following the critical shortening of telomeres, and functions primarily to help maintain, not induce, the senescent state. The exact mechanism by which $p16^{ink4a}$ is activated in senescent cells remains to be fully elucidated; however, recent research suggests that one mechanism may be via the p38 MAP kinase pathway, allowing the ETS-mediated transcription of the $p16^{ink4a}$ gene (Ohtani et al. 2001), and/or by inhibition of Polycomb group proteins, in particular Bmi1, since repression of Bmi1 by shRNA targeting has been shown to cause upregulation of $p16^{ink4a}$ and cause premature senescence of HDF (Itahana et al. 2003; see Sharpless, this volume). In any event, transcriptional upregulation of p21, mediated by active p53, appears to be the primary means by which cell cycle arrest is initiated by telomere-induced senescence because, as has been shown for some types of primary human cell cultures, $p16^{ink4a}$ is not always present at replicative senescence (Beausejour et al. 2003), knocking down $p16^{ink4a}$ has not been shown to lead to a significantly extended replicative life span (Bond et al. 2004), and some cell strains in which the $p16^{ink4a}$ gene has been mutated still undergo normal replicative senescence (Brookes et al. 2002).

A number of recent studies have also demonstrated that, at the advent of telomere-induced senescence of primary human cells, critically short telomeres are initially sensed by the DNA damage response pathway (Fig. 3.3; see Chang, this volume, and Rudolph, this volume). The DNA damage sensor encoded by the ataxia-telangiectasia mutated (ATM) gene appears to play a particularly important role in initiating the DNA damage response signal, since HDF which lack functional ATM are able to bypass the normal G1/S phase checkpoint at senescence, but then subsequently undergo cell cycle arrest via activation of a G2 phase DNA damage response pathway mediated by the sensor ATM and Rad3-related (ATR) kinase (Herbig et al. 2004). Other elegant studies have shown, utilizing immunofluorescence and chromatin immunoprecipitation (CHIP) assays, that a number of DNA response factors, including ATM, the histone variant gamma-H2AX, MRE11, the p53 binding protein 53BP1, and the checkpoint kinases chk1 and chk2 co-localize with telomeric DNA to form foci in senescent HDF (d'Adda di Fagagna et al. 2003, Herbig et al. 2004). A small number of these foci (1–10) are present in nearly all cells that have undergone telomere-induced replicative senescence, whereas only a very small fraction of early passage cells have detectable foci. These foci are likely directly responsible for activation of p53 and transcriptional activation of the p21 gene in telomere-induced senescence, because p53 is a primary target of these DNA response factors in the presence of damaged DNA (e.g., broken chromosomes). Moreover, targeted inhibition of the chk2 kinase, an important downstream target of ATM, prevents p53 activation and extends the replicative life span of HDF (Gire et al. 2004).

6 Telomere-Induced Replicative Senescence and Organismal Aging

A major question that the biology of aging field currently faces is to what extent does telomere-induced replicative senescence contribute to organismal aging? This question remains a mystery at present, in part because of the technical difficulty in assessing the frequency and localization of senescent cells in the body.

However, an exciting new study by Choudhury et al. has shown that the deletion of the p21 gene in telomerase-deficient mice prolongs the life span of these mice and improves the proliferative capacity of intestinal epithelial and hematopoietic stem cells (2007; see also Gutierrez and Ju, this volume). This study strongly emphasizes the potential importance of telomere-induced replicative senescence to in vivo aging in humans. Nevertheless, it is still uncertain as to whether senescent cells accumulate in the body during aging in humans. The development of markers, such as SA-βgal (Dimri et al. 1995), to allow detection of senescent cells in situ should help address this issue. In fact, the frequency of SA-βgal positive cells in humans has been shown to accumulate in various tissues during aging (Dimri et al. 1995). However, SA-βgal does not distinguish between different types of senescent cells, and therefore in this study, it is unclear whether the senescent state was telomere-induced, or brought about prematurely by other stressors, such as oxidative or genotoxic stress. It will be important for future efforts in the development of new markers for detection of senescent cells in situ to distinguish between telomere-induced senescence and other telomere-independent mechanisms that initiate cell senescence.

Another way to assess to what degree telomere-induced senescence contributes to organismal aging is to measure the replicative capacity of primary human cell cultures established from donors at various ages. A number of studies have shown that telomeres shorten in different tissues during organismal aging, and therefore replicative capacity would be predicted to be diminished for primary strains that are derived from these tissues. Indeed, replicative capacity has been shown to decline as a function of donor age for a number of primary human cell strains, including HDF (Schneider and Mitsui, 1976, Allsopp et al. 1992), smooth muscle cells (Bierman 1978), RPE (Tassin et al. 1979), muscle satellite cells (Decary et al. 1997), and lymphocytes (Tice et al. 1979). However, a more recent study by Pignolo et al. has challenged these observations, since they observed no correlation between donor age and replicative capacity for HDF (Pignolo et al. 1998). Thus it is possible the attrition of telomeres as a function of donor age that has been observed in skin biopsies (Lindsey et al. 1991) may be accounted for by the proliferation of cells other than HDF, such as keratinocytes, melanocytes, and epithelial cells. Further work needs to be done to resolve the discrepancy between the study by Pignolo et al. and previous studies by others. Analysis of the correlation between donor age and replicative capacity of human somatic cells is certainly an open area for future research. Of particular interest will be cell and tissue types for which an age-related decline in telomere length has already been demonstrated in vivo, for example, the hematopoietic system, including hematopoietic stem cells (Hastie et al. 1990, Vaziri et al. 1994) and regions of the vasculature exposed to high levels of hemodynamic stress (Chang and Harley 1995).

Regarding the role of telomere-induced senescence in organismal aging, it is important to also consider the relationship between telomere-induced senescence and cancer, given the high proportion of telomerase-positive tumors and the increasing risk for cancer during aging (see Rudolph, this volume). There are a number of reasons why telomere-induced replicative senescence may be viewed as a mechanism to protect against cancer. First, the most obvious one is that if a cell senesces, then it will not be able to proliferate further and develop into a tumor. Second, studies with cells that fail to induce the M1 checkpoint, either cells from telomerase-deficient mice (Artandi et al. 2001) or transformed primary human cell cultures (Counter et al. 1992), have shown that when telomeres are allowed to get very short, telomere dysfunction results, notably end-to-end chromosomal fusions, which can drive karyotype instability, specifically, repetitive chromosome breakage-fusion-bridge cycles, which in turn increases the probability of developing a mutation that will predispose the cell to a malignant state. Third, the tumor suppressors p53 and Rb are required for inducing cell senescence, including telomere-induced replicative senescence, and are also mutated in the majority of human tumors. Though all of these arguments in support of telomere-induced senescence playing a protective role against cancer are now well established and conceptually sound, they are all intrinsic to the senescent cell per se. Studies on the extrinsic affects of senescent cells in vitro and in vivo point toward possible cancer-promoting effects of senescent cells. For example, recent studies have shown that senescent HDF secrete epithelial growth factors, which in turn could cause aberrant mitogenic stimulation of nearby epithelial cells (Parrinello et al. 2005). In addition, the plethora of matrix-modifying proteins that are secreted from senescent cells are thought to cause abnormal remodeling or restructuring of the surrounding tissue, which in turn could promote or allow the metastasis of cancer cells. To what degree telomere-induced senescence is cancer promoting or cancer inhibiting in humans remains to be answered.

Finally, if future research demonstrates that telomere-induced replicative senescence contributes to organismal aging, whether in specific tissues or *in toto*, or in specific disease states or the elderly in general, then it will be of interest to develop telomerase-based mechanisms to prevent telomere shortening. Indeed, some biotech industries are already working in this area. The future development of telomerase-based therapies to prevent replicative senescence may be aided by gaining a better understanding of how telomerase is regulated in different cells and tissues of the body. Most cells of the human body contain little or no active telomerase, with the notable exception of the hematopoietic cells and the germ line. Human germ cells, at least male germ cells, not only express telomerase throughout life, but also maintain telomeres during organismal aging. Thus understanding how telomerase is regulated in germ cells, at the cell and molecular levels, may lead to the development of novel methods to reactivate telomerase in cells that are approaching telomere-induced senescence, and thereby extend the replicative life span.

References

Afshari CA, Vojta PJ, Annab LA, Futreal PA, Willard TB, Barrett JC. (1993) Investigation of the role of G1/S cell cycle mediators in cellular senescence. Exp Cell Res 209: 231–7.

Alcorta DA, Xiong Y, Phelps D, Hannon G, Beach D, Barrett JC (1996) Involvement of the cyclin-dependent kinase inhibitor p16 (INK4a) in replicative senescence of normal human fibroblasts. Proc Natl Acad Sci USA 93: 13742–7.

Allsopp RC, Vaziri H, Patterson C, Goldstein S, Younglai EV, Futcher AB, Greider CW, Harley C (1992) Telomere length predicts replicative capacity of human fibroblasts. Proc Natl Acad Sci USA 89: 10114–8.

Allsopp RC, Chang E, Kashefi-Aazam M, Rogaev EI, Piatyszek MA, Shay JW, Harley CB (1995) Telomere shortening is associated with cell division *in vitro* and *in vivo*. Exp Cell Res 220: 194–200.

Allsopp RC, Cheshier S, Weissman IL (2002) Telomerase activation and rejuvenation of telomere length in stimulated T cells derived from serially transplanted hematopoietic stem cells. J Exp Med 196: 1427–33.

Allsopp RC, Morin GB, DePinho R, Harley CB, Weissman IL (2003) Telomerase is required to slow telomere shortening and extend replicative lifespan of HSCs during serial transplantation. Blood 102: 517–20.

Artandi SE, Chnag S, Lee SL, Alson S, Gottlieb GJ, Chin L, DePinho RA (2000) Telomere dysfunction promotes non-reciprocal translocations and epithelial cancers in mice. Nature 406(6796): 641–454.

Bandyopadhyay D, Timchenko N, Suwa T, Hornsby PJ, Campisi J, Medrano EE (2001) The human melanocyte: a model system to study the complexity of cellular aging and transformation in non-fibroblastic cells. Exp Gerontol 36: 1265–75.

Benn P (1976) Specific chromosome aberrations in senescent fibroblast cell lines derived from human embryos. Am J Hum Genet 28: 465–73.

Beausejour CM, Krtolica A, Galimi F, Narita M, Lowe SW, Yaswen P, Campisi J (2003) Reversal of human cellular senescence: roles of the p53 and p16 pathways. EMBO J 22: 4212–22.

Bierman, J (1978) The effect of donor age on the *in vitro* life span of cultured human arterial smooth-muscle cells. In vitro 14: 951–5.

Bodnar AG, Ouellette M, Frolkis M, Holt SE, Chiu CP, Morin GB, Harley CB, Shay JW, Lichtsteiner S, Wright WE (1998) Extension of life-span by introduction of telomerase into normal human cells. Science 279: 349–52.

Bond JA, Wyllie FS, Wynford-Thomas D (1994) Escape from senescence in human diploid fibroblasts induced directly by mutant p53. Oncogene 9: 1885–9.

Bond J, Jones C, Haughton M, DeMicco C, Kipling D, Wynford-Thomas D (2004) Direct evidence from siRNA-directed "knock down" that p16(INK4a) is required for human fibroblast senescence and for limiting ras-induced epithelial cell proliferation. Exp Cell Res 292: 151–6.

Broccoli D, Young JW, de Lange T (1995) Telomerase activity in normal and malignant hematopoietic cells. Proc Natl Acad Sci USA 92: 9082–6.

Brookes S, Rowe J, Ruas M, Llanos S, Clark PA, Lomax M, James MC, Vatcheva R, Bates S, Vousden KH, Parry D, Gruis N, Smit N, Bergman W, Peters G (2002) INK4a-deficient human diploid fibroblasts are resistant to RAS-induced senescence. EMBO J 21: 2936–45.

Brown JP, Wei W, Sedivy JM (1997) Bypass of senescence after disruption of p21CIP1/WAF1 gene in normal diploid human fibroblasts. Science 277: 831–4.

Bryan TM, Englezou A, Gupta J, Bacchetti S, Reddel RR (1995) Telomere elongation in immortal human cells without detectable telomerase activity. EMBO J 14: 4240–8.

Chang E, Harley, H (1995) Telomere length and replicative aging in human vascular tissues. Proc Natl Acad Sci USA 92: 11190–4.

Chai W, Du Q, Shay JW, Wright WE (2006) Human telomeres have different overhang sizes at leading versus lagging strands. Mol Cell 21 427–35.

Chiu CP, Dragowska W, Kim NW, Vaziri H, Yui J, Thomas TE, Harley CB, Lansdorp PM (1996) Differential expression of telomerase activity in hematopoietic progenitors from adult human bone marrow. Stem Cells 14: 239–48.

Choudhury AR, Ju Z, Djojosubroto MW, Schienke A, Lechel A, Schaetzlein S, Jiang H, Stepczynska A, Wang C, Buer J, Lee HW, von Zglinicki T, Ganser A, Schirmacher P, Nakauchi H, Rudolph KL (2007) Cdkn1a deletion improves stem cell function and lifespan of mice with dysfunctional telomeres without accelerating cancer formation. Nat Genet 39: 99–105.

Cooke H, Smith B (1986) Variability at the telomeres of the human X/Y pseudoautosomal region. Cold Spring Harb Symp Quant Biol 51: 213–9.

Counter CM, Avilion AA, LeFeuvre CE, Stewart NG, Greider CW, Harley CB, Bacchetti S (1992) Telomere shortening associated with chromosome instability is arrested in immortal cells which express telomerase activity. EMBO J 11: 1921–9.

Counter CM, Hahn WC, Wei W, Caddle SD, Beijersbergen RL, Lansdorp PM, Sedivy JM, Weinberg RA (1998) Dissociation among *in vitro* telomerase activity, telomere maintenance, and cellular immortalization. Proc Natl Acad Sci USA 95: 14723–8.

Cristofalo VJ, Allen RG, Pignolo RJ, Martin BG, Beck JC (1998) Relationship between donor age and the replicative lifespan of human cells in culture: a reevaluation. Proc Natl Acad Sci USA 95: 10614–9.

Cristofalo VJ, Lorenzini A, Allen RG, Torres C, Tresini M (2004) Replicative senescence: a critical review. Mech Ageing Dev 125: 827–48.

d'Adda di Fagagna F, Reaper PM, Clay-Farrace L, Fiegler H, Carr P, Von Zglinicki T, Saretzki G, Carter NP, Jackson SP (2003) A DNA damage checkpoint response in telomere-initiated senescence. Nature 426: 194–8.

De Lange T, Shiue L, Myers RM, Cox DR, Naylor SL, Killery AM, Varmus HE (1990) Structure and variability of human chromosome ends. Mol Cell Biol 10: 518–27.

De Lange T (2005) Shelterin: the protein complex that shapes and safeguards human telomeres. Genes Dev 19: 2100–10.

Decary S, Mouly V, Hamida CB, Sautet A, Barbet JP, Butler-Browne GS (1997) Replicative potential and telomere length in human skeletal muscle: implications for satellite cell-mediated gene therapy. Hum Gene Ther 8: 1429–38.

Dimri GP, Hara E, Campisi J (1994) Regulation of two E2F-related genes in presenescent and senescent human fibroblasts. J Biol Chem 269: 16180–6.

Dimri GP, Lee X, Basile G, Acosta M, Scott G, Roskelley C, Medrano EE, Linskens M, Rubelj I, Pereira- Smith O, et al. (1995) A biomarker that identifies senescent human cells in culture and in aging skin in vivo. Proc Natl Acad Sci USA 92: 9363–7.

Dulic V, Beney GE, Frebourg G, Drullinger LF, Stein GH (2000) Uncoupling between phenotypic senescence and cell cycle arrest in aging p21-deficient fibroblasts. Mol Cell Biol 20: 6741–54.

Ferbeyre G, de Stanchina E, Querido E, Baptiste N, Prives C, Lowe SW (2000) PML is induced by oncogenic ras and promotes premature senescence. Genes Dev 14: 2015–27.

Finkel, T, Holbrook, N (2000) Oxidants, oxidative stress and the biology of ageing. Nature 408: 239–47.

Fukami J, Anno K, Ueda K, Takahashi T, Ide T (1995) Enhanced expression of cycling D1 in senescent human fibroblasts. Mech Ageing Dev 81: 139–57.

Gire V, Roux P, Wynford-Thomas D, Brondello JM, Dulic V (2004) DNA damage checkpoint kinase Chk2 triggers replicative senescence. EMBO J 23: 2554–63.

Griffith JD, Comeau L, Rosenfield S, Stansel RM, Bianchi A, Moss H, de Lange T (1999) Mammalian telomeres end in a large duplex loop. Cell 97: 503–14.

Goldstein, S (1990) Replicative senescence: the human fibroblast comes of age. Science 249: 1129–1134.

Gorman, S, Cristofalo, V (1985) Reinitiation of cellular DNA synthesis in BrdU-selected nondividing senescent WI-38 cells by simian virus 40 infection. J Cell Physiol 125: 122–6.

Greider C, Blackburn, E (1985) Identification of a specific telomere terminal transferase activity in Tetrahymena extracts. Cell 43: 405–13.

Hara E, Tsurui H, Shinozaki A, Nakada S, Oda K (1991) Cooperative effect of antisense-Rb and antisense-p53 oligomers on the extension of life span in human diploid fibroblasts, TIG-1. Biochem Biophys Res Commun 179: 528–34.

Harley CB, Futcher AB, Greider CW (1990) Telomeres shorten during ageing of human fibroblasts. Nature 345: 458–60.

Harley, C (1991) Telomere loss: mitotic clock or genetic time bomb? Mutat Res 256: 271–82.

Harley CB, Vaziri H, Counter CM, Allsopp RC (1992) The telomere hypothesis of cellular aging. Exp Gerontol 27: 375–82.

Harley C (2001) Telomerase is not an oncogene. Oncogen*e* 21: 494–502.

Hastie ND, Dempster M, Dunlop MG, Thompson AM, Green DK, Allshire RC (1990) Telomere reduction in human colorectal carcinoma and with ageing. Nature 346: 866–8.

Hayflick, L, Moorhead, P (1961) The serial cultivation of human diploid cell strains. Exp Cell Res 25: 585–621.

Hemann MT, Strong MA, Hao LY, Greider CW (2001) The shortest telomere, not average telomere length, is critical for cell viability and chromosome stability. Cell 107: 67–77.

Henderson S, Allsopp R, Spector D, Wang SS, Harley C (1996) In situ analysis of changes in telomere size during replicative aging and cell transformation. J Cell Biol 134: 1–12.

Herbig U, Jobling WA, Chen BP, Chen DJ, Sedivy JM (2004) Telomere shortening triggers senescence of human cells through a pathway involving ATM, p53, and p21(CIP1), but not p16(INK4a). Mol Cell 14: 501–13.

Herbig U, Sedivy J. (2006) Regulation of growth arrest in senescence: telomere damage is not the end of the story. Mech Ageing Dev 127: 16–24.

Ide T, Tsuji Y, Ishibashi S, Mitsui Y. (1983). Reinitiation of host DNA synthesis in senescent human diploid cells by infection with Simian virus 40. *Exp Cell Res*. **143**, 343–9.

Itahana K, Zou Y, Itahana Y, Martinez JL, Beausejour C, Jacobs JJ, Van Lohuizen M, Band V, Campisi J, Dimri GP (2003) Control of the replicative life span of human fibroblasts by p16 and the polycomb protein Bmi-1. Mol Cell Biol 23: 389–401.

Kim NW, Piatyszek MA, Prowse KR, Harley CB, West MD, Ho PL, Coviello GM, Wright WE, Weinrich SL, Shay JW (1994) Specific association of human telomerase activity with immortal cells and cancer. Science 66: 2011–5.

Klingelhutz AJ, Barber SA, Smith PP, Dyer K, McDougall JK (1994) Restoration of telomeres in human papillomavirus-immortalized human anogenital epithelial cells. Mol Cell Biol 14: 961–9.

Kortlever RM, Higgins PJ, Bernards R (2006) Plasminogen activator inhibitor-1 is a critical downstream target of p53 in the induction of replicative senescence. Nat Cell Biol 8: 877–84

Lansdorp PM, Verwoerd NP, van de Rijke FM, Dragowska V, Little MT, Dirks RW, Raap AK, Tanke HJ (1996) Heterogeneity in telomere length of human chromosomes. Hum Mol Genet 5: 685–91.

Liu K, Hodes RJ, Weng NP (2001) Cutting edge: telomerase activation in human T lymphocytes does not require increase in telomerase reverse transcriptase (hTERT) protein but is associated with hTERT phosphorylation and nuclear translocation. J Immunol 166: 4826–30.

Lindsey J, McGill NI, Lindsey LA, Green DK, Cooke HJ (1991) *In vivo* loss of telomeric repeats with age in humans. Mutat Res 256: 45–8.

Matsumura T, Zerrudo Z, Hayflick L (1979) Senescent human diploid cells in culture: survival, DNA synthesis and morphology. J Gerontol 34: 328–34.

McClintock B (1941) The stability of broken ends of chromosomes in Zea mays. Genetics 26: 234–282.

Meyerson M, Counter CM, Eaton EN, Ellisen LW, Steiner P, Caddle SD, Ziaugra L, Beijersbergen RL, Davidoff MJ, Liu Q, Bacchetti S, Haber DA, Weinberg RA (1998) hEST2, the putative human telomerase catalytic subunit gene, is up-regulated in tumor cells and during immortalization. Cell 22: 785–95.

Minamino T, Mitsialis SA, Kourembanas S (2001) Hypoxia extends the life span of vascular smooth muscle cells through telomerase activation. Mol Cell Biol 21: 3336–42.

Morin G (1989) The human telomere terminal transferase enzyme is a ribonucleoprotein that synthesizes TTAGGG repeats. Cell 59: 521–9.

Moyzis RK, Buckingham JM, Cram LS, Dani M, Deaven LL, Jones MD, Meyne J, Ratliff RL, Wu JR (1988) A highly conserved repetitive DNA sequence, (TTAGGG)n, present at the telomeres of human chromosomes. Proc Natl Acad Sci USA 85: 6622–6.

Mueller SN, Rosen EM, Levine EM (1980) Cellular senescence in a cloned strain of bovine fetal aortic endothelial cells. Science 207: 889–91.

Muntoni A, Reddel R (2005) The first molecular details of ALT in human tumor cells. Hum Mol Genet 14: R191–6.

Nakamura TM, Morin GB, Chapman KB, Weinrich SL, Andrews WH, Lingner J, Harley CB, Cech TR (1997) Telomerase catalytic subunit homologs from fission yeast and human. Science 277: 955–9.

Noda A, Ning Y, Venable SF, Pereira-Smith OM, Smith JR (1994) Cloning of senescent cell-derived inhibitors of DNA synthesis using an expression screen. Exp Cell Res 211: 90–8.

Norwood TH, Pendergrass WR, Sprague CA, Martin GM (1974) Dominance of the senescent phenotype in heterokaryons between replicative and post-replicative human fibroblast-like cells. Proc Natl Acad Sci SA 71: 2231–5.

Ohtani N, Zebedee Z, Huot TJ, Stinson JA, Sugimoto M, Ohashi Y, Sharrocks AD, Peters G, Hara E (2001) Opposing effects of Ets and Id proteins on p16INK4a expression during cellular senescence. Nature 409: 1067–70.

Olovnikov A (1973) A theory of marginotomy. The incomplete copying of template margin in enzymic synthesis of polynucleotides and biological significance of the phenomenon. J Theor Biol 14: 181–90.

Parrinello S, Coppe JP, Krtolica A, Campisi J (2005) Stromal-epithelial interactions in aging and cancer: senescent fibroblasts alter epithelial cell differentiation. J Cell Sci 118: 485–96.

Pereira-Smith O, Smith J (1982) Phenotype of low proliferative potential is dominant in hybrids of normal human fibroblasts. Somatic Cell Genet 8:731–42.

Ramirez RD, Morales CP, Herbert BS, Rohde JM, Passons C, Shay JW, Wright WE (2001) Putative telomere-independent mechanisms of replicative aging reflect inadequate growth conditions. Genes Dev 15: 398–403.

Rheinwald, J, Green, H (1975) Serial cultivation of strains of human epidermal keratinocytes: the formation of keratinizing colonies from single cells. Cell 6: 331–43.

Ridley AJ, Paterson HF, Noble M, Land H (1988) Ras-mediated cell cycle arrest is altered by nuclear oncogenes to induce Schwann cell transformation. EMBO J 7: 1635–45.

Romanov SR, Kozakiewicz BK, Holst CR, Stampfer MR, Haupt LM, Tlsty TD (2001) Normal human mammary epithelial cells spontaneously escape senescence and acquire genomic changes. Nature 409: 633–7.

Rudolph KL, Chang S, Lee HW, Blasco M, Gottlieb GJ, Greider C, DePinho RA (1999) Longevity, stress response, and cancer in aging telomerase- deficient mice. Cell 96: 701–12.

Samper E, Fernandez P, Eguia R, Martin-Rivera L, Bernad A, Blasco MA, Aracil M (2002) Long-term repopulating ability of telomerase-deficient murine hematopoietic stem cells. Blood 99: 2767–75.

Schneider E, Mitsui Y (1976) The relationship between *in vitro* cellular aging and *in vivo* human age. Proc Natl Acad Sci USA 73: 3584–8.

Serrano M, Lin AW, McCurrach ME, Beach D, Lowe SW (1997) Oncogenic ras provokes premature cell senescence associated with accumulation of p53 and p16INK4a. Cell 88: 593–602.

Shay J, Wright W (1989) Quantitation of the frequency of immortalization of normal human diploid fibroblasts by SV40 large T-antigen. Exp Cell Res 184: 109–18.

Shay JW, Pereira-Smith OM, Wright WE (1991) A role for both RB and p53 in the regulation of human cellular senescence. Exp Cell Res 196: 33–9.

Shelton DN, Chang E, Whittier PS, Choi D, Funk WD (1999) Microarray analysis of replicative senescence. *I* 9: 939–45.

Sherwood SW, Rush D, Ellsworth JL, Schimke RT (1988) Defining cellular senescence in IMR-90 cells: a flow cytometric analysis. Proc Natl Acad Sci USA 85: 9086–90.

Stampfer MR, Bodnar A, Garbe J, Wong M, Pan A, Villeponteau B, Yaswen P (1997) Gradual phenotypic conversion associated with immortalization of cultured human mammary epithelial cells. Mol Biol Cell 8: 2391–405.

Stein GH, Beeson M, Gordon L (1990) Failure to phosphorylate the retinoblastoma gene product in senescent human fibroblasts. Science 249: 666–9.

Takada T, Hayashi T, Arakawa M, Kominami R (1992) Telomere elongation frequently observed during tumor metastasis. Jpn J Cancer Res 83: 1124–7.

Tassin J, Malaise E, Courtois Y (1979) Human lens cells have an *in vitro* proliferative capacity inversely proportional to the donor age. Exp Cell Res 123: 388–92.

Thomson JA, Itskovitz-Eldor J, Shapiro SS, Waknitz MA, Swiergiel JJ, Marshall VS, Jones JM (1998) Embryonic stem cell lines derived from human blastocysts. Science 282: 1145–7.

Tice RR, Schneider EL, Kram D, Thorne P (1979) Cytokinetic analysis of the impaired proliferative response of peripheral lymphocytes from aged humans to phytohemagglutinin. J Exp Med 149: 1029–41.

Vaziri H, Dragowska W, Allsopp RC, Thomas TE, Harley CB, Lansdorp PM (1994) Evidence for a mitotic clock in human hematopoietic stem cells: loss of telomeric DNA with age. Proc Natl Acad Sci USA. 91: 9857–60.

Vulliamy T, Marrone A, Goldman F, Dearlove A, Bessler M, Mason PJ, Dokal I (2001) The RNA component of telomerase is mutated in autosomal dominant dyskeratosis congenita. Nature 413: 432–5.

Wang JC, Warner JK, Erdmann N, Lansdorp PM, Harrington L, Dick JE (2005) Dissociation of telomerase activity and telomere length maintenance in primitive human hematopoietic cells. Proc Natl Acad Sci USA 102: 14398–403.

Watson, J (1972) Origin of concatemeric T7 DNA. Nat New Biol. 239: 197–20.

Webley K, Bond JA, Jones CJ, Blaydes JP, Craig A, Hupp T, Wynford-Thomas D (2000) Posttranslational modifications of p53 in replicative senescence overlapping but distinct from those induced by DNA damage. Mol Cell Biol 20: 2803–8.

Weng NP, Levine BL, June CH, Hodes RJ (1995) Human naive and memory T lymphocytes differ in telomeric length and replicative potential. Proc Natl Acad Sci USA 92: 11091–4.

Wistrom C, Villeponteau B (1992) Cloning and expression of SAG: a novel marker of cellular senescence. Exp Cell Res 199: 355–62.

Wright WE, Piatyszek MA, Rainey WE, Byrd W, Shay JW (1995) Telomerase activity in human germline and embryonic tissues and cells. Dev Genet 18: 173–9.

von Zglinicki T, Saretzki G, Docke W, Lotze C (1995). Mild hyperoxia shortens telomeres and inhibits proliferation of fibroblasts: a model for senescence? Exp Cell Res 220: 186–93.

Chapter 3
Telomeres, Senescence, Oxidative Stress, and Heterogeneity

João F. Passos, Glyn Nelson, and Thomas von Zglinicki(✉)

Abstract Life span heterogeneity is a hallmark of the ageing process. It is also a characteristic feature of telomere-dependent replicative senescence. We review here evidence showing that telomere shortening is heterogeneous between telomeres and between cells, and relate this to cell-to-cell variation in mitochondrial function and production of reactive oxygen species (ROS). Telomere shortening is to a large extent governed by ROS-mediated telomeric DNA damage, and we show here that apparently stochastic variation in mitochondrial ROS production can account for vast differences in replicative potential between individual cell lineages.

1 Heterogeneity: The Hallmark of Ageing

Ageing is characterized by random accumulation of unrepaired cellular and molecular damage. The mechanisms involved are inherently stochastic, i.e., driven by chance (Kirkwood 2005). In fact, several studies have demonstrated stochastic heterogeneity as an important feature of the ageing process:

Examining the effect of ageing on cell integrity in different tissues of *Caenorhabditis elegans* by electron microscopy (Herndon et al. 2002), it was found that stochastic factors are clearly involved in the ageing process and that different cell types deteriorate at different rates. Apart from heterogeneity on a tissue and cellular level, the wide interindividual variation in life span of *C.elegans* nematodes, even in isogenic populations under very homogeneous environmental conditions, indicates the intrinsic randomness of the ageing process (Kirkwood and Finch 2002). This heterogeneity in life span variation of isogenic *C. elegans* populations has been attributed to chance variation in expression of gene *hsp-16.2* (Rea et al. 2005).

Thomas von Zglinicki
Henry Wellcome Laboratory for Biogerontology Research, Institute for Ageing and Health, Center for Integrated Systems Biology of Ageing and Nutrition, University of Newcastle upon Tyne, NE4 6BE, United Kingdom.
e-mail: T.Vonzglinicki@newcastle.ac.uk

K.L. Rudolph (ed.), *Telomeres and Telomerase in Ageing, Disease, and Cancer.*

It has been shown by dissociating single cardiomyocytes from fresh heart samples of both young and old mice that gene expression heterogeneity increased significantly with age (Bahar et al. 2006). In an attempt to attribute a cause to this cell-to-cell variation, the authors tested the involvement of oxidative stress in the process by exposing mouse embryonic fibroblasts to hydrogen peroxide. They found that gene expression variation increased after hydrogen peroxide treatment.

The free radical theory of ageing provided a first approach toward mechanistic insight into the concept of random damage in ageing. Harman's work (Harman 1956) suggested that the random nature of free radical damage could explain the marked heterogeneity in the decline of cells, tissues, and individuals in ageing. Although a large amount of data has been gathered in general support of this idea, the majority of it is still merely correlative and fails to rigidly establish causal connections. So far, mice studies have proven to be inconclusive. For instance, mice carrying a heterozygous deletion of the mitochondrial superoxide dismutase, an enzyme able to convert superoxide anion to water and hydrogen peroxide, showed indications of increased oxidative stress and high cancer incidence but not accelerated ageing (Van Remmen et al. 2003). However, targeting catalase to mitochondria increased the life span of transgenic mice (Schriner et al. 2005) and overexpression of human thioredoxin (TRX) in mice led to increased resistance to oxidative stress and extended median and maximum life spans (Mitsui et al. 2002).

In the model of replicative senescence, i.e., the irreversible loss of division potential of somatic cells after a more or less constant number of cell divisions, the role of oxidative stress is under intense debate. Recently, reports have suggested that oxidative stress might have a role as a tumor suppressor mechanism (Ramsey and Sharpless 2006), and it has been established that oxidative stress has a significant impact on telomere shortening (von Zglinicki 2002). In the remainder of this chapter we will review the role of oxidative stress in replicative senescence, its influence on telomeres, and its contribution to cell-to-cell heterogeneity.

2 Replicative Senescence and the Telomere Hypothesis

In the 19th century, August Weismann, a German zoologist, had the idea that ageing could be a consequence of somatic cells having a finite life span (Kirkwood and Cremer 1982). Eighty years were necessary to prove Weismann's prediction experimentally. Hayflick and Moorhead showed that embryo-derived fibroblasts can divide 50±10 times before arresting irreversibly when cultured in vitro (Hayflick and Moorhead 1961). The potential number of divisions became known as the "Hayflick limit," and the phenomenon was termed replicative senescence.

Early on, it was suggested that the shortening of telomeres, the ends of chromosomes, might function as a replicometer (counting the finite number of cell divisions) and as a trigger of replicative senescence in normal diploid cells (Olovnikov 1971, Watson 1972). Later, this idea was tested experimentally (Harley et al. 1990). The best evidence in support of a telomere-dependent senescence mechanism is the observation that ectopic expression of the catalytic subunit of telomerase, an enzyme able to counteract telomere shortening, can lead to cell immortalization on its own (Bodnar et al. 1998). It is believed that telomere shortening destabilizes telomeric loops (Griffith et al. 1999) and as a consequence increases the probability of telomere uncapping.

These findings led to a generalized notion of cellular senescence as result of a “biological clock” that counts the number of cell divisions through telomere shortening.

3 Heterogeneity in Senescence

One of the problems with the idea of telomeres working as a clock is that the Hayflick limit is a fixed number only with respect to large populations, that is, in a stochastic sense. In contrast, individual cell lineages vary greatly in their replicative potential.

Smith and Hayflick (1974) showed large variations in the population doubling potential of fibroblast clones isolated from WI-38 and WI-26 cell cultures (Smith and Hayflick 1974). This finding raised concerns over studies involving mass cultures due to heterogeneous populations containing cells at different population doubling levels.

Later, Smith and Whitney, in a classic experiment, showed heterogeneity in doubling potential of individual cells from clonally derived populations (Smith and Whitney 1980). Thus it was suggested that heterogeneity could not be explained solely by the differences between individual cells’ life histories. Rather, stochastic factors must be important to determine the replicative potential of cells.

Other studies showed that the fraction of cells able to divide decreased progressively with increasing population doublings, using BrdU labeling (Kill et al. 1994), Ki67 staining (Thomas et al. 1997), and p53-reporter assay (Bond et al. 1994). It was also shown that the percentage of cells stained positive for γ-H2AX, a marker for senescence-associated DNA damage foci that might be formed in response to functionally uncapped telomeres, increased with population doubling level (Fagagna et al. 2003).

Thus, life span heterogeneity can be characterized by the fraction of cells displaying a senescent phenotype. However, the question remains, what causes this heterogeneity?

It is often assumed that cells that lose their division capacity early do so by a process termed *premature senescence*, which is a telomere-independent cellular response to a variety of stresses. Premature senescence can in fact be induced by a variety of experimental conditions:

Human epithelial cells encounter a telomere-independent, p16-dependent growth arrest in response to suboptimal culture conditions (Stampfer and Yaswen 2003). Overexpression of oncogenes, such as activated RAS or RAF, can induce a senescence-like arrest in primary human or mouse cells (Ferbeyre et al. 2002, Lin et al. 1998, Serrano et al. 1997), and this is dependent on p16 (Benanti and

Galloway 2004). Modification of chromatin by inhibitors of histone deacetylases also induces a senescent phenotype (Ogryzko et al. 1996). Oxidative stress, ionizing radiation, and other DNA-damaging agents can also induce the senescent phenotype without any detectable telomere shortening (Dumont et al. 2000, Herskind and Rodemann 2000, Jeyapalan et al. 2004, Robles and Adami 1998). Therefore, it is sometimes assumed that all early-occurring senescence might be both stress-induced and telomere-independent and thus premature. In other words, in any given cell culture experiment, most cells would senesce by a process that is altogether different from telomere-driven senescence, which would happen only in the last surviving clone. However, telomere shortening is by itself distinctly stress-dependent (von Zglinicki 2002), suggesting an alternative explanation for cell-to-cell heterogeneity in replicative senescence.

4 Heterogeneity and Telomeres

One of the earlier suggestions to explain cell-to-cell heterogeneity was that telomere uncapping, the process that leads to telomeres being recognized as inducers of a DNA damage response triggering growth arrest, is itself stochastic (Blackburn 2000). While some random variation in uncapping remains a possibility, it is unclear that such an effect is sufficient to explain the observed heterogeneity in both telomere lengths and division capacity within cultures of cells.

In fact, there is good experimental evidence for large heterogeneity in telomere length, both between chromosome ends within individual cells and between cells. Interchromosomal variation in telomere length has been observed in cultured human hematopoietic cells by Q-FISH. Telomere fluorescence intensity values were found to vary up to six-fold per metaphase (Lansdorp et al. 1996). Zou et al. found that a specific subgroup of chromosomes displayed the shortest telomeres in near-senescent cells and that this subgroup induced a DNA damage response and replicative senescence (Zou et al. 2004). This is indicative that shortening of individual telomeres is not uniform in cells. It is, however, not clear yet whether such a subgroup is in itself a stochastic selection or whether there are factors that modulate telomere shortening in a chromosome-specific way.

The latter possibility has been suggested by Q-FISH analyses. Some studies indicated that intracellular, interchromosomal heterogeneity might be cell line- and donor-specific and heritable (Graakjaer et al. 2004). More studies using various methodological approaches and larger numbers of cells will be necessary to prove or disprove this point. Using STELA, a method able to measure telomere lengths from individual chromosomes, extensive cell-to-cell variation of telomere lengths on individual chromosome arms (e.g., the Xp/Yp chromosome) has been found (Baird et al. 2003), which seems to suggest a strong influence of nonheritable (chance) factors on telomere length heterogeneity. At present, the relative impact of heritable and randomly acting factors on inter- and intrachromosomal telomere length distributions cannot yet be conclusively estimated.

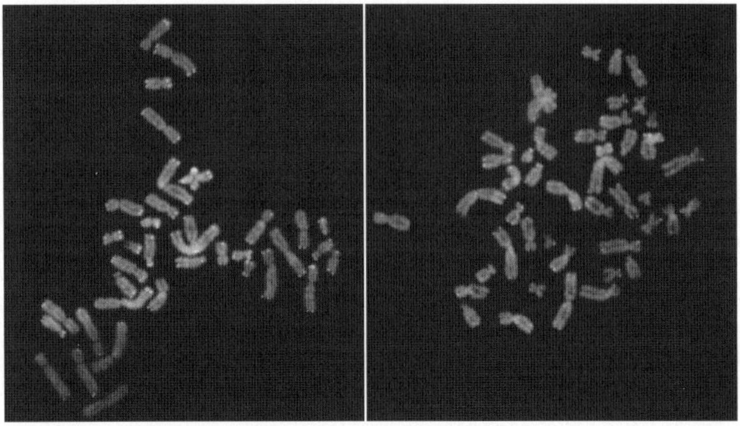

Fig. 3.1 Heterogeneity in telomere Q-FISH signals between metaphases from young proliferating fibroblast cultures (PD13). Most metaphases show strong telomeric signals (left panel), but some display signals as weak as in near-senescent cells (right panel) (*See Color Plate*)

Intercellular heterogeneity in telomere length has also been observed by Flow-FISH analysis of mouse and human cell lines, showing the existence of subpopulations of cells with varying telomere lengths (Cabuy et al. 2004). Moreover, Q-FISH analysis of metaphases shows that even in very "young" cultures, cells with long telomeres (Fig. 3.1, left panel) coexist with cells containing very short telomeres, similar to those in senescent cells (Fig. 3.1, right panel) (Passos et al. 2007).

However, the question remains: what triggers this heterogeneity? Why does telomere shortening vary from cell to cell and from chromosome end to chromosome end? One important contributing factor might be oxidative damage to the telomeres. It has been shown that a telomere-specific repair deficiency causes stress-dependent accumulation of single-strand breaks (Petersen et al. 1998), which accelerates telomere shortening during DNA replication (Sitte et al. 1998).

There is good evidence for the involvement of environmental stress in telomere shortening. A number of independent studies have shown that exposure to mild oxidative stress leads to faster telomere shortening and that maintenance of telomere length improves under low oxidative stress (Passos and von Zglinicki 2005).

The minimal rate of telomere shortening predicted from the end replication problem is approximately 3 bp/end/cell division; however, human cells lacking telomerase lose on average 50–300 bp/end/cell division (de Lange 2006). However, in conditions where cells have been exposed to lower oxidative stress, it has been observed that telomere shortening can be as slow as predicted solely by the end replication problem (Passos et al. 2007, Serra et al. 2003).

Kurz et al. showed that cultured human umbilical vein endothelial cells, when exposed to chronic oxidative stress, showed faster telomere shortening, but also increased interchromosomal telomere length heterogeneity (Kurz et al. 2004). Population studies have revealed a correlation between environmental stress levels and white blood cell telomere length in humans, suggesting that oxidative stress might be an important factor for telomere maintenance in vivo as well (Epel et al. 2004).

Together, these data raise the possibility that the intercellular variation between telomeres occurs as the result of random intrinsic variation in production of free oxygen radicals and/or defense against them. This heavily implies mitochondria, the major intracellular source of oxygen free radicals. In the next section we review data concerning the role of mitochondria in replicative senescence and their possible contribution to cell-to-cell heterogeneity.

5 Heterogeneity and Mitochondria

There is a good experimental basis to consider a possible involvement of mitochondria in ageing, mainly from data obtained in postmitotic tissues. Mitochondria are the major source of oxygen-free radicals in cells, which by their very nature are highly unstable molecules with an unpaired electron in the outer orbital, which can react indiscriminately with any organic molecules present in a cell.

Moreover, one of the key observations when accessing mitochondrial function and levels of mtDNA mutations in tissues is that both vary considerably from cell to cell. Thus, changes in mitochondrial function might be the missing link explaining cell-to-cell heterogeneity in telomeres and, consequently, in cellular senescence.

The main cause for increased mitochondrial production of free radicals is still unclear. One suggestion has been accumulation of mutations in the mitochondrial (mt) DNA (Harman 1972). Mitochondria, like telomeres, are highly susceptible to oxidative damage – and for good reasons. There is close physical proximity between the sites of reactive oxygen species production and the mtDNA. Moreover, there is relative inefficiency of the repair mechanisms in mitochondria and, unlike the nuclear genome, the mitochondrial genome is not protected by histones. Accordingly, mtDNA damage was found to be more extensive and persistent than nuclear DNA damage in cultured human fibroblasts following treatment with hydrogen peroxide (Yakes and Van Houten 1997). This damage might have functional consequences, since mtDNA damage in immortalized normal human fibroblasts (NHF) after hydrogen peroxide exposure has been shown to correlate with decreased mitochondrial membrane potential (Santos et al. 2003).

MtDNA mutations are responsible for deficient activity of respiratory chain enzymes such as cytochrome C oxidase, which is a mitochondrial membrane–bound enzyme composed of subunits that are encoded in both the mitochondria (COX subunits I, II, and III) and the nucleus (all others). Muller-Hocker showed that mutant mitochondria deficient in cytochrome c oxidase (COX) are distributed in the muscle in a few fiber segments, where the surrounding fibers are normal (Muller-Hocker 1989). The frequency of COX-deficient fibers was shown to increase significantly with age, but remained always below 1%. In situ hybridization studies using mtDNA probes from various regions of the mitochondrial genome showed accumulation of mtDNA deletions in COX-negative cells (Lee et al. 1998). The availability of microdissection techniques facilitated the direct analysis of single cells in histological sections. Using this technique (Cao et al. 2001, Wanagat et al. 2001), it was shown that all electron transport chain-deficient fibers in rat skeletal muscle contained mtDNA deletion mutations. In single neurons from substantia nigra of humans showing COX deficiency, an age-dependent increase in mtDNA deletions was found in two independent studies (Bender et al. 2006, Kraytsberg et al. 2006). Interestingly, these data reveal that interneuronal variation (measured as median absolute deviation) in the percentage of mtDNA deletions increased significantly with age.

Other recent data showed that extensive, stochastic variability in the presence and level of mitochondrial DNA mutations occurs in human colon stem cells, one of the most actively proliferating cell types in the body (Taylor et al. 2003). This result demonstrates that heterogeneity in mitochondrial mutation load is not exclusive to postmitotic cells.

Nonetheless, recent work has questioned the association between mtDNA mutation load and mitochondrial dysfunction as a major cause of aging. It has been shown that homozygous knock-in mice that express a proof-reading-deficient version of the nucleus-encoded catalytic subunit of mtDNA polymerase γ (PolgA) accumulated very high levels of mtDNA mutations and deletions and showed a significant decrease in life span (Trifunovic et al. 2004). However, the mechanism for this phenomenon remains unknown, since no increased markers of oxidative stress, mitochondrial dysfunction or defects in cellular proliferation were found in these animals (Kujoth et al. 2005, Trifunovic et al. 2005). Importantly, mice that are heterozygous for PolgA function show no significant reduction in life span despite a mtDNA mutation burden 30 times higher than in old wild-type animals (Vermulst et al. 2007). These studies suggest that mtDNA mutation load does not limit life span of wild-type mice and that mtDNA mutations, even at very high levels, do not necessarily lead to increased mitochondrial ROS generation. Mitochondrial dysfunction and generation of reactive oxygen species might be consequences of other still unexplored factors occurring with cell ageing.

With regard to cell senescence in culture, one can hypothesize that variation in mitochondrial function between cells could determine changes in their replicative potential. It is clear in mass populations of fibroblasts that mitochondrial function varies significantly between cells and determines their replicative life span, as we shall discuss in the next section (Passos et al. 2007 and see Fig. 3.2).

6 Mitochondrial Dysfunction Causes Telomere Dysfunction and Thus Heterogeneity in Replicative Senescence

Are mitochondria at all causally involved in the process of replicative senescence? There is increasing evidence that they do play an important role. Using fluorescent dyes which measure cellular peroxides, some studies showed that senescent cells

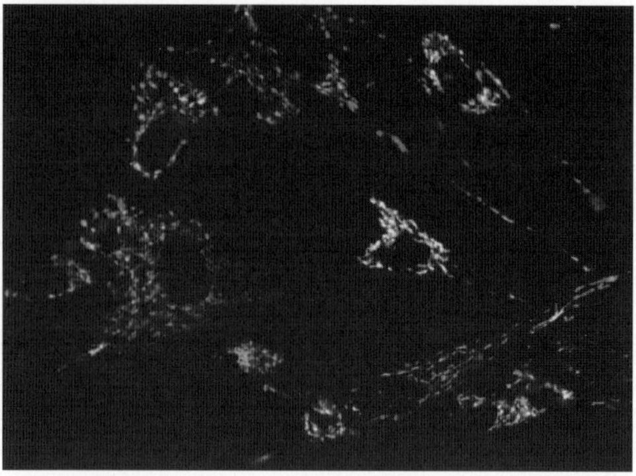

Fig. 3.2 Confocal image of live young fibroblasts staining with JC-1. The dye emits green fluorescence at low concentrations and red fluorescence at high density. It is transported into mitochondria in a membrane potential (MMP)-dependent fashion, so that green indicates low MMP, while mitochondria stained red have high MMP. Note the cell-to-cell heterogeneity in JC-1 ratios, as well as intracellular heterogeneity (*See Color Plate*)

were associated with high levels of endogenous reactive oxygen species (Hutter et al. 2002, Allen et al. 1999). Accumulation of oxidation products, such as protein carbonyls and lipofuscin, occurs in senescent fibroblasts grown in vitro (Sitte et al. 2000, Sitte et al. 2001). More recently, we have shown that senescent cells display mitochondrial dysfunction, characterized by lower mitochondrial membrane potential, mtDNA damage, and increased superoxide production (Passos et al. 2007). This was in accordance with earlier reports showing that senescent fibroblasts have impaired metabolism, with strong reduction of ATP and other nucleotide triphosphates (Zwerschke et al. 2003).

Beyond mere correlation, there is evidence supporting a causal role for mitochondrial dysfunction in telomere-dependent senescence. In one study, mitochondria from fibroblasts of 21 individuals between 20 weeks and 103 years of age were introduced into human mtDNA-less cells. The authors found both a slight age-dependent decrease in growth rate and a decline in respiratory rate in the transformants, suggesting a causal role of mitochondria in the senescence process (Laderman et al. 1996). Selective targeting of antioxidants directly to the mitochondria counteracted telomere shortening and increased life span in fibroblasts under mild oxidative stress (Saretzki et al. 2003). Continuous treatment with nicotinamide, which led to lower ROS generation and changes in mitochondrial function, has been reported to extend life span (a remarkable 1.6-fold increase) and decelerate telomere shortening (Kang et al. 2006). Also, pharmacological mild chronic uncoupling of mitochondria that reduced the production of superoxide anion improved telomere maintenance and extended telomere-dependent life span (Passos et al. 2007).

Accordingly, mitochondrial dysfunction generated by severe mitochondrial depolarization using an uncoupling agent (FCCP) led to an increased production of ROS, telomere attrition, telomere loss, and chromosome fusion in mouse embryos (Liu et al. 2002). Moreover, Oexle and Zwirner (1997) showed that patients with mitochondrial diseases MELAS and LHON had, on average, 1.5 kb shorter telomeres than those of age-matched controls.

Thus, there is good evidence for a causal role of mitochondria in telomere-dependent senescence. Based on these data, it is reasonable to assume that cell-to-cell variation in mitochondrial function would impact telomere length and might explain heterogeneity in the replicative potential of cells. A stochastic network model, combining telomere attrition with other mechanisms such as oxidative stress and mitochondrial and nuclear mutations, was able to show good agreement with published data on heterogeneity in division potential (Sozou and Kirkwood 2001).

In a series of FACS sorting experiments, we have recently proven the interconnection between mitochondrial dysfunction, ROS generation, and accelerated telomere shortening as cause for cell-to-cell heterogeneity of division potential of human fibroblasts (Martin-Ruiz et al. 2004, Passos et al. 2007, von Zglinicki et al. 2003).

Senescent cells present in a culture of early passage fibroblasts have been shown to have higher levels of reactive oxygen species, together with decreased mitochondrial membrane potential, which is indicative of metabolic inefficiency. These sorted cells showed high frequency of telomeric γ-H2A.X foci and shorter telomeres than the remainder of the population, indicating that mitochondrial ROS generation could be targeting telomeres specifically. Additionally, sorting of cells at early passage according to mitochondrial superoxide production (fibroblasts were stained with MitoSOX) showed that the cells which produced higher levels of superoxide were also the ones which contained the highest frequency of telomeric γ-H2A.X foci (Passos et al. 2007).

These results support a model in which telomeres are sensors of mitochondrial function/dysfunction (Passos and von Zglinicki 2005) and replicative life span of individual cells is determined by a network of processes involving mitochondrial dysfunction, oxidative stress, and telomere shortening (Sozou and Kirkwood 2001). It is clear that stress levels (external and/or internal) will determine the relative importance of the different components of the network for induction of senescence. Thus, at very high levels of stress (for instance, caused by ionizing radiation), sufficient unrepairable DNA damage will occur to arrest cells in telomere-independent senescence. Under relatively mild stress conditions (typically cell culture under 20%–40% oxygen partial pressure) accelerated telomere shortening due to mitochondrially derived ROS appears to be the major route toward senescence and the main cause for its heterogeneity. Under constant low oxidative stress conditions or improved mitochondrial function, the "end replication problem" becomes more important as a cause of telomere shortening and there is less heterogeneity in the life span of cells (Passos et al. 2007).

However, the proximal cause for mitochondrial dysfunction in senescence and its heterogeneity remains still unclear. Is it to do with random accumulation of mtDNA mutations, mtDNA damage, or other changes (like differential gene expression) that can affect mitochondrial function and assembly? This is an important area of research which has yet to be further explored.

7 Complexities Beyond

Recent evidence suggests that the relation between telomeres and mitochondria might be vastly more complex than that between producers of free radicals (mitochondria) and targets of free radical damage (telomeres). Several reports have suggested that similar mechanisms that protect telomeres might have effects on mitochondrial DNA. One example is the Pif1p family of DNA helicases, which are conserved from yeast to man. In budding yeast ScPif1p affects both repair of the mitochondrial genome and telomere shortening (Boule and Zakian 2006).

Another example relates to telomerase. Overexpression of *hTERT*, the catalytic subunit of telomerase, counteracts telomere shortening. In many cell strains, this is sufficient to greatly extend the replicative potential and to prevent replicative senescence (Bodnar et al. 1998). There is also evidence that telomerase promotes cell survival and stress resistance by mechanisms which appear to be largely independent of its effect on telomere length maintenance (Flores et al. 2005, Kang et al. 2004, Ludwig et al. 2001, Oh et al. 2003, Sarin et al. 2005, Zhang et al. 2003). More interestingly, recent reports show that *hTERT* is imported into mitochondria under acute oxidative stress and has an effect on mtDNA integrity (Santos et al. 2006). However, there is still uncertainty regarding its role in mitochondria (Passos et al. 2006).

Acknowledgements We would like to thank Dr. Conor Lawless for his analysis of median absolute deviation (data extracted from Kraytsberg et al. 2006) and helpful discussions. Work leading to this chapter has been supported by Research into Ageing UK and BBSRC/EPSRC (CISBAN).

References

Allen RG, Tresini M, Keogh BP, Doggett DL. Cristofalo VJ (1999) Differences in electron transport potential, antioxidant defenses, and oxidant generation in young and senescent fetal lung fibroblasts (WI-38). J Cell Physiol 180 (1): 114–22

Bahar R, Hartmann CH, Rodriguez KA, Denny AD, Busuttil RA, Dolle MET, Calder RB, Chisholm GB, Pollock BH, Klein CA, Vijg J (2006) Increased cell-to-cell variation in gene expression in ageing mouse heart. Nature 441 (7096): 1011–14

Baird DM, Rowson J, Wynford-Thomas D. Kipling D (2003) Extensive allelic variation and ultrashort telomeres in senescent human cells. Nat Genet 33(2): 203–07

Benanti JA, Galloway DA (2004) Normal human fibroblasts are resistant to RAS-induced senescence Mol Cell Biol 24(7): 2842–52

Bender A, Krishnan KJ, Morris CM, Taylor GA, Reeve AK, Perry RH, Jaros E, Hersheson JS, Betts J, Klopstock T, Taylor RW, Turnbull DM (2006) High levels of mitochondrial DNA deletions in substantia nigra neurons in aging and Parkinson disease. Nat Genet 38(5): 515–17

Blackburn EH (2000) Telomere states and cell fates. Nature 408(6808): 53–56

Bodnar AG, Ouellette M, Frolkis M, Holt SE, Chiu CP, Morin GB, Harley CB, Shay JW, Lichtsteiner S, Wright WE (1998) Extension of life-span by introduction of telomerase into normal human cells. Science 279(5349): 349–52

Bond, JA, Wyllie FS. and Wynford-Thomas, D. (1994) 'Escape from senescence in human diploid fibroblasts induced directly by mutant p53', *Oncogene,* 9, (7), pp. 1885–89

Boule, J-B. and Zakian, VA. (2006) 'Roles of Pif1-like helicases in the maintenance of genomic stability', *Nucl. Acids Res.,* 34, (15), pp. 4147–4153.

Cabuy, E, Newton C, Roberts, T, Newbold, R. and Slijepcevic, P. (2004) 'Identification of subpopulations of cells with differing telomere lengths in mouse and human cell lines by flow FISH', *Cytometry Part A,* 62A, (2), pp. 150–161.

Cao, Z, Wanagat, J, McKiernan, SH. and Aiken, JM. (2001) 'Mitochondrial DNA deletion mutations are concomitant with ragged red regions of individual, aged muscle fibers: analysis by laser-capture microdissection', *Nucleic Acids Res,* 29, (21), pp. 4502–8

de Lange, T. (2006) 'Mammalian Telomeres', in de lange, T., Lundbard, V. and Blackburn, E. H.(eds) *Telomeres.* Cold Spring Harbour Laboratory Press: New York, pp. 387–431.

Dumont P, Burton M, Chen QM, Gonos ES, Frippiat C, Mazarati, JB, Eliaers, F, Remacle, J. and Toussaint, O. (2000) 'Induction of replicative senescence biomarkers by sublethal oxidative stresses in normal human fibroblast', *Free Radical Biology and Medicine,* 28, (3), pp. 361–373

Epel, E. S., Blackburn, E. H., Lin, J., Dhabhar, F. S., Adler, N. E., Morrow, J. D. and Cawthon, R. M. (2004) 'Accelerated telomere shortening in response to life stress', *PNAS,* 101, (49), pp. 17312–17315.

Fagagna Fd Ad, Reaper PM, Clay-Farrace L, Fiegler H, Carr P, von Zglinicki T, Saretzki G, Carter NP, Jackson SP (2003) A DNA damage checkpoint response in telomere-initiated senescence. Nature 426(6963): 194–98

Ferbeyre G, de Stanchina E, Lin AW, Querido E, McCurrach ME, Hannon GJ, Lowe SW (2002) Oncogenic ras and p53 cooperate to induce cellular senescence. Mol Cell Biol 22(10): 3497–3508

Flores I, Cayuela ML, Blasco MA (2005) Effects of telomerase and telomere length on epidermal stem cell behavior. Science 309(5738): 1253–56

Graakjaer J, Pascoe L, Der-Sarkissian H, Thomas G, Kolvraa S, Christensen K, Londono-Vallejo J.-A (2004) The relative lengths of individual telomeres are defined in the zygote and strictly maintained during life. Aging Cell 3(3): 97–102

Griffith JD, Comeau L, Rosenfield S, Stansel RM, Bianchi A, Moss H, de Lange T (1999) Mammalian telomeres end in a large duplex loop. Cell 97(4): 503–14

Harley CB, Futcher AB, Greider CW (1990) Telomeres shorten during ageing of human fibroblasts. Nature 345(6274): 458–60

Harman D (1956) Aging: A theory based on free radical and radiation chemistry. J Gerontol 11: 298–300

Harman D (1972) The biologic clock: the mitochondria? J Am Geriatr Soc 20(4): 145–47

Hayflick L, Moorhead PS (1961) The serial cultivation of human diploid cell strains, Exp Cell Res 25: 585–621

Herndon LA, Schmeissner PJ, Dudaronek JM, Brown PA, Listner KM, Sakano Y, Paupard MC, Hall DH, Driscoll M (2002) Stochastic and genetic factors influence tissue-specific decline in ageing C. elegans. Nature 419(6909): 808–14

Herskind C, Rodemann HP (2000) Spontaneous and radiation-induced differentiation of fibroblasts. Exp Gerontol 35(6–7): 747–55

Hutter E, Unterluggauer H, Uberall F, Schramek H, and Jansen-Durr P (2002) Replicative senescence of human fibroblasts: the role of Ras-dependent signaling and oxidative stress. Exp Gerontol 37(10–11): 1165–74

Jeyapalan J, Leake A, Ahmed S, Saretzki G, Tilby M, von Zglinicki T (2004) The role of telomeres in Etoposide induced tumor cell death. Cell Cycle 3(9): 1169–76

Kang HJ, Choi YS, Hong S-B, Kim K-W, Woo R-S, Won SJ, Kim EJ, Jeon HK, Jo S-Y, Kim TK, Bachoo R, Reynolds I J, Gwag BJ, Lee H-W (2004) Ectopic expression of the catalytic subunit of telomerase protects against brain injury resulting from ischemia and NMDA-induced neurotoxicity. J Neurosci 24(6): 1280–87

Kang HT, Lee HI, Hwang ES (2006) Nicotinamide extends replicative life span of human cells. Aging Cell 5(5): 423–36

Kill IR, Faragher RG, Lawrence K, Shall S (1994) The expression of proliferation-dependent antigens during the life span of normal and progeroid human fibroblasts in culture. J Cell Sci 107: 571–79

Kirkwood T, Cremer T (1982) Cytogerontology since 1881: a reappraisal of August Weissmann and a review of modern progress. Hum Genet 60: 101–21

Kirkwood TB, Finch CE (2002) Ageing: the old worm turns more slowly. Nature 419(6909): 794–95

Kirkwood TBL (2005) Understanding the odd science of aging. Cell 120(4): 437–447

Kraytsberg Y, Kudryavtseva E, McKee AC, Geula C, Kowall NW, Khrapko K (2006) Mitochondrial DNA deletions are abundant and cause functional impairment in aged human substantia nigra neurons. Nat Genet 38(5): 518–20

Kurz DJ, Decary S, Hong Y, Trivier E, Akhmedov A, Erusalimsky JD (2004) Chronic oxidative stress compromises telomere integrity and accelerates the onset of senescence in human endothelial cells. J Cell Sci 117(11): 2417–26

Laderman KA, Penny JR, Mazzucchelli F, Bresolin N, Scarlato G, Attardi G (1996) Aging-dependent functional alterations of mitochondrial DNA (mtDNA) from human fibroblasts transferred into mtDNA-less cells. J Biol Chem 271(27): 15891–97

Lansdorp PM, Verwoerd NP, van de Rijke FM, Dragowska V, Little MT, Dirks RW, Raap AK, Tanke HJ (1996) Heterogeneity in telomere length of human chromosomes. Hum Mol Genet 5(5): 685–91

Lee CM, Lopez ME, Weindruch R, Aiken JM (1998) Association of age-related mitochondrial abnormalities with skeletal muscle fiber atrophy. Free Rad Biol Med 25(8): 964–72

Lin AW, Barradas M, Stone JC, van Aelst L, Serrano M, Lowe SW (1998) Premature senescence involving p53 and p16 is activated in response to constitutive MEK/MAPK mitogenic signaling. Genes Dev 12(19): 3008–19

Liu L, Trimarchi JR, Smith PJ, Keefe DL (2002) Mitochondrial dysfunction leads to telomere attrition and genomic instability. Aging Cell 1(1): 40–46

Ludwig A, Saretzki G, Holm PS, Tiemann F, Lorenz M, Emrich T, Harley CB, von Zglinicki T (2001) Ribozyme cleavage of telomerase mRNA sensitizes breast epithelial cells to inhibitors of topoisomerase. Cancer Res 61(7): 3053–61

Martin-Ruiz C, Saretzki G, Petrie J, Ladhoff J, Jeyapalan J, Wei W, Sedivy J, von Zglinicki T (2004) Stochastic variation in telomere shortening rate causes heterogeneity of human fibroblast replicative life span. J Biol Chem 279(17): 17826–33

Mitsui A, Hamuro J, Nakamura H, Kondo N, Hirabayashi Y, Ishizaki-Koizumi S, Hirakawa T, Inoue T, Yodoi J (2002) Overexpression of human thioredoxin in transgenic mice controls oxidative stress and life span. Antioxidants Redox Signal 4(4): 693–96

Muller-Hocker J (1989) Cytochrome-c-oxidase deficient cardiomyocytes in the human heart–an age-related phenomenon. A histochemical ultracytochemical study. Am J Pathol 134(5): 1167–73

Oexle K, Zwirner A (1997) Advanced telomere shortening in respiratory chain disorders. Hum Mol Genet 6(6): 905–8

Ogryzko V, Hirai T, Russanova V, Barbie D, Howard B (1996) Human fibroblast commitment to a senescence-like state in response to histone deacetylase inhibitors is cell cycle dependent. Mol Cell Biol 16(9): 5210–18

Oh H, Bradfute SB, Gallardo TD, Nakamura T, Gaussin V, Mishina Y, Pocius J, Michael LH, Behringer RR, Garry D J, Entman ML, Schneider MD (2003) Cardiac progenitor cells from adult myocardium: homing, differentiation, and fusion after infarction. PNAS 100(21): 12313–18

Olovnikov AM (1971) [Principle of marginotomy in template synthesis of polynucleotides]. Dokl Akad Nauk SSSR 201(6): 1496–99

Passos JF, Saretzki G, Ahmed S, Nelson G, Richter T, Peters H, Wappler I, Birkett M, Harold G, Schaeuble K, Birch-Machin M, Kirkwood T, von Zglinicki T (2007) Mitochondrial dysfunction accounts for the stochastic heterogeneity in telomere-dependent senescence. PLoS Biol 5(5): e 110

Passos JF, von Zglinicki T (2005) Mitochondria, telomeres and cell senescence. Exp Gerontol 40(6): 466–72

Passos JF, von Zglinicki T, Saretzki G (2006) Mitochondrial dysfunction and cell senescence: cause or consequence? Rejuv Res 9(1); 64–68

Petersen S, Saretzki G, von Zglinicki T (1998) Preferential accumulation of single-stranded regions in telomeres of human fibroblasts. Exp Cell Res 239(1): 152–60

Ramsey M, Sharpless N (2006) ROS as a tumour suppressor? Nat Cell Biol. 8(11): 1213–15

Rea SL, Wu D, Cypser JR, Vaupel JW, Johnson TE (2005) A stress-sensitive reporter predicts longevity in isogenic populations of Caenorhabditis elegans. Nat Genet 37(8): 894–98

Robles S, Adami G (1998) Agents that cause DNA double strand breaks lead to p16INK4a enrichment and the premature senescence of normal fibroblasts.' Oncogene 16(9): 1113–23

Santos JH, Hunakova L, Chen Y, Bortner C, Van Houten B (2003) Cell sorting experiments link persistent mitochondrial DNA damage with loss of mitochondrial membrane potential and apoptotic cell death. J Biol Chem 278 (3): 1728–34

Santos JH, Meyer JN, Van Houten B (2006) Mitochondrial localization of telomerase as a determinant for hydrogen peroxide-induced mitochondrial DNA damage and apoptosis. Hum Mol Genet 15(11): 1757–68

Saretzki G, Murphy MP, von Zglinicki T (2003) MitoQ counteracts telomere shortening and elongates life span of fibroblasts under mild oxidative stress. Aging Cell 2(2): 141–43

Sarin KY, Cheung P, Gilison D, Lee E, Tennen RI, Wang E, Artandi MK, Oro AE, Artandi SE (2005) Conditional telomerase induction causes proliferation of hair follicle stem cells. 436(7053): 1048–52

Schriner SE, Linford NJ, Martin GM, Treuting P, Ogburn CE, Emond M, Coskun PE, Ladiges W, Wolf N, Van Remmen H, Wallace DC, Rabinovitch PS (2005) Extension of murine life span by overexpression of catalase targeted to mitochondria. Science 308(5730): 1909–11

Serra V, von Zglinicki T, Lorenz M, Saretzki G (2003) Extracellular superoxide dismutase is a major antioxidant in human fibroblasts and slows telomere shortening. J Biol Chem 278(9): 6824–30

Serrano M, Lin AW, McCurrach ME, Beach D, Lowe SW (1997) Oncogenic ras provokes premature cell senescence associated with accumulation of p53 and p16INK4a. Cell 88(5): 593–602

Sitte N, Merker K, Grune T. von Zglinicki T (2001) Lipofuscin accumulation in proliferating fibroblasts in vitro: an indicator of oxidative stress. Exper Gerontol 36(3): 475–86

Sitte N, Merker K, von Zglinicki T, Grune T (2000) Protein oxidation and degradation during proliferative senescence of human MRC-5 fibroblasts. Free Rad Biol Med 28(5): 701–08

Sitte N, Saretzki G, von Zglinicki T (1998) Accelerated telomere shortening in fibroblasts after extended periods of confluency. Free Rad Biol Med 24(6): 885–93

Smith JR, Hayflick L (1974) Variation in the life-span of clones derived from human diploid cell strains. J Cell Biol 62(1): 48–53

Smith JR, Whitney RG (1980) Intraclonal variation in proliferative potential of human diploid fibroblasts: stochastic mechanism for cellular aging. Science 207(4426): 82–84

Sozou PD, Kirkwood T (2001) A stochastic model of cell replicative senescence based on telomere shortening, oxidative stress, and somatic mutations in nuclear and mitochondrial DNA. J Theor Biol 213: 573–86

Stampfer MR, Yaswen P (2003) Human epithelial cell immortalization as a step in carcinogenesis. Cancer Lett 194(2): 199–208

Taylor RW, Barron MJ, Borthwick GM, Gospel A, Chinnery PF, Samuels DC, Taylor GA, Plusa SM, Needham SJ, Greaves LC, Kirkwood T, Turnbull DM (2003) Mitochondrial DNA mutations in human colonic crypt stem cells. J Clin Inv 112(9): 1351–60

Thomas E, Al-Baker E, Dropcova S, Denyer S, Ostad N, Lloyd A, Kill IR, Faragher RGA (1997) Different kinetics of senescence in human fibroblasts and peritoneal mesothelial cells. Exper Cell Res 236(1): 355–58

Van Remmen H, Ikeno Y, Hamilton M, Pahlavani M, Wolf N, Thorpe SR, Alderson NL, Baynes JW, Epstein CJ., Huang TT, Nelson J, Strong R, Richardson A (2003) Life-long reduction in MnSOD activity results in increased DNA damage and higher incidence of cancer but does not accelerate aging. Physiol Genom 16: 29–37

von Zglinicki T (2002) Oxidative stress shortens telomeres. Trends Biochem Sci 27(7): 339–44

von Zglinicki T, Petrie J, Kirkwood TB (2003) Telomere-driven replicative senescence is a stress response. Nat Biotechnol 21(3): 229–30

Wanagat J, Cao Z, Pathare P, Aiken JM (2001) Mitochondrial DNA deletion mutations colocalize with segmental electron transport system abnormalities, muscle fiber atrophy, fiber splitting, and oxidative damage in sarcopenia. Faseb J 15 (2): 322–32

Watson JD (1972) Origin of concatemeric T7 DNA. Nat New Biol 239(94): 197–201

Yakes FM, Van Houten B (1997) Mitochondrial DNA damage is more extensive and persists longer than nuclear DNA damage in human cells following oxidative stress. Proc Natl Acad Sci U S A 94(2): 514–49

Zhang P, Chan SL, Fu W, Mendoza M, Mattson MP (2003) TERT suppresses apoptosis at a premitochondrial step by a mechanism requiring reverse transcriptase activity and 14-3-3 protein binding ability. FASEB J pp. 02–0603

Zou Y, Sfeir A, Gryaznov SM, Shay JW, Wright WE (2004) Does a sentinel or a subset of short telomeres determine replicative senescence? Mol Biol Cell 15(8): 3709–18

Zwerschke W, Mazurek S, Stockl P, Hutter E, Eigenbrodt E, Jansen-Durr P (2003) Metabolic analysis of senescent human fibroblasts reveals a role for AMP in cellular senescence. Biochem J 376 (Pt 2): 403–11

Chapter 4
Initiation of Genomic Instability, Cellular Senescence, and Organismal Aging by Dysfunctional Telomeres

Sandy Chang

Abstract Telomeres are ribonucleoprotein structures that protect the end of linear chromosomes from recognition as DNA double-stranded breaks and activation of a DNA damage response. Telomere-associated proteins also regulate telomerase, the protein responsible for maintaining telomere length. Loss of telomere function results from either alteration in the telosome/shelterin complex that exerts capping function at telomeres, or from progressive loss of telomeric repeats necessary to maintain proper telomeric structure. Dysfunctional telomeres activate p53 to initiate cell cycle arrest/ cellular senescence and may play an important role in aging. Here I describe the use of mouse models to probe the impact of dysfunctional telomeres upon organismal aging.

1 Telomeres and Telomerase

Telomeres are G-rich, simple repeat sequences that comprise the physical termini of chromosomes (reviewed by Greider 1996). The synthesis and maintenance of telomeric repeats are mediated by telomerase, a specialized ribonucleoprotein complex consisting of an essential telomerase RNA (Terc) that serves as a template for the addition of telomere repeats (reviewed by Blackburn 2001) and a protein component, the telomerase reverse transcriptase catalytic subunit (Tert) (Counter et al. 1997, Harrington et al. 1997, Meyerson et al. 1997, Nakamura et al. 1997, Nakayama et al. 1997, Greenberg et al. 1998). In addition, the telomere ends in a 3′ G-rich overhang, which serves as the substrate to which telomeric repeats are added. Electron microscopic analysis revealed that telomeres could fold back into a structure termed the t-loop, in which the 3′ overhang invades a duplex region of the telomere to generate a displaced d-loop that sequesters the overhang (Griffith et al. 1999). This structure likely prevents telomeres from

Sandy Chang
Dept. of Cancer Genetics and Hematopathology, The M.D. Anderson Cancer Center, Houston, Texas, United States.
e-mail: schang@mdanderson.org

being recognized as a double-strand DNA damage signal by DNA repair enzymes. Several telomere-binding proteins regulate t-loop formation and influence telomere length (see below).

The early work of Muller (Muller 1938) and McClintock (McClintock 1941) first led to the concept that telomeres play a critical role in the maintenance of chromosomal integrity and cell viability. The inability of DNA polymerase to fully synthesize terminal ends of the lagging DNA strand (the end replication problem) leads to progressive telomere shortening with each round of replication. This is countered by the enzyme telomerase, a specialized ribonucleoprotein complex composed of a RNA subunit (Terc) that provides a template for addition of telomeric repeats, and a protein catalytic subunit (Tert). Deletion of either Terc or Tert genes results in progressive telomere attrition and, after a lag period, to loss of cell viability (Yu et al. 1990, Singer and Gottschling 1994, McEachern and Blackburn 1996). The lag period comprises cell divisions during which telomeres shorten before they reach a critical length that is the threshold for chromosome instability and cell death. A link between cell viability and telomere length was also confirmed in human fibroblasts in vitro wherein telomere shortening was shown to accompany a reduction of cell growth (Harley et al. 1990). Tert expression is repressed in human somatic cells but weakly expressed in some stem cells with self-renewal capacities, e.g., hematopoietic stem cells and intestinal stem cells. In human somatic tissues, the strict downregulation of telomerase accounts for the age-dependent decline in telomere lengths (Harley et al. 1990, Allsopp et al. 1992). Several studies have documented a decrease in telomere length in several human epithelial cell types, ranging from 50–100 bp per population doubling, for a total lifetime loss of approximately 2–4 kb (Allsopp et al. 1992, Harley et al. 1994). This rate of telomere length attrition would be significant in long-lived organisms such as humans.

Cells from most multicellular organisms have finite replicative life spans manifested by an eventual cessation of cell division. This phenomenon, termed replicative senescence, is thought to be important in human aging by the following observations: (1) there is a correlation between in vitro replicative life span and donor age (Martin et al. 1970), (2) a correlation exists between in vitro lifespan with the average life expectancy of the species (Rohme 1981), (3) individuals with premature aging syndromes have cells with a diminished in vitro replicative life span (Faragher et al. 1993), and (4) senescent cells accumulate in aged individuals (Dimri et al. 1995). In addition, there is evidence that senescent cells do not persist in vivo but are eliminated through apoptosis and immune responses (Choudhury et al. 2007, Cosme-Blanco et al. 2007, Xue et al. 2007). Thus the impact of telomere shortening and cellular senescence on organismal aging might be underestimated by simply examining cells positive for senescence markers (see also Gutierrez and Ju, this volume).

Mounting evidence supports the hypothesis that replicative senescence in cell culture is due to telomere attrition (reviewed in Campisi 1997, Sedivy 1998). Most primary human somatic cells lack telomerase activity and exhibit progressive telomere shortening with each cell division in culture (Harley et al. 1990). This telomere shortening eventually triggers entry into replicative senescence, also termed the Hayflick limit (Hayflick 1961) or mortality stage 1 (M1) (Wright and Shay 1992; see Allsopp, this volume). It has thus been proposed that telomere shortening could be a molecular clock that signals replicative senescence. Correspondingly, inactivation of p53 and pRb by antisense neutralization (Hara et al. 1991) or by viral oncoproteins (Shay et al. 1991) can extend replicative potential, driving further telomere erosion. This degree of telomere erosion triggers a new block to replication, termed mortality stage 2 (M2) or crisis, a period characterized by genomic instability, chromosomal fusions, and massive cell death (Counter et al. 1992; Wright and Shay 1992). Virally transformed human cells escape crisis at extremely low frequencies (Shay et al. 1993), and in order for M2-arrested human cells to emerge from the crisis period as immortalized cancer cells, telomerase must be upregulated or alternative telomere maintenance mechanisms (ALT) need to be activated to stabilize telomere length (see below) (Bryan et al. 1997).

Confirmation that telomere shortening indeed signals replicative senescence in vitro came from studies in which telomerase was inappropriately activated in normal human cells (Bodnar et al. 1998, Vaziri and Benchimol 1998). Activation of telomerase by enforced expression of human Tert in many different primary human cell types resulted in telomere elongation and a bypass of senescence. Instead of senescencing after a well-defined number of cell divisions, these clones exceeded their normal life span while maintaining both a youthful phenotype and a normal karyotype. These results clearly demonstrate that telomere length is the molecular clock that establishes the Hayflick limit, and that restoration of telomerase activity (hence telomere length) increases cellular life span.

2 A Six-Protein Core Complex Mediates Telomere End Protection

This telomeric t-loop is bound and stabilized by a number of telomere-specific binding proteins that form a six-protein complex, termed telosome (Liu et al. 2004) or shelterin (de Lange 2005). It is postulated that this highly regulated telomere structure serves three essential functions: (1) protecting natural chromosomal DNA ends from being inappropriately recognized as double-stranded breaks (DSBs) and therefore initiating an inappropriate DNA damage response (DDR), (2) protecting chromosomal ends from inappropriate enzymatic degradation, and (3) preventing chromosomal end-to-end fusion. Three sequence-specific DNA binding proteins are recruited to chromosomal ends: the duplex telomere-binding proteins TRF1 and TRF2 and the single-stranded TTAGGG repeat binding protein POT1 (Protection Of Telomeres 1). These proteins are interconnected by three additional proteins: TIN2, TPP1, and RAP1. TPP1 heterodimerizes with POT1 to augment its association with telomeres (Wang et al. 2007, Xin et al. 2007).

Mounting evidence suggests that POT1 is critical for telomere maintenance. It is highly conserved among eukaryotes and is the only protein in the complex that recognizes telomeric single-stranded DNA. All POT proteins examined to date contain two highly conserved oligonucleotide/oligosaccharide-binding folds (OB folds) that bind to the 3′ terminus of the single-stranded G-rich telomeric overhang (Baumann and Cech 2001, Loayza and de Lange 2003, Wu et al. 2006). In accord with its end protective function, deletion of POT1 in a large variety of organisms results in chromosomal end-to-end fusions and activation of a DNA damage response. In addition, POT1 is also required to protect telomeres from participating in aberrant homologous recombination (HR) at telomeres (Wu et al. 2006). Since t-loops resemble Holliday junctions, substrates that readily undergo HR, POT1 functions to prevent telomeres from engaging in inappropriate recombination reactions. Inappropriate HR at telomeres could result in rapid telomere shortening and engagement of a p53-dependent senescence response (see below).

The POT1-TPP1 complex appears to play a major role in telomere length regulation. The POT1-TPP1 complex functions to both positively and negatively control telomere length. As a positive length regulator, the POT1-TPP1 complex binds telomerase to increase its activity and processitivity on telomeres (Wang et al. 2007, Xin et al. 2007). As a negative length regulator, POT1 inhibits telomerase activity by limiting its access to the terminal G residue of telomeres (Kelleher et al. 2005, Lei et al. 2005). Therefore, the POT1-TPP1 complex might modulate telomere length by recruiting telomerase to dysfunctional telomeric ends to promote regulated telomere extension.

3 Telomere Dysfunction Results in Genomic Instability

Cytogenetic analysis by McClintock on breakage and fusion of maize chromosomes provided the first evidence that proper maintenance of telomeres is important for chromosomal integrity and genome stability. Dysfunctional telomeric ends are highly recombinogenic, leading to the improper chromosomal fusions, some of which may end up as dicentric chromosomes. During anaphase, dicentric chromosomes may be pulled to opposite ends of the cell, forming an anaphase bridge if the two kinetochores from the same chromatid are attached to the opposite spindle poles. Subsequent random chromosome breakage will generate broken chromosomes in the daughter cells, leading to the breakage-fusion-bridge (BFB) cycle proposed by McClintock. Since the break could occur anywhere along the anaphase bridge, this type of genomic instability may lead to loss of heterozygocity (LOH) in one daughter cell while the other daughter gains extra genetic information: regional chromosome amplification occurs if information from a homologous chromosome is gained or nonreciprocal translocations (NRTs) occur if information is from a different chromosome. Therefore, the occasional formation of dicentric chromosomes fueled by telomere dysfunction may trigger an endless chain of BFB events, constantly generating novel chromosomal variants with each cell division.

The importance of telomere length in regulating genome stability has been extended across many species. Fission yeast engineered without telomerase die after extensive passage, with rare survivors emerging with recombined (circularized) chromosomes to eliminate the need for telomeres entirely (Nakamura et al. 1997) or activate the ALT pathway to regenerate functional telomeres (Lundblad and Blackburn 1993). Elevated genomic instability characterized by the formation of NRTs via the BFB cycle was observed in telomerase-null *Sachromyces cerevisiae* possessing dysfunctional telomeres (Hackett et al. 2001). In late-generation mTerc-/- mice, chromosomes isolated from lymphocytes and embryonic fibroblasts exhibited p-arm-to-p-arm chromosomal fusions, a hallmark of telomere dysfunction (Blasco et al. 1997, Lee et al. 1998, Rudolph et al. 1999). Increased telomere loss leads to the formation of complex cytogenetic rearrangements, including dicentric chromosomes and the generation of NRTs (reviewed in Maser and DePinho 2002, Hande et al. 1999, Artandi et al. 2000, Chang et al. 2003). The decline in telomere length, rather than the absence of telomerase activity per se, appears to be the most important parameter dictating chromosomal integrity, since early-generation mTerc-/- mice, which still possess long telomeres, are cytogenetically and biologically normal. These results further highlight the importance of telomerase in preserving genome stability by preventing telomere dysfunction, thus inhibiting the initiation of BFB cycles and abrogation of their adverse effects.

4 Telomere Dysfunction Initiates p53-Dependent Replicative Senescence

Primary human cells have a limited replicative potential due to progressive telomere shortening, eventually resulting in the onset of replicative senescence, a state of permanent cell cycle arrest in which cells remain metabolically active, exhibit characteristic morphological changes, and stain for senescence-associated β-galactosidase (SA-β-gal) activity (Hayflick 1961, Harley et al. 1990, Allsopp et al. 1992, Wright and Shay 1992, Dimri et al. 1995; see also Allsopp, this volume). Replicative senescence has been shown to be due to dysfunctional telomeres engaging the p53 and pRb-dependent DDR (reviewed in Campisi 2005). In response to genotoxic stress, p53 initiates cell-cycle arrest, cellular senescence, or apoptosis to eliminate genomically unstable cells and suppress tumorigenesis (Schmitt 2003). Senescent human fibroblasts display molecular markers characteristic of cells bearing DNA double-strand breaks (reviewed in Jackson 2003). These markers include the DNA damage response proteins phosphorylated γH2AX, 53BP1, NBS1, and the DNA damage checkpoint kinase CHK2 (d'Adda di Fagagna et al. 2003, Taki et al. 2003). Many of these proteins localize directly to dysfunctional telomeres, and their inactivation in senescent cells restores cell cycle progression into S-phase (d'Adda di Fagagna et al. 2003). These results suggest that dysfunctional telomeres are recognized by the DNA damage machinery as double-strand DNA breaks, which signal entry into replicative senescence. DNA damage response proteins may therefore be excellent biochemical markers for the detection of replicative senescence in vivo, a hypothesis that can be tested in premature aging mouse models and in human progeric syndromes.

Although a definitive connection between telomere dynamics and normal aging in humans has yet to be established, accumulating evidence has strengthened the view that accelerated telomere attrition contributes directly to acquired and inherited degenerative conditions and premature aging syndromes such as Werner Syndrome (see below and also Davis and Kipling, this volume). In support of this notion, analysis of telomere lengths in cells derived from peripheral blood of humans over age 60 revealed that individuals possessing shorter telomeres than age-matched controls had significantly poorer survival rates, which was attributed in part to an elevated mortality rate from heart and infectious diseases (Cawthon et al. 2003). Increased telomere shortening also correlates with increased incidence of coronary heart disease and elevated mortality (Brouilette et al. 2007). While it is somewhat surprising that the function of a postmitotic tissue such as the heart might be adversely impacted upon by dysfunctional telomeres, it is possible that cardiac stem cells with long telomeres are required to confer long-term cardiac protection.

It is postulated that senescent cells accumulate with normal aging and may contribute to age-related pathologies by inhibiting tissue regenerative capacities (Krtolica and Campisi 2002, Itahana et al. 2001). Increasing evidence suggests that senescent cells do accumulate in aging human tissues. For example, SA-β-gal-positive senescent cells have been identified in aged liver, atherosclerotic plaques, and skin (Dimri et al. 1995, Vasile et al. 2001). Perhaps the best evidence that senescent cells exist in vivo comes from studies of aging baboon skin: ~15% of aged baboon skin fibroblasts possess prominent foci of DNA damage markers, including γ-H2AX, 53-BP1, and phosphorylated ATM kinase that co-localize with telomeres, indicating that telomeres are dysfunctional in aging skin (Herbig et al. 2006). These results suggest that in old primates, telomere dysfunction activates the ATM-dependent DNA-damage signaling pathway to initiate cellular senescence in tissues, likely negatively impacting upon tissue physiology and function.

5 Consequences of Telomerase Loss In Vivo: The Telomerase Knockout Mouse

Telomeric repeats in human cells are typically between 10 and 15 kb in length. Furthermore, telomerase activity is stringently regulated in human tissues, i.e., undetectable in many somatic tissues but present in germ cells, activated leukocytes, and stem cells from a variety of organs (Harley et al. 1994, Wright et al. 1996, Newbold 1997, Weng et al. 1997). This regulation is achieved almost entirely through stringent downregulation of hTert (Meyerson et al. 1997). In contrast to humans, the laboratory mouse species *Mus musculus* possesses much longer telomeres, ranging in size from 40 to 80 kb, although some chromosomes harbor significantly shorter telomeres (Blasco et al. 1997, Zijlmans et al. 1997). In addition, telomerase activity and gene expression of mouse Tert (mTert) appear to be less stringently regulated in murine somatic cells (Kipling and Cooke 1990, Prowse et al. 1993, Prowse and Greider 1995, Greenberg et al. 1998, Martin-Rivera et al. 1998). Despite these species-specific differences, mTert expression appears to be the critical determinant of telomerase activity in mouse cells, since mTert expression correlates well with the distribution of telomerase activity in mouse tissues and cells (Greenberg et al. 1998). Although murine telomerase is regulated less stringently, murine telomeres do exhibit telomere shortening during development and somatic growth in some tissues, implying that the low levels of telomerase present may be inadequate or that other factors allowing telomere access may be limiting (Prowse and Greider 1995, Rudolph et al. 1999).

Work from a number of laboratories has shown that the mouse is an excellent system for studying telomere biology (reviewed in Chang et al. 2001, Goytisolo and Blasco 2002). The telomerase-deficient mouse lacks the critical RNA subunit mTerc, is viable, fertile, and has no significant morphological abnormalities (Blasco et al. 1997). Telomere length measurements revealed that, as in yeast and human cells lacking telomerase, telomeres shortened by approximately 120 bp per cell division. Given the long telomeres in mice, successive generations of telomerase-deficient mice were produced to achieve sufficient telomere attrition. By the sixth generation (G6), defects were revealed in multiple tissue compartments with high proliferative histories, including hematopoietic and reproductive systems. Defects in the hematopoietic system were manifested by a marked decline in hematopoietic precursor cell numbers, reduced proliferation of T and B lymphocytes upon mitogenic stimulation, and splenic atrophy, with a reduction in germinal center function (Lee et al. 1998, Herrera et al. 2000). Decreased fecundity was first noted in G4 mice, culminating in sterility in late-generation (G6) animals. Both male and female reproductive functions were severely compromised at this point, with the males displaying marked testicular atrophy accompanied by germ cell depletion, leading to a complete absence of spermatogenesis. In the females, a decrease in the number of oocytes upon ovulation was observed (Lee et al. 1998). Importantly, reconstitution of telomerase activity in late-generation mTerc-/- mice increased telomere lengths preferentially at the shortest telomeres and rescued many of the pathologies associated with telomere dysfunction (Hemann et al. 2001, Samper et al. 2001). Taken together, these results strongly support the view that the long-term renewal of germ and hematopoietic cells is dependent upon telomere function, and that telomerase expression is vital in long-term cellular survival and organ homeostasis (see also Zimmermann and Martens, this volume, Gutierrez and Ju, this volume).

Owing to its anti-proliferative effects, replicative senescence has long been postulated to be a potent tumor suppression mechanism in vivo (reviewed by Campisi 2005; see also Rudolph, this volume, and Blasco, this volume). Mouse models bearing dysfunctional telomeres have been generated to address this hypothesis. The *INK4a*$^{-/-}$ mouse lacks both *p16INK4a* and *p19ARF*, which have important roles in both the pRb and p53 pathways. Importantly, the p53-dependent DDR is intact in the *INK4a*$^{-/-}$ mouse. Treatment of early-generation *mTerc*$^{-/-}$ *INK4a*$^{-/-}$ mice with DMBA and UVB revealed that these mice are highly cancer prone (Greenberg et al. 1999).

However, similar treatment of late-generation *mTerc*$^{-/-}$*INK4a*$^{-/-}$ mice with short dysfunctional telomeres yielded a reduction in tumor incidence and much longer survival. This result suggests that in the setting of telomere dysfunction and an intact *p53* checkpoint, tumor formation is impaired. A similar finding was observed in a skin carcinogenesis model, in which late-generation *mTerc*$^{-/-}$ mice produced significantly fewer skin tumors upon chemical carcinogenesis of the skin compared to wild-type controls with long telomeres (Gonzalez-Suarez et al. 2000). Upregulation of p53 was detected in late-generation *mTerc*$^{-/-}$ papillomas, most likely due to dysfunctional telomeres being sensed as DSBs, resulting in induction of p53 levels.

However, given that the aforementioned mouse models all possess wild-type p53, it is not clear how dysfunctional telomeres limit neoplastic growth in vivo. Which function of p53 (apoptosis versus cell cycle arrest/senescence) is more important for telomere-dependent tumor suppression? The generation of *mTerc*$^{-/-}$ mice with dysfunctional telomeres in the setting of a p53 knock-in mutation (p53^{R172P} allele) that is incapable of initiating a p53-dependent apoptosis but is competent to execute p53-mediated cell cycle arrest/replicative senescence allowed this question to be addressed directly (Cosme-Blanco et al. 2007). In this system, dysfunctional telomeres induce p53-mediated replicative senescence to suppress spontaneous tumorigenesis, while p53-dependent apoptosis appears dispensable for tumor suppression. Dysfunctional telomere-induced senescence was accompanied by robust increase in p53, p21, and SA-β-gal activity in all tissue compartments examined, suggesting that a telomere-dependent DDR is activated in vivo. These observations parallel a recent study indicating that p53 is dispensable for tumor suppression induced by radiation damage (a process mediated mainly through p53's apoptotic functions), while suppression of radiation-induced lymphomagenesis requires an intact p19 pathway (presumably to initiate oncogene-mediated cellular senescence) (Christophorou et al. 2006). Since the DDR is activated at the earliest stages in many human carcinomas (Bartkova et al. 2005), these results predict that activation of an intact DDR pathway by dysfunctional telomeres would promote p53-dependent senescence, suppressing further tumor progression. The increased genome instability in nascent tumor cells that stochastically inactivate the DDR and/or the p53-p21 dependent senescence pathways would promote tumor progression. However, recent studies in *mTERC*$^{-/-}$*p21*$^{-/-}$ mice revealed that deletion of p21 improved stem cell function, organ maintenance, and life span of mice possessing dysfunctional telomeres without accelerating tumor formation (Choudhury et al. 2007; see also Rudolph, this volume). The impact of cell cycle checkpoints and apoptosis for tumor suppression in response to telomere shortening thus needs to be better defined in future studies.

6 Telomere Shortening and Organismal Aging

Genomic instability has long been recognized as a common characteristic of both aging and cancer cells. Accumulation of genomic changes such as random point mutations has been proposed as a cause of aging (Dolle et al. 1997). In support of this hypothesis, increasing evidence links human premature aging syndromes and mouse models of premature aging with DNA damage–induced genome dysfunction. Mutant mice unable to repair accumulated DNA damages exhibit symptoms of premature aging and die early (Rudolph et al. 1999, Tyner et al. 2002, Cao et al. 2003, de Boer et al. 2002, Wong et al. 2003), suggesting that proper maintenance of genomic integrity is essential for longevity. Late-generation telomerase null animals display a subset of aging phenotypes, including alopecia, hair greying, a reduced stress response, and decreased longevity (Rudolph et al. 1999). These animals also exhibit elevated genomic instability, manifested as chromosomal fusions and increased cancer incidence. Enhanced sensitivity to genotoxic agents such as ionizing radiation and alkylating agents is also noted in late generation *mTerc*$^{-/-}$ mice and derivative cell lines (Wong et al. 2000, Goytisolo et al. 2000, Lee et al. 2001), suggesting that activation of the DNA damage response pathways cooperates with dysfunctional telomeres to further compromise cellular functions in vivo. In support of this notion, inactivating p53 in late-generation mTerc$^{-/-}$ mice rescues many of the associated premature aging phenotypes (Chin et al. 1999).

Although telomeres shorten in aged individuals, and telomere shortening triggers replicative senescence in cultured human cells (Allsopp et al. 1992), the causal contribution of telomere shortening to organismal aging remained unclear until the advent of the telomerase knockout mouse (Rudolph et al. 1999, Herrera et al. 2000). The generation of additional mouse models possessing dysfunctional telomeres therefore promised to further illuminate the importance of telomere attrition during organismal aging. The ATM kinase plays a critically important role in sensing double-strand DNA breaks and in the coordination of how this signal is relayed to downstream DNA damage checkpoint proteins, in particular p53 (reviewed in Shiloh 2003). Late-generation *mTerc*$^{-/-}$*ATM*$^{-/-}$ mice experience increased telomere attrition, elevated genomic instability, and generalized proliferation defects across all cellular compartments examined, leading to the onset of premature aging phenotypes and early death (Wong et al. 2003). Interestingly, late-generation *mTerc*$^{-/-}$*ATM*$^{-/-}$ neuronal stem cells failed to proliferate and differentiate into viable neurons, possibly explaining the neurodegenerative phenotype observed in AT patients. An unexpected finding is the near total suppression of lymphomas in late-generation *mTerc*$^{-/-}$*ATM*$^{-/-}$ mice (Qi et al. 2003, Wong et al. 2003), in marked contrast to the increased cancer incidence observed in the *p53*$^{-/-}$*mTerc*$^{-/-}$ mouse (Chin et al. 1999). These contrasting phenotypes could be explained by the fact that p53 can still be constitutively activated in *mTerc*$^{-/-}$*ATM*$^{-/-}$ mice by dysfunctional telomeres, leading to tumor suppression. Indeed, constitutive activation of a hyperfunctional mutant p53 in the mouse germ line has been shown to suppress tumorigenesis while promoting the onset of premature aging phenotypes in mice (Tyner et al. 2002). These results suggest that inappropriate activation of p53 by dysfunctional telomeres in the setting of ATM deficiency may lead to the manifestation of aging phenotypes observed in *mTerc*$^{-/-}$*ATM*$^{-/-}$ mice. Support for this hypothesis comes from the observation that in late-generation *mTerc*$^{-/-}$ mice, dysfunctional telomeres upregulate p53 in diverse cellular compartments including the skin and heart (Gonzalez-Suarez et al. 2000, Leri et al. 2003).

Perhaps the most direct evidence linking telomere dynamics and age-related degenerative conditions in humans comes from the investigations of the human disease, Dyskeratosis congenita (DC) (Dokal 2001; see also Du et al., this volume). DC is a multisystem disorder characterized by a triad of skin abnormalities, including abnormal skin pigmentation, nail dystrophy, and mucosal leukoplakia. DC patients also experience increased cancer risk and most often succumb to bone marrow failure by the fourth decade of life. These phenotypes bear a striking resemblance to those observed in the *mTerc* null mouse (Blasco et al. 1997, Lee et al. 1998), and in a satisfying confirmation of the relevance of the telomerase knockout mouse in modeling human telomere biology, the autosomal-dominant form of DC was found to be linked to mutations in genes governing *Terc* stability (Mitchell et al. 1999) or in the human *Terc* gene itself (Vulliamy et al. 2001). These mutations result in diminished telomerase activity in critical stem cell compartments, leading to telomere dysfunction, premature bone marrow depletion, and ultimately bone marrow failure. Similar to the generational effects seen in the *mTerc* null mouse, patients with DC are more severely affected in later generations, most likely due to the inheritance of progressively shorter telomeres (Vulliamy et al. 2004). This disease anticipation, in which the onset of disease occurs at progressively younger ages in subsequent generations, is observed only in siblings inheriting a defective copy of *hTerc*. These genetic observations strongly implicate critical telomere shortening in successive generations as a mechanism for disease anticipation, and support the notion that a steady increase in the level of telomere dysfunction could contribute to age-related disease processes in normal elderly individuals. Finally, the concordant biological features between the telomerase knockout mouse and DC patients provided a measure of validation that the mouse can serve as a relevant model organism to dissect the complex roles of telomeres in human pathobiology including normal aging, age-related disorders, and inherited premature aging syndromes.

7 The $mTerc^{-/-}$ $Wrn^{-/-}$ Compound Mutant Mouse as a Model of Human Werner Syndrome

Although a definitive connection between telomere dynamics and normal aging in humans has yet to be established, accumulating evidence has strengthened the view that accelerated telomere attrition contributes directly to acquired and inherited degenerative conditions and premature aging syndromes such as Werner Syndrome (WS; see also Davis and Kipling, this volume). WS is characterized by early onset of age-related pathologies and cancer. The protein mutated in WS, WRN, appears to play a major role in genome stability, particularly during DNA replication and telomere metabolism. Much of the pathophysiology associated with WS, including the rapid onset of cellular senescence, early cancer onset, and premature aging, can be attributed to a defect in telomere maintenance. Recent genetic evidence from the telomerase-Wrn double knockout mouse demonstrates that mice with critically shortened telomeres display aging phenotypes reminiscent of human WS, further reinforcing the notion that telomere dysfunction is required for the manifestation of aging pathophysiologies in the setting of *WRN* deficiency (Chang et al. 2004, Du et al. 2004).

Extensive physiological analyses of telomerase-knockout mice revealed that late-generation *mTerc*$^{-/-}$ animals that have short, dysfunctional telomeres exhibit decreased lifespan, a range of aging phenotypes including alopecia, hair graying, a reduced capacity to cope with acute and chronic stress, and a modestly increased incidence of cancer (Rudolph et al. 1999). Age-matched animals at earlier generations that possess longer telomeres do not manifest these phenotypes, which suggests that telomere dysfunction is important for the emergence of premature aging phenotypes. However, only a subset of these human aging phenotypes is present in mice with dysfunctional telomeres, suggesting that other factors contribute to human aging.

Interestingly, despite the wide-ranging activities of the WRN helicase in human cells, *Wrn* null mice do not exhibit any cellular or clinical phenotypes characteristic of human WS (Lombard et al. 2000). The observations that WRN is required for telomere maintenance, and that laboratory mice normally possess very long telomeres, suggest that manifestation of premature aging phenotypes in *Wrn*-null mice may require the presence of critically short telomeres. This hypothesis has since received genetic support from studies of mice lacking both telomerase and *Wrn* (Chang et al. 2004, Du et al. 2004). Strikingly, these late-generation *mTerc*$^{-/-}$*Wrn*$^{-/-}$ mice display clinical symptoms resembling human WS, including early onset of age-related disorders such as impaired wound repair, osteoporosis and skeletal fractures, hypogonadism, cataract formation, type II diabetes, and premature death. These mice also exhibit an elevated incidence of mesenchymal cancers compared with control mice (Chang et al. 2004, Du et al. 2004). Compared to age-matched controls, quantitative telomere-FISH revealed that chromosomes from late-generation *mTerc*$^{-/-}$*Wrn*$^{-/-}$ mice displayed elevated loss in overall telomere length that correlated with increased chromosomal fusions and nonreciprocal translocations (NRTs) (Chang et al. 2004, Du et al. 2004). Late-generation *mTerc*$^{-/-}$*Wrn*$^{-/-}$ mouse embryo fibroblasts (MEFs) exhibited prominent DNA damage markers such as 53BP1 and γ-H2AX and profound reduction in replicative life span (Chang et al. 2004), which is similar to growth defects observed in human WS fibroblasts. Dysfunctional telomeres thus appear to cooperate with *WRN* deficiency to activate the DNA damage response and subsequent entry into replicative senescence. It is important to emphasize that the phenotype observed in the *mTerc*$^{-/-}$*Wrn*$^{-/-}$ double mutant is not simply a worsening of aging phenotypes observed in the telomerase-null mouse, but a recapitulation of specific phenotypes encountered in WS patients that are not observed in late-generation *mTerc*$^{-/-}$ mice. Data from these mouse models therefore strongly support the hypothesis that manifestation of the diverse pathophysiological phenotypes observed in WS patients requires both the presence of dysfunctional telomeres as well as WRN deficiency.

Given the elevated number of chromosomal aberrations and robust DNA damage response observed in organs and cells derived from late-generation *mTerc*$^{-/-}$*Wrn*$^{-/-}$ mice, it is tempting to speculate that a key factor underlying WS may be the inability of WS cells to suppress elevated genomic instability. When p53 is lost, the genomic instability induced by dysfunctional telomeres can provide a mutator mechanism enabling would-be cancer cells to achieve a threshold of cancer-promoting changes required for tumorigenesis (Artandi et al. 2000, Rudolph et al. 1999, O'Hagan et al. 2002). To understand the impact of *Wrn* deficiency on telomere maintenance, chromosomal stability, and tumorigenesis, G5 *mTerc*$^{-/-}$*Wrn*$^{-/-}$ mouse cell lines were allowed to undergo spontaneous cellular senescence. Surprisingly, loss of *Wrn* in this setting promoted escape from cellular senescence and the emergence of immortalized clones in which the p53 pathway is compromised (Laud et al. 2005). These clones continue to shorten their telomeres and exhibit marked chromosomal aberrations. However, telomere sister chromatid exchange (T-SCE), an indicator of homologous recombination (HR) between sister telomeres, specifically increases in these cells. Restoration of wild-type, but not helicase dead, WRN to G5 *mTerc*$^{-/-}$*Wrn*$^{-/-}$ cell lines decreased T-SCE levels (Laud et al. 2005), which suggests that the helicase domain of WRN is required to repress inappropriate telomeric recombination. Sgs1, the yeast homolog of WRN, represses recombination at homeologous sequences (Myung et al. 2001, Sugawara et al. 2004), and this observation is consistent with a role for WRN in repressing HR at homeologous telomeric sequences. The observation that human ALT cell lines also display elevated T-SCE (Bechter et al. 2004, Londono-Vallejo et al. 2004) suggests that HR-mediated telomere recombination mechanisms are important prerequisites for transition to the ALT phenotype.

When immortalized G5 *mTerc*$^{-/-}$*Wrn*$^{-/-}$ clones are injected subcutaneously into SCID mice, aberrant telomeric HR, coupled with the strong selective pressure to maintain telomere length in the absence of telomerase activity, results in the activation of ALT and rapid formation of tumors (Laud et al. 2005). T-SCE levels in ALT tumor cell lines are not repressed when telomerase is expressed in these cells, which is consistent with previous observations indicating that once ALT is activated, it is not normally repressible by telomerase expression (Cerone et al. 2001, Perrem et al. 2001, Londono-Vallejo et al. 2004). These results indicate that the increased incidence of chromosomal instability and cancer observed in WS patients may be due to aberrant recombination at telomeres, which promotes the activation of ALT that maintains telomeres through a mechanism analogous to the type I survival pathway observed in *sgs1*$^{-/-}$, telomerase-null yeasts. While not all Wrn, telomerase-null human cells maintain telomeres by this mechanism, the recent discovery of a Wrn, telomerase-null SV40-immortalized ALT cell line that maintains telomeres in a manner reminiscent of the type I survival pathway supports this hypothesis (Fasching et al. 2005, Marciniak et al. 2005). The telomerase-Werner compound knockout mouse thus offers a unique system to discover novel genes involved in ALT activation during tumor formation.

8 Conclusions

The telomerase-deficient mouse has been invaluable in providing insights into basic questions pertaining to consequences of telomere dysfunction during aging and cancer in the context of the mammalian organism. Studies in this mouse model have demonstrated that cellular responses to telomere dysfunction are fundamentally conserved in both humans and mice, and that p53 plays a key role in sensing and mediating responses to dysfunctional telomeres Figure 4.1. Telomere dysfunction leads to genomic instability and activation of the DNA damage response checkpoints. In the setting of an intact p53 pathway, telomere dysfunction in late-generation *mTerc-/-* mice activates p53, leading to tumor suppression as a result of increased cellular senescence (Cosme-Blanco et al. 2007). However, inappropriate activation of p53 also promotes premature organismal aging (Rudolph et al. 1999, Herrera et al. 2000). On the other hand, loss of p53 function cooperates with dysfunctional telomeres to promote carcinogenesis at the expense of premature aging (Chin et al. 1999). The conclusion suggested by these contrasting results argue that

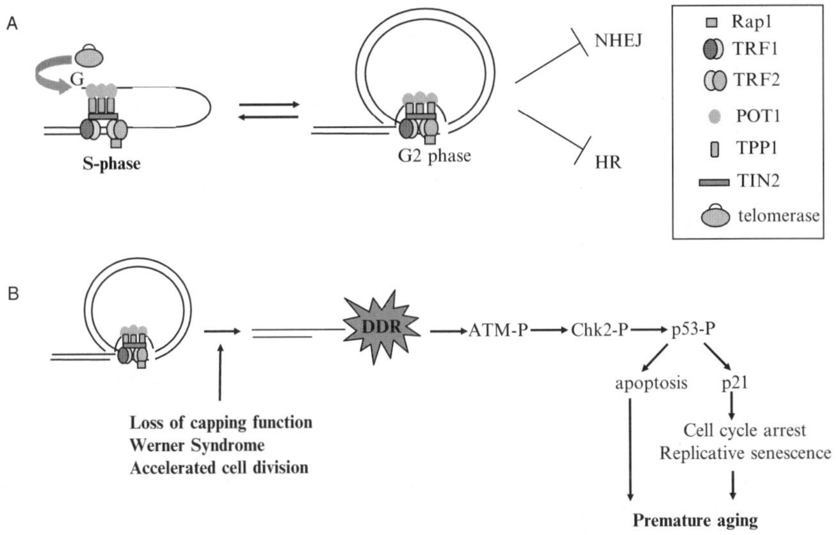

Fig. 4.1 **(A)** Telomere structure. The telomere folds back onto itself to form a double-stranded t-loop and a single-stranded d-loop. This complex protects telomeres at the G2 phase of the cell cycle from inappropriate nonhomologous end joining (NHEJ) and homologous recombination (HR)–mediated repair of telomeric DNA. The six-component core protein complex is shown schematically. Telomere extension by telomerase requires translocation of POT1 to an internal site, permitting telomerase to access the terminal G-residue. B) Telomere dysfunction elicits a DNA damage response at telomeres and activation of ATM, resulting in p53-meduated cell cycle arrest/replicative senescence or apoptosis to trigger the onset of premature aging symptoms (*See Color Plate*)

how telomere length is sensed by p53 in an organism may be one important factor in mediating the balance between predisposition to aging or cancer.

With respect to WS, important clues regarding its function have emerged from mouse models. For example, premature entry into replicative senescence, a hallmark of WS fibroblasts, is likely to be due in part to dysfunctional telomeres activating a p53-dependent DNA damage response, culminating in premature entry into cellular senescence. The demonstration that WS cells with dysfunctional telomeres accumulate markers of DNA damage (Chang et al. 2004, Szekely et al. 2005) and the intriguing observation that aged primate fibroblasts also bear identical markers of telomere dysfunction (Herbig et al. 2006) suggest a causal link between dysfunctional telomere-induced cellular senescence and organismal aging. This link is further supported by genetic evidence from the *mTerc*$^{-/-}$*Wrn*$^{-/-}$ mouse model demonstrating that only mice with critically shortened telomeres manifest organismal aging phenotypes. Accumulation of prematurely senescent cells in vivo probably has profound adverse physiological consequences. I speculate that the transition from normalcy at birth to the onset of premature aging phenotypes observed at the second decade of life in WS patients relates to the erosion of a population of telomeres to a critically short threshold that is capable of initiating a DNA damage response. This view is in line with the emerging consensus that telomere maintenance, DNA repair, and DNA metabolism pathways are intertwined and function either to suppress or precipitate aging and cancer phenotypes, depending upon the status of the DNA damage pathway.

References

Allsopp RC, Vaziri H, Patterson C, Goldstein S, Younglai EV, Futcher AB, Greider CW, Harley CB (1992) Telomere length predicts replicative capacity of human fibroblasts. Proc Natal Aced Sci U S A 89: 10114–18

Artandi S, Chang S, Lee S, Alson, S, Gottlieb G, Chin L, and Depinho RA (2000) Telomere dysfunction promotes non-reciprocal translocations and epithelial cancers in mice. Nature 406: 641–45

Artandi S, Alson S, Tietze M, Sharpless N, Ye S, Greenberg RA, Castrillon DH, Horner JW, Weiler S, Carrasco R, DePinho, RA (2002) Constitutive telomerase expression promotes mammary carcinomas in aging mice. Proc Natl Acad Sci U S A 99: 8191–94

Bartkova J, Horejsi Z, Koed K, Kramer A, Tort F, Zieger K, Guldberg P, Sehested M, Nesland JM, Lukas C, Orntoft T, Lukas J, Bartek J (2005) DNA damage response as a candidate anti-cancer barrier in early human tumorigenesis. Nature 434: 864–70

Baumann P, Cech TR (2001) Pot1, the putative telomere end-binding protein in fission yeast and humans. Science 292: 1171–75

Blackburn EH (2001) Switching and signaling at the telomere. Cell 106: 661–73

Blasco MA, Lee HW, Hande MP, Samper E, Lansdorp PM, DePinho RA., Greider CW (1997) Telomere shortening and tumor formation by mouse cells lacking telomerase RNA. Cell 91: 25–34

Bodnar AG, Ouellette M, Frolkis M, Holt SE, Chiu CP, Morin GB, Harley CB, Shay JW, Lichtsteiner S, Wright WE (1998) Extension of life-span by introduction of telomerase into normal human cells [see comments]. Science 279: 349–52

Brouilette SW, Moore JS, McMahon AD, Thompson JR, Ford I, Shepherd J, Packard CJ, Samani NJ; West of Scotland Coronary Prevention Study Group (2007) Telomere length, risk of coronary heart disease, and statin treatment in the West of Scotland Primary Prevention Study: a nested case-control study. Lancet 369: 107–14

Bryan TM, Englezou A, Dalla-Pozza L, Dunham MA, Reddel RR (1997) Evidence for an alternative mechanism for maintaining telomere length in human tumors and tumor-derived cell lines. Nat Med 3: 1271–74

Campisi J (1997) The biology of replicative senescence. Eur J Cancer 33: 703–09

Campisi J (2005) Senescent cells, tumor suppression, and organismal aging: good citizens, bad neighbors. Cell 120: 513–22

Cao L, Li W, Kim S, Brodie SG, Deng CX (2003) Senescence, aging, and malignant transformation mediated by p53 in mice lacking the Brca1 full-length isoform. Genes Dev 17: 201–13

Cawthon RM, Smith KR, O'Brien E, Sivatchenko A, Kerber RA (2003) Association between telomere length in blood and mortality in people aged 60 years or older. Lancet 361: 393–95

Chang S, Khoo C, DePinho RA (2001) Modeling chromosomal instability and epithelial carcinogenesis in the telomerase deficient mouse. Sem Oncol 11: 227–38

Chang S, Khoo C, Naylor M, Maser R, DePinho RA (2003) Telomere-based crisis: functional differences between telomerase activation and ALT in tumor progression. Genes Dev 17: 88–100

Chang S, Multani A S, Cabrera NG, Naylor MI, Laud P, Lombard, D, Pathak S, Guarente L, DePinho RA (2004) Essential role of limiting telomeres in the [athogenesis of Werner Syndrome. Nature Gen 36: 877–82

Chin L, Artandi S, ShenQ, Tam A, Lee S, Gottlieb G, Greider CW, DePinho RA (1999) p53 deficiency rescues the adverse effects of telomere loss and cooperates with telomere dysfunction to accelerate carcinogenesis. Cell 97: 527–38

Choudhury AR, Ju Z, Djojosubroto MW, Schienke A, Lechel A, Schaetzlein S, Jiang H, Stepczynska A, Wang C, Buer J, Lee HW, von Zglinicki T, Ganser A, Schirmacher P, Nakauchi H, Rudolph KL (2007) Cdkn1a deletion improves stem cell function and lifespan of mice with dysfunctional telomeres without accelerating cancer formation. Nat Genet 39: 99–105

Christophorou MA, Martin-Zanca D, Soucek L, Lawlor ER, Brown-Swigart L, Verschuren EW, Evan GI (2006) The pathological response to DNA damage does not contribute to p53-mediated tumour suppression. Nature 443: 214–17

Cosme-Blanco W, Shen W, Lazar A, Pathak S, Lozano G, Multani AS, and Chang S (2007) Telomere dysfunction suppresses spontaneous tumorigenesis *in vivo* by activating p53-mediated cellular senescence. EMBO Reports, in press.

Counter CM, Avilion AA, LeFeuvre CE, Stewart NG, Greider CW, Harley CB, Bacchetti S (1992) Telomere shortening associated with chromosome instability is arrested in immortal cells which express telomerase activity. EMBO J 11: 1921–29

Counter CM, Meyerson M, Eaton EN, Weinberg RA (1997) The catalytic subunit of yeast telomerase. Proc Natl Acad Sci USA 94: 9202–07

d'Adda di Fagagna F, Reaper PM, Clay-Farrace L, Fiegler H, Carr P, Von Zglinicki T, Saretzki G, Carter NP, Jackson SP (2003) A DNA damage checkpoint response in telomere-initiated senescence. Nature 426: 194–98

de Boer J, Andressoo JO, de Wit J, Huijmans J, Beems RB, van Steeg H, Weeda G, van der Horst GT, van Leeuwen W, Themmen AP, Meradji M, Hoeijmakers JH (2002) Premature aging in mice deficient in DNA repair and transcription. Science 296: 1276–79

de Lange, T (2005) Shelterin: the protein complex that shapes and safeguards human telomeres. Genes Dev 19: 2100–10

Dimri GP, Lee X, Basile G, Acosta M, Scott G, Roskelley C, Medrano EE, Linskens M, Rubelj I, Pereira-Smith O (1995) A biomarker that identifies senescent human cells in culture and in aging skin *in vivo*. Proc Natl Acad Sci USA 92: 9363–67

Dokal I (2001) Dyskeratosis congenita. A disease of premature ageing. Lancet 358 Suppl, S27.

Dolle ME, Giese H, Hopkins CL, Martus HJ, Hausdorff JM, Vijg J (1997) Rapid accumulation of genome rearrangements in liver but not in brain of old mice. Nat Genet 17: 431–34

Du X, Shen J, Kugan N, Furth E, Lombard D, Cheung C, Pak S, Luo G, Pignolo R, DePinho RA, Guarente L, Johnson FB (2004) Telomere shortening exposes functions for the Werner and Bloom syndrome genes in mice. Mol Cell Biol. 19: 8437–46

Faragher RG, Kill IR, Hunter JA, Pope FM, Tannock C, Shall S (1993) The gene responsible for Werner syndrome may be a cell division "counting" gene. Proc Natl Acad Sci U S A 90: 12030–34

Gonzalez-Suarez E, Samper E, Flores JM, Blasco, MA (2000) Telomerase-deficient mice with short telomeres are resistant to skin tumorigenesis. Nat Genet 26: 114–17

Goytisolo FA, Samper E, Martin-Caballero J, Finnon P, Herrera E, Flores JM, Bouffler SD, Blasco MA (2000) Short telomeres result in organismal hypersensitivity to ionizing radiation in mammals. J Exp Med. 192: 1625–36

Goytisolo FA Blasco MA (2002) Many ways to telomere dysfunction: in vitro studies using mouse models. Oncogene 21: 584–91

Greenberg RA, Allsopp RC, Chin L, Morin GB, DePinho RA (1998) Expression of mouse telomerase reverse transcriptase during development, differentiation and proliferation. Oncogene 16: 1723–30

Greenberg RA, Chin L, Femino A, Lee KH, Gottlieb GJ, Singer RH, Greider CW and DePinho, RA (1999) Short dysfunctional telomeres impair tumorigenesis in the INK4a(delta2/3) cancer-prone mouse. Cell 97: 515–25

Greider CW (1996) Telomere length regulation. Annu Rev Biochem 65: 337–65

Griffith JD, Comeau L, Rosenfield S, Stansel RM, Bianchi A, Moss H, and de Lange T (1999) Mammalian telomeres end in a large duplex loop. Cell 97: 503–14

Hackett JA, Feldser DM, Greider CW (2001) Telomere dysfunction increases mutation rate and genomic instability. Cell 106: 275–86

Hande, MP, Samper, E, Lansdorp, P Blasco MA (1999) Telomere length dynamics and chromosomal instability in cells derived from telomerase null mice. J Cell Biol 144: 589–601

Hara E, Tsurui, H, Shinozaki, A, Nakada, S Oda, S (1991) Cooperative effect of antisense-Rb and antisense-p53 oligomers on the extension of life span in human diploid fibroblasts, TIG-1. Biochem Biophys Res Commun 179: 528–34

Harley CB, Futcher AB, Greider CW (1990) Telomeres shorten during ageing of human fibroblasts. Nature 345: 458–60

Harley CB, Kim NW, Prowse KR, Weinrich SL, Hirsch KS, West MD, BacchettiS, Hirte HW, Counter CM, Greider CW (1994). Telomerase, cell immortality, and cancer. Cold Spring Harb Symp Quant Biol 59: 307–15

Harrington L, Zhou W, McPhail T, Oulton R, Yeung DS, Mar V, Bass MB, and Robinson MO (1997) Human telomerase contains evolutionarily conserved catalytic and structural subunits. Genes Dev 11: 3109–15

Hayflick L (1961) The serial cultivation of human diploid cell strains. Exp. Cell Res. 25: 585–621

Hemann MT, Strong, MA, Hao, LY, Greider, CW (2001) The shortest telomere, not average telomere length, is critical for cell viability and chromosome stability. Cell 107: 67–77

Herbig U, Ferreira M, Condel L, Carey D, Sedivy JM (2006) Cellular senescence in aging primates. Science 311: 1257

Herrera E, Martinez AC, Blasco MA (2000) Impaired germinal center reaction in mice with short telomeres. EMBO J 19: 472–81

Itahana K, Dimri G, Campisi J (2001) Regulation of cellular senescence by p53. Eur J Biochem 268: 2784–91

Kelleher C, Kurth I, Lingner J (2005) Human protection of telomeres 1 (POT1) is a negative regulator of telomerase activity in vitro. Mol Cell Biol. 25: 808–18

Kipling D, Cooke HJ (1990) Hypervariable ultra-long telomeres in mice. Nature 347: 400–02

Krtolica A, Campisi J (2002) Cancer and aging: a model for the cancer promoting effects of the aging stroma. Int J Biochem Cell Biol 34: 1401–14

Laud PR, Multani AS, Bailey SM, Wu L, Ma J, Kingsley C, Lebel M, Pathak S, DePinho RA, Chang S (2005) Elevated telomere-telomere recombination in WRN-deficient, telomere dysfunctional cells promotes escape from senescence and engagement of the ALT pathway. Genes Dev 19: 2560–70

Lee HW, Blasco MA, Gottlieb GJ, Horner JW, Greider CW, DePinho R A (1998) Essential role of mouse telomerase in highly proliferative organs. Nature 392: 569–74

Lei M, Zaug AJ, Podell ER, Cech TR (2005) Switching human telomerase on and off with hPOT1 protein in vitro. J Biol Chem. 280: 20449–56

Liu D, O'Connor MS, Qin J, Songyang Z (2004) Telosome, a mammalian telomere-associated complex formed by multiple telomeric proteins. J Biol Chem. 279: 51338–42

Lombard DB, Beard C, Johnson B, Marciniak RA, Dausman J, Bronson R, Buhlmann JE, Lipman R, Curry R, Sharpe A, Jaenisch R, Guarente L (2000) Mutations in the WRN gene in mice accelerate mortality in a p53-null background. Mol Cell Biol 9: 3286–91

Loayza D, de Lange (2003) POT1 as a terminal transducer of TRF1 telomere length control. Nature 423: 1013–18

Lundblad V, Blackburn EH (1993) An alternative pathway for yeast telomere maintenance rescues est1- senescence. Cell 73: 347–60

Martin GM, Sprague CA, Epstein CJ (1970) Replicative life-span of cultivated human cells. Effects of donor's age, tissue, and genotype. Lab Invest 23: 86–92

Martin-Rivera L, Herrera E, Albar JP, Blasco MA (1998) Expression of mouse telomerase catalytic subunit in embryos and adult tissues. Proc Natl Acad Sci USA 95: 10471–76

Maser RS, DePinho RA (2002) Connecting chromosomes, crisis, and cancer. Science 297: 565–69

McClintock B (1941) The stability of broken ends of chromosomes in Zea mays. Genetics 26: 234–82

McEachern MJ, Blackburn EH (1996) Cap-prevented recombination between terminal telomeric repeat arrays (telomere CPR) maintains telomeres in Kluyveromyces lactis lacking telomerase. Genes Dev 10: 1822–34

Meyerson M, Counter CM, Eaton EN, Ellisen LW, Steiner P, Caddle SD, Ziaugra L, Beijersbergen RL, Davidoff MJ, Liu Q, Bacchetti S, Haber DA, Weinberg RA (1997) hEST2, the putative human telomerase catalytic subunit gene, is up- regulated in tumor cells and during immortalization. Cell 90: 785–95

Mitchell JR, Cheng J, Collins K (1999) A box H/ACA small nucleolar RNA-like domain at the human telomerase RNA 3′ end. Mol Cell Biol 19: 567–76

Muller HJ (1938) The remaking of chromosomes. Collecting net 8: 182–95

Nakamura TM, Morin GB, Chapman KB, Weinrich SL, Andrews WH, Lingner J, Harley CB, Cech TR (1997) Telomerase catalytic subunit homologs from fission yeast and human. Science 277: 955–59

Nakayama J, Saito M, Nakamura H, Matsuura A, Ishikawa F (1997) TLP1: a gene encoding a protein component of mammalian telomerase is a novel member of WD repeats family. Cell 88: 875–84

Newbold RF (1997) Genetic control of telomerase and replicative senescence in human and rodent cells. Ciba Found Symp 211: 177–89

O'Hagan R, Chang S, Maser R, Mohan R, Artandi S, Chin L, DePinho RA (2002) Telomere dysfunction provokes regional amplification and deletion in cancer genomes. Cancer Cell 2: 149–57

Prowse KR, Avilion AA, Greider CW (1993) Identification of a nonprocessive telomerase activity from mouse cells. Proc Natl Acad Sci U S A 90: 1493–97

Prowse KR, Greider CW (1995) Developmental and tissue-specific regulation of mouse telomerase and telomere length. Proc Natl Acad Sci USA 92: 4818–22

Qi L, Strong MA, Karim BO, Armanios M, Huso DL, Greider CW (2003) Short telomeres and ataxia-telangiectasia mutated deficiency cooperatively increase telomere dysfunction and suppress tumorigenesis. Cancer Res. 63: 8188–96

Rohme D (1981) Evidence for a relationship between longevity of mammalian species and life spans of normal fibroblasts *in vitro* and erythrocytes *in vivo*. Proc Natl Acad Sci U S A 78: 5009–13

Rudolph KL, Chang S, Lee HW, Blasco M, Gottlieb GJ, Greider C, DePinho RA (1999) Longevity, stress response, and cancer in aging telomerase-deficient mice. Cell 96: 701–12
Samper E, Flores JM, Blasco MA (2001) Restoration of telomerase activity rescues chromosomal instability and premature aging in Terc–/– mice with short telomeres. EMBO Rep 23: 23–30
Schmitt CA (2003) Senescence, apoptosis and therapy–cutting the lifelines of cancer Nat Rev Cancer 3: 286–95
Sedivy JM (1998) Can ends justify the means: telomeres and the mechanisms of replicative senescence and immortalization in mammalian cells. Proc Natl Acad Sci U S A 95: 9078–81
Shay JW, Pereira-Smith OM, Wright WE (1991) A role for both RB and p53 in the regulation of human cellular senescence. Exp Cell Res 196: 33–39
Shay JW, Van Der Haegen BA, Ying Y, Wright, WE (1993) The frequency of immortalization of human fibroblasts and mammary epithelial cells transfected with SV40 large T-antigen. Exp Cell Res 209: 45–52
Shiloh Y (2003) ATM and related protein kinases: safeguarding genome integrity. Nat Rev Cancer 3: 155–68
Singer MS, Gottschling DE (1994) TLC1: template RNA component of Saccharomyces cerevisiae telomerase. Science 266: 404–09
Szekely AM, Bleichert F, Numann A, Van Komen S, Manasanch E, Ben Nasr A, Canaan A, Weissman SM (2005) Werner protein protects nonproliferating cells from oxidative DNA damage. Mol Cell Biol. 25: 10492–506
Tyner SD. et al. (2002) p53 mutant mice that display early ageing-associated phenotypes. Nature 415: 45–53
Vasile E, Tomita Y, Brown LF, Kocher O, Dvorak HF (2001) Differential expression of thymosin beta-10 by early passage and senescent vascular endothelium is modulated by VPF/VEGF: evidence for senescent endothelial cells in vivo at sites of atherosclerosis. FASEB J 15: 458–66
Vaziri H, Benchimol S (1998) Reconstitution of telomerase activity in normal human cells leads to elongation of telomeres and extended replicative life span. Curr Biol 8: 279–82
Vulliamy T, Marrone A, Goldman F, Dearlove A, Bessler M, Mason PJ, Dokal I (2001) The RNA component of telomerase is mutated in autosomal dominant dyskeratosis congenita. Nature 413: 432–35
Vulliamy T, Marrone A, Szydlo R, Walne A, Mason PJ, Dokal I (2004) Disease anticipation is associated with progressive telomere shortening in families with dyskeratosis congenita due to mutations in TERC. Nat Genet 36: 447–49
Wang F, Podell ER, Zaug AJ, Yang Y, Baciu P, Cech TR, Lei M (2007) The POT1-TPP1 telomere complex is a telomerase processivity factor. Nature 445: 506–10
Weng NP, Granger L, Hodes RJ (1997) Telomere lengthening and telomerase activation during human B cell differentiation. Proc Natl Acad Sci U S A 94: 10827–32
Wong K, Chang S, Weiler SR, Ganesan S, Chaudhuri J, Zhu C, Artandi SE, Rudolph KL, Gottlieb GJ, Chin L, Alt FW, DePinho RA (2000) Telomere dysfunction impairs DNA repair and enhances sensitivity to ionizing radiation. Nat Genet 26: 85–88.
Wong KK, Maser RS, Bachoo RM, Menon J, Carrasco DR, Gu Y, Alt FW, DePinho RA (2003) Telomere dysfunction and ATM deficiency compromises organ homeostasis and accelerates ageing. Nature 421: 643–48
Wright WE, Shay JW (1992) The two-stage mechanism controlling cellular senescence and immortalization. Exp Gerontol 27: 383–89
Wright WE, Piatyszek MA, Rainey WE, Byrd W, Shay JW (1996) Telomerase activity in human germline and embryonic tissues and cells. Dev Genet 18: 173–79
Wu L, Multani AS, He H, Cosme-Blanco W, Deng Y, Deng JM, Bachilo O, Pathak S, Tahara H, Bailey SM, Deng Y, Behringer RR, Chang S (2006) *Pot1a* deficiency initiates DNA damage checkpoint activation and aberrant homologous recombination at telomeres. Cell 126: 49–62
Xin H, Liu D, Wan M, Safari A, Kim H, Sun W, O'Connor MS, Songyang Z. (2007) TPP1 is a homologue of ciliate TEBP-beta and interacts with POT1 to recruit telomerase. Nature 445: 559–62

Xue W, Zender L, Miething C, Dickins RA, Hernando E, Krizhanovsky V, Cordon-Cardo C, Lowe SW (2007) Senescence and tumour clearance is triggered by p53 restoration in murine liver carcinomas. Nature 445: 656–60

Yu GL, Bradley JD, Attardi LD, Blackburn EH (1990) *In vivo* alteration of telomere sequences and senescence caused by mutated Tetrahymena telomerase RNAs. Nature 344: 126–32

Zijlmans JM, Martens UM, Poon SS, Raap AK, Tanke HJ, Ward RK, Lansdorp PM (1997) Telomeres in the mouse have large inter-chromosomal variations in the number of T2AG3 repeats. Proc Natl Acad Sci USA 94: 7423–28

Chapter 5
Telomerase Mutations and Premature Ageing in Humans

Hong-Yan Du, Monica Bessler, and Philip J. Mason(✉)

Abstract Dyskeratosis congenita (DC) is a rare inherited bone marrow failure syndrome which gets its name from the classic diagnostic triad of abnormal skin pigmentation, nail dystrophy, and leukoplakia. Most DC patients have bone marrow failure, and this is the most common cause of premature death. DC is genetically heterogeneous. The most common X-linked form is due to mutations in the *DKC1* gene encoding dyskerin. Autosomal-dominant DC is caused by mutations in either TERT or TERC; no genes have so far been identified that cause the Autosomal-dominant form of the disease. That all three known genes underlying DC are components of the telomerase complex suggests strongly that DC is caused by defective telomerase, and this suggestion is supported by the fact that all DC patients have very short telomeres. Thus the study of DC offers a unique opportunity to observe the effects of defective telomerase and failure to maintain telomere length on the human organism. The tissues affected in DC, skin, blood, lung, and the gut lining are those which need to be constantly renewed by stem cell activity and can be considered to undergo a kind of tissue-specific accelerated ageing. In this chapter, we propose a model of the pathogenesis of DC. Mutations in the telomerase complex result in excessive telomere shortening either by decreasing the stability of telomerase complex, reducing the accumulation, or directly affecting the enzymatic activity of telomerase. Telomere shortening leads to cell cycle arrest (senescence) or cell death (apoptosis) in rapidly dividing cells, including bone marrow cells. To fulfill the requirement for normal function, these cells have to increase the recruitment of more quiescent stem cells into the cell cycle and subsequent proliferation, which in turn accelerates telomere shortening, senescence, and apoptosis. When the stem cell compartments are exhausted, clinical manifestations of bone marrow failure occur. Meanwhile, cells surviving the telomere crisis accumulate genomic instability and become potential malignant cells.

Philip J. Mason
Department of Internal Medicine, Washington University School of Medicine, 660 S Euclid Avenue, St. Louis, Missouri, United States.
e-mail: PMason@im.wustl.edu

K.L. Rudolph (ed.), *Telomeres and Telomerase in Ageing, Disease, and Cancer.*

1 Introduction

Telomeres are DNA/protein structures at the ends of chromosomes, needed to protect chromosome ends from degradation and distinguish them from double-stranded breaks (de Lange 2002). Human telomere DNA consists of thousands of repeats of TTAGGG.

DNA polymerase I cannot copy the extreme end of a DNA strand (Watson 1972), so telomeres would tend to get shorter with each cell division. In germ cells (Wright et al. 1996), some stem cells (Chiu et al. 1996, Harle-Bachor and Boukamp 1996) and most cancer cells (Kim et al. 1994, Broccoli et al. 1995), telomeres are maintained by the action of the enzyme telomerase (Greider and Blackburn 1985), which contains two core components, a reverse transcriptase TERT (telomerase reverse transcriptase), and an RNA molecule, *TERC* (telomerase RNA component), that acts as a template for the synthesis of the telomere repeats. Telomerase is active in HSCs, but there is shortening of telomeres during ageing (Vaziri et al. 1994), indicating that the level of telomerase activity is not sufficient to stabilize telomeres in ageing stem cells (see also Zimmermann and Martens, this volume).

In the absence of telomerase in most somatic cells, telomeres do shorten with each cell division (Harley et al. 1990). In most tissues, therefore, telomere length decreases with age (Vaziri et al. 1994), reflecting the history of cell division.

2 Telomerase and Ageing

Because somatic cells that lose telomeres during cell division have a limited replicative life span, it has been hypothesized that telomere shortening is associated with the physiological process of ageing. Initial speculation about the relationship between telomere shortening and cellular senescence came from the observation that telomeres capping the long arms of human sex chromosomes were strikingly shorter in somatic cells than in germ line cells (Cooke and Smith 1986). Other evidence supporting this idea includes the correlation between the proliferative potential of primary human cells and their telomere length (Allsopp et al. 1992) and the fact that telomeres shorten during cellular ageing in culture (Harley et al. 1990) and ageing of human tissue in vivo (Allsopp et al. 1992). Furthermore, the finding that telomerase is activated in the majority of human tumors, which otherwise require an alternative mechanism to maintain telomeric DNA, substantiates the importance of telomere maintenance in the proliferation of human cells (Kim et al. 1994).

Most somatic cells do not express telomerase, and therefore telomeres undergo gradual shortening until a critical minimal telomere length is reached, when cells stop dividing and enter into a quiescent viable state called replicative senescence, which is proposed to prevent cells from excessive cell division and genetic instability (Harley 1991). Telomeres were observed to shorten in human tissues, including peripheral blood cells, liver, kidney, spleen, dermal fibroblasts, and mucosal keratinocytes (Broccoli et al. 1995, Hiyama et al. 1995, Harle-Bachor and Boukamp 1996, Yasumoto et al. 1996). On the other hand, a low level of telomerase activity is detected in cells that undergo rapid expansion, such as germ cells, certain populations of stem cells and their progeny, activated lymphocytes, keratinocytes, and the basal layer of the epidermis (Masutomi et al. 2003, Harrington 2004, Flores et al. 2005). To test the role of telomerase in the proliferative potential of primary human cells, Bodnar et al. reconstituted telomerase activity by expressing TERT in human cells, including retinal epithelial cells, foreskin fibroblasts, and vascular endothelial cells. It was found that activation of telomerase resulted in elongation of telomeres and bypass of senescence in these cells (Bodnar et al. 1998).

While accumulating evidence has supported the notion that telomerase insufficiency leads to telomere shortening and replicative senescence of cultured cells, the role of telomere maintenance in the ageing process has not been clear. Ageing as a physiological process or phenomenon is characterized by impaired response to stress, unbalanced homeostasis, and increased risk of disease (Weinert and Timiras 2003). The relevance of telomere shortening induced by dysfunctional telomerase to physiological ageing was first demonstrated by the generation of a mouse model devoid of telomerase activity (Blasco et al. 1997). It was observed that late-generation mice deficient for telomerase had a shorter life span than wild-type mice. In addition, although early generations of *mTR-/-* mice did not show abnormalities (laboratory mice have a much longer telomere length than humans or wild mice, and telomeres only become critically short after several generations with no telomerase (Herrera et al. 1999, Rudolph et al. 1999) defects in cell viability of highly proliferative tissues were observed in later generations. In human, a severe, inherited bone marrow failure syndrome associated with telomerase deficiency, known as dyskeratosis congenita (DC), has also been considered to be related to premature ageing (Dokal 2001, Vulliamy et al. 2001a).

Late generations of *mTR* knockout mice and DC patients share similar phenotypes in terms of hair loss or early greying of hair, increased incidence of malignancy, decreased capacity for wound healing, and telomere shortening, which are common to the process of normal ageing. Furthermore, several syndromes/diseases have also been considered to be related to telomere shortening, and have been referred to as premature ageing syndromes. These include cardiovascular disease (Cawthon et al. 2003), hypertension (Serrano and Andres 2004), paroxysmal nocturnal hemoglobinuria (PNH) (Beier et al. 2005), Fanconi anemia (Franco et al. 2004), Werner syndrome (Chang et al. 2004), ataxia telangiectasia (Metcalfe et al. 1996), Bloom syndrome (Cohen and Sinclair 2001), and Nijmegen breakage syndrome (Paull and Gellert 1999). In this chapter, DC, which is caused by mutations in the components of telomerase ribonucleoprotein complex, will be discussed in order to understand the role of telomerase in a premature ageing syndrome.

3 Dyskeratosis Congenita as a Premature Ageing Syndrome

As we will discuss below, DC is genetically heterogeneous, but the two known genetic causes of DC both affect components of the telomerase complex and lead to decreased telomerase levels (Heiss et al. 1998, Mitchell et al. 1999b, Vulliamy et al. 2001a). Moreover, all DC patients have very short telomeres (Vulliamy et al. 2001b). Thus DC offers us the opportunity to observe the effects of telomerase deficiency on the whole organism and constitutes a stringent test of the importance of telomerase in the process of ageing.

3.1 *Clinical Presentation of Dyskeratosis Congenita*

DC is also known as Zinsser-Engman-Cole syndrome (Zinsser 1906, Engman 1926, Cole et al. 1930). Classically DC is characterized by the triad of abnormal skin pigmentation, nail dystrophy, and mucosal leukoplakia. A variety of other noncutaneous abnormalities have also been reported since its first description (Table 5.1), including dental caries, tooth loss, esophageal webs, epiphora, short stature, developmental delay, hypogonadism, osteoporosis, pulmonary disease, premature grey hair, and hair loss (Trowbridge et al. 1977, Womer et al. 1983, Wiedemann et al. 1984, Zijlmans et al. 1997, Knight et al. 1998, Solder et al. 1998, Dokal 2000, Mason et al. 2005).

The principal cause of mortality in DC patients is bone marrow failure; pulmonary complications and malignancies are also frequent. The prevalence of DC is approximately 1 in 1,000,000 individuals, with death occurring at a median age of 16 (Drachtman and Alter 1995). The clinical phenotype of this disorder is highly heterogeneous, as shown by the variable features of age of onset, severity of

Table 5.1 Clinical Features of DC

	Review of 224 Patients (%) (Drachtman and Alter 1995)	118 Males from DKC Registry London (%) (Dokal 2000)
Classic		
Nail changes/dystrophy	83	88
Skin pigmentation	88	89
Leukoplakia, changes of the buccal mucosa	64	78
Bone marrow failure	100	85.5
Non-classic		
Epiphora and eye changes	45	30.5
Chromosomal breaks	13	NA
Cancer and leukemia	10	7.6
Pulmonary fibrosis/ abnormalities	0.4	20.3
Hyperhiderosis	12	15.3
Teeth abnormalities, osteoporosis	17	16.9
Hair loss/early graying	17	16
Liver cirrhosis	NA	7.3
Short stature	15	19.5
Developmental delay, mental retardation		25.4
Low birth weight	16	7.6
Hypogonadism	4	6
Others (urogenital, cadiopulmonary)	8	10

somatic or hematological disorders, and the age of death. Greater than 90% of patients develop at least a single cytopenia by the age of 40, and 50% develop severe aplastic anemia. A very severe form of DC is known as Hoyeraal-Hreidarsson Syndrome (HHS) and is characterized by cerebellar ataxia, intrauterine growth failure, immunodeficiency, and aplastic anemia. Patients with HHS usually present symptoms within the first two years of life and die before the age of 5 (Hoyeraal et al. 1970, Hreidarsson et al. 1988, Dokal 2001).

Despite highly variable manifestation, typical patients with DC appear healthy at birth, develop mucocutaneous features and bone marrow failure in adolescence, and die from complications in the second and third decades. Three distinct genetic forms of DC exist: X-Linked recessive (MIM # 305000), Autosomal-dominant (AD, MIM # 127550), and Autosomal-recessive (AR, MIM # 224230). The severity of the mucocutaneous and noncutaneous features are generally associated with the mode of inheritance, although there are exceptions. Compared to the X-linked form, the manifestation of the Autosomal-dominant form is less severe. On the other hand, Autosomal-recessive cases include more and less severe forms.

3.2 Similarities and Differences Between DC and Ageing

An important feature of human ageing is increased risk of disease, such as cancer and a decrease in maintenance and function of almost all organs. Greying of hair is another obvious feature of the ageing population. In accordance with these, both DC patients and late generations of *mTR* knockout mice were observed to have increased incidence of malignancy and hair greying. The occurrence of cancer in these situations probably arises from chromosomal instability due to critically short telomeres (Counter et al. 1992, Dokal et al. 1992), whereas in ageing, additional factors also contribute to carcinogenesis. Such factors include free-radical-induced DNA damage, the accumulation of DNA mutations, changes in metabolic rate, and oxidative stress (Sohal et al. 2002, Frisard and Ravussin 2006). Another distinction between DC patients and the ageing population is the occurrence of hematological disorders. While hematological disorders are usually seen in DC patients and *mTR* knockout mice, severe dysfunction of the hematological system does not necessarily occur in natural ageing, though the proliferative capacity of the hematopoietic system in response to stress or environmental insult, such as drugs and viruses, does decrease with age; age-related anemia and lymphopenia are frequently seen. Early studies have shown that while the hematological disorder is manifested as bone marrow failure in DC patients (Dokal 2001, Vulliamy et al. 2001a), *mTR* knockout mice showed lymphocytopenia and inhibited response to hematopoietic ablation (Rudolph et Al. 1999, Herrera et al. 1999). Furthermore, unlike patients with some progeria syndromes, DC patients do not look older than they are, probably reflecting the fact that DC affects some but not all tissues.

Most clinical presentations of DC are associated with impaired proliferative capacity of tissues (Luzzatto and Karadimitris 1998). One example is that skin fibroblasts from DC patients had shorter telomeres and had decreased population doubling and abnormal chromosomal rearrangements in in vitro culture (Dokal et al. 1992). In addition, the number of hematopoietic progenitor cells was decreased (Marsh et al. 1992). As a consequence, tissues requiring constant renewal, i.e., skin, oral mucosa, and bone marrow are affected most frequently in DC.

As mentioned above, DC patients have a median life span of 16 years, while some can survive up to 50 years old and milder cases even longer. The shorter life span in DC patients could be inferred to be the result of dysfunctional telomerase. The death of DC patients is usually caused by complications of bone marrow failure (immune deficiency or bleeding), pulmonary fibrosis, or malignancies. These are extreme examples of some of the organ or tissue dysfunction that takes place in normal ageing. Perhaps a key feature of the disease process in DC is that it involves the failure of some tissues, principally highly proliferative tissues that are dependent on renewal by stem cell activity. Ageing, however, can affect the entire organism. Nevertheless, in the first generation of mice carrying a deletion of *mTR,* when no hematopoietic disorders were apparent, the longevity was reduced in comparison to wild-type mice (Geserick and Blasco 2006). This observation supports a role of telomerase in overall life span of organisms.

4 Dyskeratosis Congenita and Telomerase Mutations

The most common form of DC is the X-linked form that is associated with mutations in *DKC1* gene encoding dyskerin. In this mode of inheritance, the patients are males with brothers, maternal uncles, or male cousins who are affected. In some cases, the heterozygous females may have subtle, mainly cutaneous, manifestations. AD-DC is a rare form of the disease that was found to be caused by mutations in the *TERC* gene located on chromosome 3q26 (Vulliamy et al. 2001a). Recently, it was shown that the *TERT* gene on 5p15.3 might also be mutated in some cases of AD-DC (Armanios et al. 2005, Vulliamy et al. 2005, Yamaguchi et al. 2005). No genes have been found to be responsible for the AR-DC yet, and a high percentage of DC patients are sporadic cases with males more frequently diagnosed than females (Vulliamy et al. 2006).

4.1 X-linked DC and Mutations in DKC1

The *DKC1* gene located on Xq28 that encodes dyskerin is responsible for the X-linked recessive form of the disorder (Heiss et al. 1998). Dyskerin is a component of a small nucleolar ribonucleoprotein (snoRNP) that is highly conserved throughout evolution. The orthologs in yeast, fly, and rat are Cbf5p (Cadwell et al. 1997), N60B/minifly (Phillips et al. 1998), and NAP57 (Meier and Blobel 1994), respectively. The ubiquitous expression of dyskerin in all tissues of the body suggests a vital housekeeping function (Meier 2005), which correlates with the multisystem phenotype of the disease. Dyskerin, together with three other proteins, GAR1, NOP10, and NHP2 (homologous to Gar1p, Nhp2p, and Nop10p, respectively, in yeast), bind to small nucleolar RNAs (snoRNAs) containing the box H and box ACA motif (Henras et al. 1998, Lafontaine et al. 1998, Henras et al. 2004, Wang and Meier 2004) (Fig. 5.1A). The snoRNAs guide the snoRNP complex to newly synthesized ribosomal RNA (rRNA) via specific residue base pairing. Subsequently, the rRNA is modified by pseudouridylation, a process that transforms uridine to pseudouridine to stabilize RNA helices and aid in the formation of protein binding sites (Newby and Greenbaum 2002). Pseudouridylation is essential for biogenesis, maturation, and assembly of ribosomes (Tollervey and Kiss 1997, Filipowicz and Pogacic 2002). It is likely that dyskerin is the active human pseudouridine synthase that catalyzes the isomerization of uridine to pseudouridine, based on the structural similarity with pseudouridine synthases in yeast and bacteria (Cadwell et al. 1997).

In addition to its involvement in RNA processing, dyskerin has been shown to be associated with *TERC*, which also contains H/ACA consensus sequence (Mitchell et al. 1999a, Mitchell et al. 1999b) (Fig. 5.1B). The interaction between dyskerin and the telomerase complex was demonstrated by coimmunoprecipitation of dyskerin with *TERC* in a telomerase-positive cell line (Pogacic et al. 2000). SnoRNA was also immunoprecipitated with dyskerin, suggesting the structural association among dyskerin, telomerase, and snoRNAs.

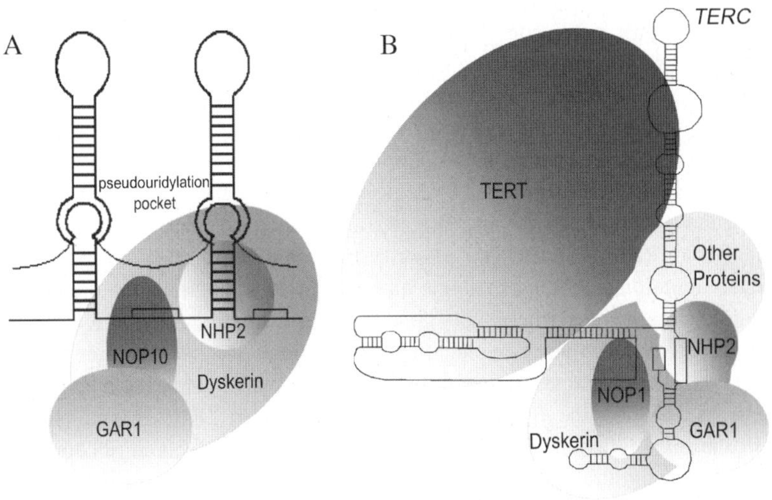

Fig. 5.1 Dyskerin is an important component of both the H/ACA small nucleolar (snoRNA) ribonucleoprotein (RNP) particles and telomerase complex. (A) Dyskerin, together with GAR1, NHP2, NOP10, and H/ACA snoRNA, comprises snoRNP particles. The H/ACA RNA-protein complex catalyzes pseudouridylation and processing of ribosomal (r) RNA. (B) *TERC* (telomerase RNA component) and TERT (telomerase reverse transcriptase) are the core components of telomerase complex. As an H/ACA snoRNA, *TERC* associates with dyskerin, GAR1, NHP2, and NOP10, which are also present in the telomerase complex (*See Color Plate*)

DC was hypothesized to be a disease resulting from defective telomere maintenance based on the discovery of interaction between dyskerin protein and *TERC*, and clinical observation of abnormally short telomeres and reduced level of *TERC* in DC individuals (Mitchell et al. 1999b, Vulliamy et al. 2001a). On the other hand, significant defects in pseudouridylation or processing of 18S or 28S rRNA were not detected (Mitchell et al. 1999b). Telomerase activity in fibroblasts from DC-affected male cells is lower than in cells from DC female carriers when transfected with TERT; and the diminished telomerase activity in DC patients could be recovered by expressing additional *TERC* (Wong and Collins 2006). A recent study showed that co-expression of TERT and *TERC* in dermal fibroblasts from DC patients carrying ΔL37 dyskerin mutation or A386T mutation could gain and stabilize telomere length, whereas telomere shortening was not inhibited by expressing TERT alone (Wong and Collins 2006). These findings indicate that dyskerin plays a role in telomere maintenance by stabilizing *TERC*.

However, the observation of the function of dyskerin in mouse models suggested that the involvement in ribosome biosynthesis cannot be ruled out. When two human mutations in the DKC1 gene were introduced into murine embryonic stem cells (A353V and G402E), defects in pseudouridylation and pre-rRNA processing were observed in cell lines with each mutation. On the other hand, only mutation A353V induced destabilization of *TERC* and reduced telomerase activity (Mochizuki et al. 2004). Furthermore, mice carrying a hypomorphic *Dkc1* mutation display some features of DC before the onset of significant telomere shortening (Ruggero et al. 2003). The Dkc1 mutant was shown to lead to a defect in IRES (internal ribosome entry site)–dependent translation, which resulted in impaired translation of mRNA containing IRES elements (Yoon et al. 2006). Therefore the relative contribution of ribosome biogenesis and telomere maintenance to the phenotype of DC is still controversial, at least in the mouse model. The importance of the two mechanisms might be dependent on the site of mutation within dyskerin.

Forty-eight mutations, including five polymorphisms, have been identified in the *DKC1* gene to date (Marrone and Dokal 2004, Kanegane et al. 2005, Mason et al. 2005) (Table 5.2; Fig. 5.2). The majority of the mutations (38) (Fig. 5.2A) are missense mutations resulting in single amino acid substitutions. One large deletion removes the last exon of the *DKC1* gene transcript; one mutation is located in the promoter region, and three in noncoding regions. The absence of frameshift mutations or nonsense mutations indicates that null mutations might be lethal to organisms as demonstrated by the observation that disruption of *Dkc1* caused embryonic lethality in mice (He et al. 2002), which could be due to defects in ribosomal RNA processing. So far, the functional consequences of the mutations are far from clear.

The recent elucidation of the crystal structure of the dyskerin-NOP10-GAR1 complex facilitates the identification of the domains in dyskerin where most mutations cluster (Li and Ye 2006; Rashid et al. 2006). Based on information from orthologs in other species, it was inferred that dyskerin has several functional domains: nuclear localization signals, a pseudouridine synthase domain (TruB/PUS), a pseudouridine synthase and archaeosine-spicific transglycosylase domain (PUA), and lysine-rich carboxy domains (Meier and Blobel 1994, Ni et al. 1997,

Table 5.2 Mutations in *DKC1*

Mutation	Exon	Amino Acid Change	Clinical Manifestation	Refs
−141C->G	Promotor	NA	DC	(Knight et al. 2001)
C5T	1	A2V	DC	(Knight et al. 1999b; Safa et al. 2001)
IVS1 592C->G	IVS1	-	DC	(Knight et al. 2001)
C29T	2	P10L	DC/HH	(Vulliamy et al. 2006)
IVS2 473C->G	IVS2	-	DC	(Knight et al. 1999b)
C91G	3	Q31E	DC	(Wong et al. 2004)
C91A	3	Q31K	DC	(Kanegane et al. 2005)
T106G	3	F36V	DC	(Heiss et al. 1998)
109–111 DCTT	3	D37L	DC	(Heiss et al. 1998; Mitchell et al. 1999b)
T113C	3	I38T	HH	(Cossu et al. 2002)
A115G	3	K39E	DC	(Knight et al. 1999b)
C119G	3	P40R	DC	(Heiss et al. 1998)
G121A	3	E41K	HH	(Knight et al. 1999b)
A127G	3	K43E	DC	(Heiss et al. 2001)
C146T	3	T49M	HH	(Knight et al. 1999b; Heiss et al. 2001; Sznajer et al. 2003)
G194C	4	R65T	DC	(Knight et al. 1999b)
A196G	4	T66A	DC	(Hassock et al. 1999; Knight et al. 1999b; Mitchell et al. 1999b)
C200T	4	T67I	DC/HH	(Vulliamy et al. 2006)
C204A	4	H68Q	DC/HH	(Vulliamy et al. 2006)
CT214-215TA	4	L72Y	DC	(Heiss et al. 1998)
A361G	5	S121G	HH	(Knight et al. 1999b)
C472T	6	R158W	DC	(Knight et al. 2001)
IVS 6 T40G	IVS6	-	polymorphism	(Marrone and Dokal 2004)
A838C	9	S280R	DC	(Knight et al. 2001)
A941G	10	K314R	DC/HH	(Vulliamy et al. 2006)
C949T	10	L317F	DC	(Rostamiani et al. 1999)
C961G	10	L321V	DC	(Knight et al. 1999b)
G965A	10	R322Q	DC	(Rostamiani et al. 1999)
T1049C	11	M350T	DC	(Knight et al. 1999b)
G1050A/C	11	M350I	DC	(Knight et al. 1999b)
C1058T	11	A353V	DC/HH	(Knight et al. 1999b; Yaghmai et al. 2000; Yoshimoto et al. 2000; Treister et al. 2004)
A1069G	11	T357A	DC	(Kanegane et al. 2005)
G1075A	11	D359N	DC	(Vulliamy et al. 2006)
C1150T	11	P384S	DC	(Rostamiani et al. 1999)
C1151T	11	P384L	DC	(Knight et al. 2001)
G1156A	12	A386T	DC/HH	(Vulliamy et al. 2006)
T1193C	12	L398P	DC	(Hiramatsu et al. 2002)
G1204A	12	G402R	DC	(Knight et al. 1999b)
G1205A	12	G402E	DC	(Heiss et al. 1998)

(continued)

Table 5.2 (continued)

Mutation	Exon	Amino Acid Change	Clinical Manifestation	Refs
C1223T	12	T408I	DC	(Vulliamy et al. 2006)
C1226T	12	P409L	DC	(Ding et al. 2004)
AG1258-1259TA	12	S420Y	DC	
C1461T	14	-	polymorphism	(Hassock et al. 1999)
IVS 14 G403T	IVS 14	-	polymorphism	(Marrone and Dokal 2004)
IVS 14 A473G	IVS14	-	DC	(Vulliamy et al. 2006)
ΔExon15	15	ΔD493–514	DC	(Heiss et al. 1998; Vulliamy et al. 1999)
1494insAAG	15	498insK	polymorphism	(Dokal 2000)
G1551A	15	-	polymorphism	(Hassock et al. 1999)

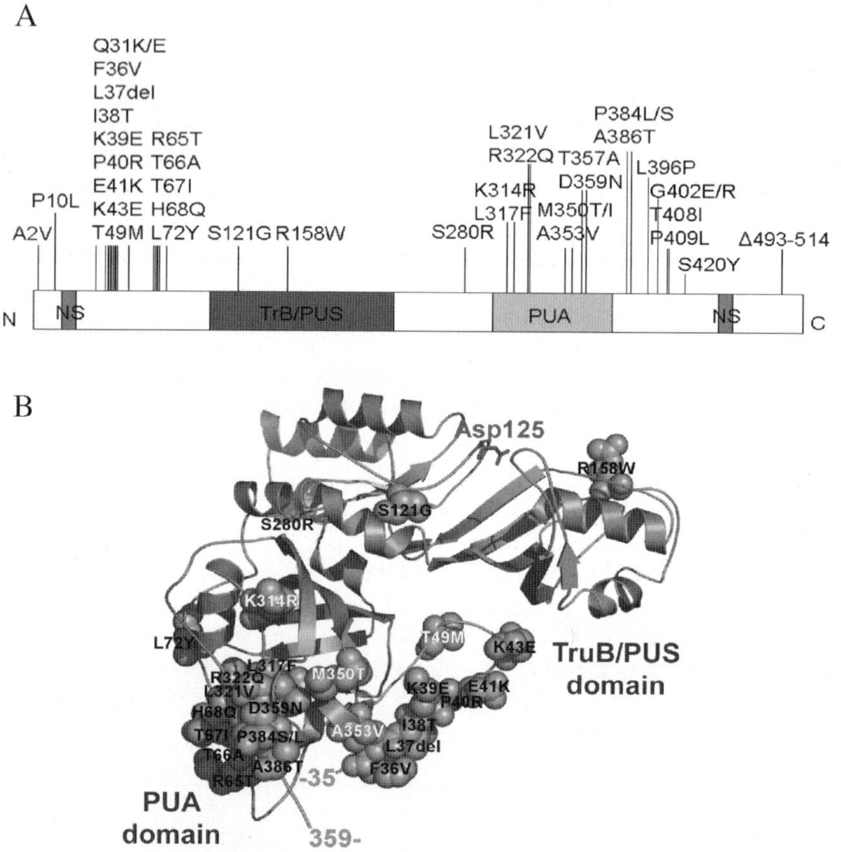

Fig. 5.2 Nonsynonymous amino acid substitutions in dyskerin affected in dyskeratosis congenita. (A) Linear representation of dyskerin protein that contains 514 amino acids. The functional domains: nuclear localization signals (NL), pseudouridine synthase domain (TruB/PUS), pseudouridine synthase, and archaeosine-spicific transglycosylase domain (PUA) are shown. Locations of nonsynonymous amino acid substitutions are labeled. (B) 3D structure of dyskerin with amino acid changes identified in dyskeratosis congenita. Mutations in the regions 1–35 and 359–513 are not shown due to the unavailability of modeling in these regions (Courtesy of Dr. Hong Li, Department of Chemistry and Biochemistry, Florida State University) (*See Color Plate*)

Youssoufian et al. 1999). Recently, the three-dimensional structure of dyskerin was modeled on the proposed crystal structure of Cbf5 from the archaebacterium *Pyrococcus furiosus* and sequence homology between dyskerin and Cbf5 (Li and Ye 2006, Rashid et al. 2006). In the primary sequence of dyskerin, most mutations found in DC are clustered in the N-terminal 100 residues and PUA domain (Fig. 5.2B). In the crystal three-dimension structure, the PUA domain is encircled by the N terminus. Therefore, the regions where the mutations are concentrated fall within or near PUA domain, which is necessary for recognition and binding of H/ACA RNA and telomerase RNA (Li and Ye 2006, Rashid et al. 2006).

Amino acid substitutions in the PUA domain were predicted, though the important residues were not those that specifically affect the interaction between dyskerin and target RNA (see Li 2006). It has been observed that certain mutations lead to decreased accumulation of H/ACA RNA or telomerase RNA (Mitchell et al. 1999b, Mochizuki et al. 2004), thereby substantiating the importance of functional interaction of dyskerin and telomerase in the assembly, processing, and stability of telomerase RNA.

The A353V mutation located in the PUA domain has been found in more than 40% of individuals with DC. Only two mutations have been found in the TruB domain, one of which causes amino acid substitution S121G. This locus is only three residues from the highly conserved essential aspartic acid residue D125 required for pseudouridylation (Ramamurthy et al. 1999). A large deletion removing the last exon was identified that induces a truncated protein devoid of 22 amino acids at the C-terminus (Vulliamy et al. 1999). A C-G mutation in the promoter region at −141 is at a potential Sp1 binding site, inducing decreased transcription in an in vitro promoter assay. (Knight et al. 2001, Salowsky et al. 2002). The intronic mutation C592G has been shown to affect splicing with generation of aberrantly spliced mRNA, and G473C and IVS 14 G403T change the splice site acceptor motif (Knight et al. 2001, Vulliamy et al. 2006).

In the most severe form of DC–HH syndrome–nine mutations have been identified which are distributed in the two clusters, the N-terminal 100 residues and PUA domain, as well as in the TruB domain. The amino acid substitutions I38T and T49M have, to date, only been seen in the HH patients (Knight et al. 1999a, Sznajer et al. 2003). On the other hand, the most frequent mutation A353V leads to DC with different extents of severity, including HH (Vulliamy et al. 2006). Therefore, although there appears to be some correlation between severity and mutation, the phenotype can be affected by other genetic and environmental factors.

4.2 AD DC and Mutations in TERC and TERT

4.2.1 Mutations in *TERC*

The Autosomal-dominant form of DC is caused by mutations in the telomerase RNA component gene, *TERC*. Since the first report establishing the link between *TERC* mutations and DC (Vulliamy et al. 2001a), 26 mutations have been identified is implicated in the regulation of localization of TERT to telomeres. The N-terminus and C-terminus of telomerase both contain a DAT domain (dissociates activities of telomerase) that is important for recruiting regulatory proteins or associating with telomeres. Mutations in this region resulted in disassociation of TERT from telomeres, but did not influence telomerase activity (Counter et al. 1998, Friedman and Cech 1999, Xia et al. 2000, Armbruster et al. 2001, Banik et al. 2002). In accordance with these findings, the two mutations found in the CTE domain among patients with DC or AA, R979W, and F1127L were not shown to impact on telomerase activity (Xin et al. 2006). Another missense mutation V1090M, however, drastically reduced telomerase activity to less than 1% (Xin et al. 2006).

Investigations of *TERT* gene mutations in DC or AA patients suggest that mutations in *TERT* may be different in penetrance and in the range of phenotypes produced compared with mutations in TERC. Among the mutations identified, only one family, with members heterozygous for the K902N mutation, displays autosomal-dominant inheritance, telomere shortening, and disease anticipation (Armanios et al. 2005). In a family with an A202T mutation, although all the family members with one mutated allele showed shortened telomere length in leukocytes, only the proband presented with clinical manifestation of a hematological disorder (Yamaguchi et al. 2005). There are also mutations (e.g., F1127L) that do not segregate with disease. Interestingly, in several pedigrees inherited idiopathic pulmonary fibrosis has been found to be associated with TERT mutations. (Armanios et al. 2005, Armanios et al, 2007, Tsakiri et al, 2007).

4.3 AR DC and Mutations in Other Components

In addition to the X-linked recessive and autosomal-dominant forms, a significant fraction of DC families shows an autosomal-recessive form of inheritance. In fact, from the DC Registry of 228 families (Vulliamy et al. 2006), 8% DC patients were observed to fall within this group. Sporadic cases can also be seen, which means only one family member carries a mutation and presents the disease. Efforts have been made to survey mutations in NOP10, NHP2, and GAR1, which are associated with dyskerin and are therefore possible candidates responsible for the autosomal-recessive subtype of this disease (Marrone and Mason 2003, Yamaguchi et al. 2005). So far no diseases causing mutations have been identified. Studies on the archaeal homologs of the four core proteins (Cbf5, Nop10, L7Ae, and Gar1) revealed that Cbf5 independently interacts with the H/ACA RNA, Nop10, and Gar1, while L7Ae did not directly bind with other proteins without mediation of RNA (Baker et al. 2005, Charpentier et al. 2005). Cbf5 associates with the lower stem and ACA motif of RNA via the PUA domain, and the RNA upper stem binds to the structure formed by L7Ae, Nop10, and Cbf5 (Li and Ye 2006). Mutations in the core proteins might affect the association with RNA and the structure and function of the telomerase nucleoprotein complex.

5 Telomerase and Cancer

Both DC and ageing exhibit an increased incidence of cancer (Alter 2000, Dokal 2000). While telomerase activity is undetectable in most somatic cells, above 85% of tumors express telomerase (Counter et al. 1992, Kim et al. 1994). Reactivation of telomerase is therefore regarded as a crucial step in the development of cancer. It has been observed that expression of TERT is sufficient for immortalizing primary cells (Bodnar et al. 1998, Vaziri and Benchimol 1998). Additionally, mice deficient for telomerase exhibited resistance to tumorigenesis (Greenberg et al. 1999, Gonzalez-Suarez et al. 2000; see also Blasco, this volume). A recent study reported that a virus-encoded telomerase RNA promotes formation of T cell lymphoma in chicken, highlighting the significance of *TERC* in oncogenesis (Trapp et al. 2006). The hypothesis that elevated expression of TERT or *TERC* is positively related with cancer occurrence raises the question why there is increased cancer susceptibility in DC and ageing where telomerase is deficient or telomeres are massively shortened. The possible explanation is that progressive telomere shortening increases genome instability, which subsequently leads to malignancy (Marciniak et al. 2000, Greenberg 2005, Mason et al. 2005, Rudolph et al. 1999 and 2001). According to the model of pathogenesis of DC proposed by us (Fig. 5.5), excessive telomere shortening induced by telomerase deficiency leads to cell cycle arrest (senescence) or cell death (apoptosis) in rapidly dividing cells, including bone marrow cells. The increased requirement for replenishment results in recruitment of more quiescent stem cells into the cell cycle and subsequent proliferation, which in turn accelerates telomere shortening, senescence, and apoptosis. This will ultimately result in an early exhaustion of the stem cell compartment, causing the clinical manifestations of bone marrow failure. In some cases, few cells survive from the telomere-based crisis, possibly via mutated p53 genes which enable the cells to proliferate beyond the senescence checkpoint (Artandi and DePinho 2000). These cells have gained the property of genome instability as the eroded telomeres can no longer protect the chromosome ends from fusion and recombination, thus becoming potential malignant cells. This model coincides with the increased risk of cancer in the ageing population where accumulation of cell damage and oxidative stress over time increases the chance of genome instability (Weinstein and Ciszek 2002). In addition to the increase in chromosomal instability, telomere shortening may induce cancer progression by impairing the replicative competition of progenitor cells (Bilousova et al. 2005; see also Rudolph, this volume).

The reactivation of telomerase in the majority of tumor types underscores the importance of telomere maintenance in tumor pathogenesis. This is further supported by the fact that in those tumor cells that do not express telomerase, an alternative telomerase- independent (ALT) pathway is required to maintain telomere length (Bryan et al. 1997, Dunham et al. 2000). It is therefore plausible that telomerase contributes to cancer development by extending telomeres. However, in recent years, several studies have shown that TERT may play a role in proliferation and tumorigenesis that is independent of telomere maintenance. Transgenic expression of TERT in mice promoted the formation of carcinogen-induced skin

Lue NF, Lin YC, Mian IS (2003) A conserved telomerase motif within the catalytic domain of telomerase reverse transcriptase is specifically required for repeat addition processivity. Mol Cell Biol 23:8440–49

Luzzatto L, Karadimitris A (1998) Dyskeratosis and ribosomal rebellion. Nat Genet 19:6–7

Ly H, Blackburn EH, Parslow TG (2003) Comprehensive structure-function analysis of the core domain of human telomerase RNA. Mol Cell Biol 23:6849–56

Ly H, Calado RT, Allard P, Baerlocher GM, Lansdorp PM, Young NS, Parslow TG (2005a) Functional characterization of telomerase RNA variants found in patients with hematologic disorders. Blood 105:2332–39

Ly H, Schertzer M, Jastaniah W, Davis J, Yong SL, Ouyang Q, Blackburn EH, Parslow TG, Lansdorp PM (2005b) Identification and functional characterization of 2 variant alleles of the telomerase RNA template gene (TERC) in a patient with dyskeratosis congenita. Blood 106:1246–52

Marciniak RA, Johnson FB, Guarente L (2000) Dyskeratosis congenita, telomeres and human ageing. Trends Genet 16:193–95

Marrone A, Dokal I (2004) Dyskeratosis congenita: molecular insights into telomerase function, ageing and cancer. Expert Rev Mol Med 6:1–23

Marrone A, Mason PJ (2003) Dyskeratosis congenita. Cell Mol Life Sci 60:507–17

Marrone A, Stevens D, Vulliamy T, Dokal I, Mason PJ (2004) Heterozygous telomerase RNA mutations found in dyskeratosis congenita and aplastic anemia reduce telomerase activity via haploinsufficiency. Blood 104:3936–42

Marsh JC, Will AJ, Hows JM, Sartori P, Darbyshire PJ, Williamson PJ, Oscier DG, Dexter TM, Testa NG (1992) "Stem cell" origin of the hematopoietic defect in dyskeratosis congenita. Blood 79:3138–44

Martin-Rivera L, Blasco MA (2001) Identification of functional domains and dominant negative mutations in vertebrate telomerase RNA using an in vivo reconstitution system. J Biol Chem 276:5856–65

Mason PJ, Wilson DB, Bessler M (2005) Dyskeratosis congenita – a disease of dysfunctional telomere maintenance. Curr Mol Med 5:159–70

Masutomi K, Yu EY, Khurts S, Ben-Porath I, Currier JL, Metz GB, Brooks MW, Kaneko S, Murakami S, DeCaprio JA, Weinberg RA, Stewart SA, Hahn WC (2003) Telomerase maintains telomere structure in normal human cells. Cell 114:241–53

Meier UT (2005) The many facets of H/ACA ribonucleoproteins. Chromosoma 114:1–14

Meier UT, Blobel G (1994) NAP57, a mammalian nucleolar protein with a putative homolog in yeast and bacteria. J Cell Biol 127:1505–14

Metcalfe JA, Parkhill J, Campbell L, Stacey M, Biggs P, Byrd PJ, Taylor AM (1996) Accelerated telomere shortening in ataxia telangiectasia. Nat Genet 13:350–53

Miller MC, Liu JK, Collins K (2000) Template definition by Tetrahymena telomerase reverse transcriptase. Embo J 19:4412–22

Mitchell JR, Cheng J, Collins K (1999a) A box H/ACA small nucleolar RNA-like domain at the human telomerase RNA 3′ end. Mol Cell Biol 19:567–76

Mitchell JR, Collins K (2000) Human telomerase activation requires two independent interactions between telomerase RNA and telomerase reverse transcriptase. Mol Cell 6:361–71

Mitchell JR, Wood E, Collins K (1999b) A telomerase component is defective in the human disease dyskeratosis congenita. Nature 402:551–55

Mochizuki Y, He J, Kulkarni S, Bessler M, Mason PJ (2004) Mouse dyskerin mutations affect accumulation of telomerase RNA and small nucleolar RNA, telomerase activity, and ribosomal RNA processing. Proc Natl Acad Sci USA 101:10756–61

Moriarty TJ, Marie-Egyptienne DT, Autexier C (2005) Regulation of 5′ template usage and incorporation of noncognate nucleotides by human telomerase. Rna 11:1448–60

Newby MI, Greenbaum NL (2002) Sculpting of the spliceosomal branch site recognition motif by a conserved pseudouridine. Nat Struct Biol 9:958–65

Ni J, Tien AL, Fournier MJ (1997) Small nucleolar RNAs direct site-specific synthesis of pseudouridine in ribosomal RNA. Cell 89:565–73

Ortmann CA, Niemeyer CM, Wawer A, Ebell W, Baumann I, Kratz CP (2006) TERC mutations in children with refractory cytopenia. Haematologica 91:707–08

Paull TT, Gellert M (1999) Nbs1 potentiates ATP-driven DNA unwinding and endonuclease cleavage by the Mre11/Rad50 complex. Genes Dev 13:1276–88
Peng Y, Mian IS, Lue NF (2001) Analysis of telomerase processivity: mechanistic similarity to HIV-1 reverse transcriptase and role in telomere maintenance. Mol Cell 7:1201–11
Phillips B, Billin AN, Cadwell C, Buchholz R, Erickson C, Merriam JR, Carbon J, Poole SJ (1998) The Nop60B gene of Drosophila encodes an essential nucleolar protein that functions in yeast. Mol Gen Genet 260:20–29
Pogacic V, Dragon F, Filipowicz W (2000) Human H/ACA small nucleolar RNPs and telomerase share evolutionarily conserved proteins NHP2 and NOP10. Mol Cell Biol 20:9028–40
Ramamurthy V, Swann SL, Spedaliere CJ, Mueller EG (1999) Role of cysteine residues in pseudouridine synthases of different families. Biochemistry 38:13106–111
Rashid R, Liang B, Baker DL, Youssef OA, He Y, Phipps K, Terns RM, Terns MP, Li H (2006) Crystal structure of a Cbf5-Nop10-Gar1 complex and implications in RNA-guided pseudouridylation and dyskeratosis congenita. Mol Cell 21:249–60
Ren X, Gavory G, Li H, Ying L, Klenerman D, Balasubramanian S (2003) Identification of a new RNA.RNA interaction site for human telomerase RNA (hTR): structural implications for hTR accumulation and a dyskeratosis congenita point mutation. Nucleic Acids Res 31:6509–15
Rostamiani K, Heiss NS, Poustka A, Metzenberg AB (1999) ❑Novel missense mutations in the DKC1 gene in patients with ❑dyskeratosis congenita. Am J Hum Genet 65 [Suppl Ab] :2777
Rudolph KL, Chang S, Lee HW, Blasco M, Gottlieb GJ, Greider C, DePinho RA (1999) Longevity, stress response, and cancer in aging telomerase-deficient mice. Cell 96:701–12
Rudolph KL, Millard M, Bosenberg MW, DePinho RA (2001) Telomere dysfunction and evolution of intestinal carcinoma in mice and humans. Nat Genet. 2001 28(2):155–59.
Ruggero D, Grisendi S, Piazza F, Rego E, Mari F, Rao PH, Cordon-Cardo C, Pandolfi PP (2003) Dyskeratosis congenita and cancer in mice deficient in ribosomal RNA modification. Science 299:259–62
Safa WF, Lestringant GG, Frossard PM (2001) X-linked dyskeratosis congenita: restrictive pulmonary disease and a novel mutation. Thorax 56:891–94
Salowsky R, Heiss NS, Benner A, Wittig R, Poustka A (2002) Basal transcription activity of the dyskeratosis congenita gene is mediated by Sp1 and Sp3 and a patient mutation in a Sp1 binding site is associated with decreased promoter activity. Gene 293:9–19
Sarin KY, Cheung P, Gilison D, Lee E, Tennen RI, Wang E, Artandi MK, Oro AE, Artandi SE (2005) Conditional telomerase induction causes proliferation of hair follicle stem cells. Nature 436:1048–52
Serrano AL, Andres V (2004) Telomeres and cardiovascular disease: does size matter? Circ Res 94:575–84
Shay JW, Wright WE (2004) Telomeres in dyskeratosis congenita. Nat Genet 36:437–8
Sohal RS, Mockett RJ, Orr WC (2002) Mechanisms of aging: an appraisal of the oxidative stress hypothesis. Free Radic Biol Med 33:575–86
Solder B, Weiss M, Jager A, Belohradsky BH (1998) Dyskeratosis congenita: multisystemic disorder with special consideration of immunologic aspects. A review of the literature. Clin Pediatr (Phila) 37:521–30
Stewart SA, Hahn WC, O'Connor BF, Banner EN, Lundberg AS, Modha P, Mizuno H, Brooks MW, Fleming M, Zimonjic DB, Popescu NC, Weinberg RA (2002) Telomerase contributes to tumorigenesis by a telomere length-independent mechanism. Proc Natl Acad Sci USA 99:12606–11
Sznajer Y, Baumann C, David A, Journel H, Lacombe D, Perel Y, Blouin P, Segura JF, Cezard JP, Peuchmaur M, Vulliamy T, Dokal I, Verloes A (2003) Further delineation of the congenital form of X-linked dyskeratosis congenita (Hoyeraal-Hreidarsson syndrome). Eur J Pediatr 162:863–67
Tesmer VM, Ford LP, Holt SE, Frank BC, Yi X, Aisner DL, Ouellette M, Shay JW, Wright WE (1999) Two inactive fragments of the integral RNA cooperate to assemble active telomerase with the human protein catalytic subunit (hTERT) in vitro. Mol Cell Biol 19:6207–16
Theimer CA, Finger LD, Trantirek L, Feigon J (2003) Mutations linked to dyskeratosis congenita cause changes in the structural equilibrium in telomerase RNA. Proc Natl Acad Sci USA 100:449–54

Tollervey D, Kiss T (1997) Function and synthesis of small nucleolar RNAs. Curr Opin Cell Biol 9:337–42

Trapp S, Parcells MS, Kamil JP, Schumacher D, Tischer BK, Kumar PM, Nair VK, Osterrieder N (2006) A virus-encoded telomerase RNA promotes malignant T cell lymphomagenesis. J Exp Med 203:1307–17

Treister N, Lehmann LE, Cherrick I, Guinan EC, Woo SB (2004) Dyskeratosis congenita vs. chronic graft versus host disease: report of a case and a review of the literature. Oral Surg Oral Med Oral Pathol Oral Radiol Endod 98:566–71

Trowbridge AA, Sirinavin C, Linman JW (1977) Dyskeratosis congenita: hematologic evaluation of a sibship and review of the literature. Am J Hematol 3:143–52

Tsakiri, KD, Cronkhite, JT, Kuan, PJ, Xing, C, Raghu, G Weissler , JC, Rosenblatt, RL, Shay, JW and Garcia, CK (2007) Adult-Onset Pulmonary Fibrosis Caused by Mutations in Telomerase. Proc Natl Acad Sci USA 104:7552–7

Ueda CT, Roberts RW (2004) Analysis of a long-range interaction between conserved domains of human telomerase RNA. Rna 10:139–47

Vaziri H, Benchimol S (1998) Reconstitution of telomerase activity in normal human cells leads to elongation of telomeres and extended replicative life span. Curr Biol 8:279–82

Vaziri H, Dragowska W, Allsopp RC, Thomas TE, Harley CB, Lansdorp PM (1994) Evidence for a mitotic clock in human hematopoietic stem cells: loss of telomeric DNA with age. Proc Natl Acad Sci U S A 91:9857–60

Vulliamy T, Marrone A, Dokal I, Mason PJ (2002) Association between aplastic anaemia and mutations in telomerase RNA. Lancet 359:2168–70

Vulliamy T, Marrone A, Goldman F, Dearlove A, Bessler M, Mason PJ, Dokal I (2001a) The RNA component of telomerase is mutated in autosomal dominant dyskeratosis congenita. Nature 413:432–35

Vulliamy T, Marrone A, Szydlo R, Walne A, Mason PJ, Dokal I (2004) Disease anticipation is associated with progressive telomere shortening in families with dyskeratosis congenita due to mutations in TERC. Nat Genet 36:447–49

Vulliamy TJ, Knight SW, Heiss NS, Smith OP, Poustka A, Dokal I, Mason PJ (1999) Dyskeratosis congenita caused by a 3′ deletion: germline and somatic mosaicism in a female carrier. Blood 94:1254–60

Vulliamy TJ, Knight SW, Mason PJ, Dokal I (2001b) Very short telomeres in the peripheral blood of patients with X-linked and autosomal dyskeratosis congenita. Blood Cells Mol Dis 27:353–57

Vulliamy TJ, Marrone A, Knight SW, Walne A, Mason PJ, Dokal I (2006) Mutations in dyskeratosis congenita: their impact on telomere length and the diversity of clinical presentation. Blood 107:2680–85

Vulliamy TJ, Walne A, Baskaradas A, Mason PJ, Marrone A, Dokal I (2005) Mutations in the reverse transcriptase component of telomerase (TERT) in patients with bone marrow failure. Blood Cells Mol Dis 34:257–63

Walne AJ, Vulliamy T, Marrone A, Beswick R, Kirwan M, Masunari, Al-Qurashi FH, Aljurf M, Dokal I (2007) Genetic heterogeneity in autosomal recessive dyskeratosis congentia with one subtype due to mutuations in the telomerase-associated protein in NOP10. Hum Mol Genet 16:1619–29

Wang C, Meier UT (2004) Architecture and assembly of mammalian H/ACA small nucleolar and telomerase ribonucleoproteins. Embo J 23:1857–67

Watson JD (1972) Origin of concatemeric T7 DNA. Nat New Biol 239:197–201

Weinert BT, Timiras PS (2003) Invited review: Theories of aging. J Appl Physiol 95:1706–16

Weinstein BS, Ciszek D (2002) The reserve-capacity hypothesis: evolutionary origins and modern implications of the trade-off between tumor-suppression and tissue-repair. Exp Gerontol 37:615–27

Wiedemann HP, McGuire J, Dwyer JM, Sabetta J, Gee JB, Smith GJ, Loke J (1984) Progressive immune failure in dyskeratosis congenita. Report of an adult in whom Pneumocystis carinii and fatal disseminated candidiasis developed. Arch Intern Med 144:397–99

Wilson DB, Ivanovich J, Whelan A, Goodfellow PJ, Bessler M (2003) Human telomerase RNA mutations and bone marrow failure. Lancet 361:1993–94

Womer R, Clark JE, Wood P, Sabio H, Kelly TE (1983) Dyskeratosis congenita: two examples of this multisystem disorder. Pediatrics 71:603–09

Wong JM, Collins K (2006) Telomerase RNA level limits telomere maintenance in X-linked dyskeratosis congenita. Genes Dev 20:2848–58

Wong JM, Kyasa MJ, Hutchins L, Collins K (2004) Telomerase RNA deficiency in peripheral blood mononuclear cells in X-linked dyskeratosis congenita. Hum Genet 115:448–55

Wright WE, Piatyszek MA, Rainey WE, Byrd W, Shay JW (1996) Telomerase activity in human germline and embryonic tissues and cells. Dev Genet 18:173–79

Xia J, Peng Y, Mian IS, Lue NF (2000) Identification of functionally important domains in the N-terminal region of telomerase reverse transcriptase. Mol Cell Biol 20:5196–5207

Xin ZT, Beauchamp AD, Calado RT, Bradford JW, Regal JA, Shenoy A, Liang Y, Lansdorp PM, Young NS, Ly H (2006) Functional characterization of natural telomerase mutations found in patients with hematological disorders. Blood

Yaghmai R, Kimyai-Asadi A, Rostamiani K, Heiss NS, Poustka A, Eyaid W, Bodurtha J, Nousari HC, Hamosh A, Metzenberg A (2000) Overlap of dyskeratosis congenita with the Hoyeraal-Hreidarsson syndrome. J Pediatr 136:390–93

Yamaguchi H, Baerlocher GM, Lansdorp PM, Chanock SJ, Nunez O, Sloand E, Young NS (2003) Mutations of the human telomerase RNA gene (TERC) in aplastic anemia and myelodysplastic syndrome. Blood 102:916–18

Yamaguchi H, Calado RT, Ly H, Kajigaya S, Baerlocher GM, Chanock SJ, Lansdorp PM, Young NS (2005) Mutations in TERT, the gene for telomerase reverse transcriptase, in aplastic anemia. N Engl J Med 352:1413–24

Yasumoto S, Kunimura C, Kikuchi K, Tahara H, Ohji H, Yamamoto H, Ide T, Utakoji T (1996) Telomerase activity in normal human epithelial cells. Oncogene 13:433–39

Yoon A, Peng G, Brandenburger Y, Zollo O, Xu W, Rego E, Ruggero D (2006) Impaired control of IRES-mediated translation in X-linked dyskeratosis congenita. Science 312:902–06

Yoshimoto T, Yanabe Y, Mizukami T, Ishibashi F, Adachi N, Takaki K, Nunoi H (2000) [Genetic analysis of a patient with dyskeratosis congenita]. Rinsho Ketsueki 41:524–29

Youssoufian H, Gharibyan V, Qatanani M (1999) Analysis of epitope-tagged forms of the dyskeratosis congenital protein (dyskerin): identification of a nuclear localization signal. Blood Cells Mol Dis 25:305–09

Zijlmans JM, Martens UM, Poon SS, Raap AK, Tanke HJ, Ward RK, Lansdorp PM (1997) Telomeres in the mouse have large inter-chromosomal variations in the number of T2AG3 repeats. Proc Natl Acad Sci USA 94:7423–28

Zinsser F (1906) Atropha cutis reticularis cum pigmentatione, dystrophia ungium et leukoplakia oris. Ikonogr Dermatol (Hyoto) 5:219–23

Part II
Telomerase, Telomeres, and Stem Cell Aging

Chapter 6
Mechanisms of Stem Cell Ageing

Amanda Waterstrat, Erin Oakley, Alison Miller, Carol Swiderski, Ying Liang, and Gary Van Zant(✉)

Abstract There are many definitions of the term "ageing" and perhaps even more theories that seek to explain its causes (Balcombe and Sinclair 2001). From a physiological standpoint, ageing beyond reproductive maturity is often viewed as a progression of multisystem deficits in tissue function. In adult mammals, tissue homeostasis is maintained by stem cell populations that reside in, or migrate between, a variety of adult tissues. These stem cells ensure proper tissue function by generating new cells to replace those lost or damaged over time. Despite the presence of adult stem cells in muscle, nervous, gastrointestinal, hematopoietic, and other tissues, each of these systems exhibit functional decline with age (Edwards et al. 2002, Campisi 2003, Kondo et al. 2003, Penninx et al. 2003, Pinto et al. 2003, Linton and Dorshkind 2004, Balducci and Ershler 2005, Fulle et al. 2005, Kamminga 2005, Bauer et al. 2006, Keller 2006, Theise 2006). It is compelling to consider that functional changes in the stem cell compartment of adult tissues precedes and perhaps contributes to the manifestation of ageing phenotypes. In this chapter we will review literature describing the effects of ageing on the well-characterized stem cells of the hematopoietic system. In this system, ageing is accompanied by immune compromise, anemia, and a dramatic increase in the incidence of malignancy (Edwards et al. 2002, Campisi 2003, Penninx et al. 2003, Pinto et al. 2003, Linton and Dorshkind 2004, Balducci and Ershler 2005). Several studies support the hypothesis that these ageing phenotypes stem from functional changes in hematopoietic stem cells (HSCs).

1 Introduction

Adult HSCs reside primarily in the marrow cavity in discrete locations termed stem cell niches. In the niche, intrinsic properties of the stem cell and extrinsic signals from the local microenvironment cooperatively regulate HSC function. We will review intrinsic

Gary Van Zant
Henry Wellcome Laboratory for Biogerontology Research, Institute for Ageing and Health, Center for Integrated Systems Biology of Ageing and Nutrition, University of Newcastle upon Tyne, NE4 6BE, United Kingdom.
e-mail: gvzant1@uky.edu

K.L. Rudolph (ed.), *Telomeres and Telomerase in Ageing, Disease, and Cancer.*

and extrinsic regulatory mechanisms governing HSCs and consider the effects of ageing on both the stem cell and the microenvironment. We will conclude by reviewing some of the functional consequences of these age-related changes. Our goal is to summarize HSC regulatory mechanisms that are subject to deleterious effects of ageing and to consider evidence that suggests a compromised stem cell compartment contributes to age-related phenotypes in the entire organism and potentially limits longevity (Fig. 6.1).

2 Cell Intrinsic Regulation of HSCs

HSCs must balance self-renewal, differentiation, and cell death decisions to sustain hematopoiesis throughout the lifetime of an organism. They must also be poised for extensive proliferation upon demand in the event of trauma such as major blood loss. The normal fate of decisions of stem cells can easily be summarized as follows: a stem cell may (1) remain quiescent and not divide at all; (2) undergo symmetric cell division to generate two new self-renewing stem cells; (3) undergo symmetric cell division to generate two new differentiated progeny, thereby eliminating the stem cell; (4) undergo asymmetric cell division to give rise to a new stem cell and a differentiated cell; and finally (5) a stem cell may undergo apoptosis and no longer contribute to hematopoiesis. There is strong evidence that cell intrinsic factors tightly regulate the survival, differentiation, and proliferation of stem cells, and thus ultimately influence the functional quality and size of the HSC compartment throughout the lifetime of an organism. Not surprisingly, cellular fate decisions change during ageing and lead to a variety of ageing phenotypes that will be discussed below.

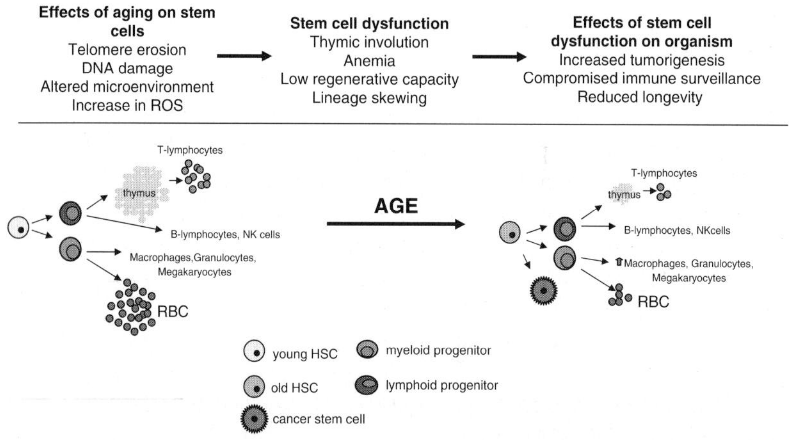

Fig. 6.1 This schematic represents the stem cell theory of ageing as it pertains to the hematopoietic system. The accumulation of damage and the series of changes that accompany ageing influence the HSC compartment in a way that results in compromised function of this system and reduced health of the entire organism. This illustrates the possible connection between adult stem cells and organismal longevity (*See Color Plate*)

2.1 HSC Cell Cycle Status

One of the most important factors in sustaining hematopoietic homeostasis and preventing exhaustion of the stem cell pool is appropriate cell cycle control. Strict regulatory mechanisms ensure that most HSCs remain quiescent and only a limited number enter cycle at any given time. Cyclin-dependent kinase inhibitors (CKIs) halt cell cycle progression and play an important role in stem cell regulation (Cheng et al. 2000b, Yuan et al. 2004, Walkley et al. 2005). The G_1 checkpoint regulator, $p21^{cip1/waf1}$ (p21) restricts HSC cycling, and thus contributes to the maintenance of stem cell quiescence and survival (Cheng et al. 2000b). p21-deficiency increases HSC proliferation and amplifies the size of the HSC pool under homeostatic conditions in the mouse (Cheng et al. 2000b). The proportion of quiescent cells in the G_0 phase of the cell cycle is reduced in $p21^{-/-}$ mice and, in agreement with these findings, $p21^{-/-}$ mice exhibit increased sensitivity to 5-fluorouracil, a cytotoxin. Hematopoietic potential is reduced so significantly by p21 deficiency that HSCs lacking p21 are almost completely exhausted after two rounds of serial bone marrow transplantation (Cheng et al. 2000b). Together these results indicate p21 as a key regulator of stem cell cycling that functions to prevent stem cell exhaustion during prolonged stress (Ezoe et al. 2004).

Unlike p21, the CKI $p27^{KIP1}$ (p27) is expressed in more mature progenitor cells rather than the primitive stem cell compartment. The number, cycling status, and self-renewal of HSCs are normal in p27-deficient mice, but the mice display an increase in the number and proliferative capacity of progenitor cells (Cheng et al. 2000a). Progenitor cells from p27-null mice expand and regenerate after serial transplantation, indicating that p27 is an important modulator of the repopulating efficiency but not overall number of HSCs (Cheng et al. 2000b). Another CKI, $p18^{INK4C}$ (p18), a member of the INK4 family of CKIs, appears to function similarly to p27. HSCs deficient in p18 display diminished stem cell exhaustion compared to wild-type cells after serial transplantation over three years (Yu et al. 2006). Interestingly, deletion of p18 overcomes the accelerated exhaustion of HSC repopulating ability induced by p21 deficiency. The advantage imparted to p18-deficient cells is thought to be due to increased self-renewal of the transplanted HSCs in vivo rather than increased donor cell proliferation (Yuan et al. 2004). These studies demonstrate the importance of CKIs in regulating HSC cell cycle kinetics and the highly divergent effects that cell cycle inhibitors can have on stem cell function.

2.2 HSC Gene Expression Profiles: Transcription Factors and Epigenetic Modification

In addition to cell cycle regulators, numerous cell intrinsic transcription factors, transcriptional repressors, and signaling molecules have been identified as important regulators of stem cell self-renewal and maintenance. We will consider some of the most well-known intrinsic regulatory models, but note that, due to space limitations, we will provide only a sampling of the many cell autonomous factors that influence HSC dynamics.

The homeobox (Hox) genes encode transcription factors that regulate segment identity in developing *Drosophila* and vertebrate embryos (Lewis 1978, Dubrulle et al. 2001). Numerous Hox family genes have also been implicated in the regulation of various stages of hematopoiesis (Lawrence et al. 1996, Abramovich and Humphries 2005, Abramovich et al. 2005). In particular, HoxB4 has been considered an important regulator of primitive HSCs because it is expressed highly in human CD34+ cells but is rapidly downregulated upon differentiation (Sauvageau et al. 1994). Overexpression of HoxB4 in the bone marrow increases HSC numbers, suggesting that it acts as a positive regulator of HSC self-renewal (Sauvageau et al. 1995, Antonchuk et al. 2002; Krosl et al. 2003; Miyake et al. 2006). Because maintenance of the HSC compartment is governed by a balance between growth-promoting signals and signals that promote senescence, differentiation, and apoptosis, one group investigated the combined effects of HoxB4 overexpression and p21 downregulation (Miyake et al. 2006). As mentioned previously, p21 acts to maintain HSCs in a quiescent state; therefore it was hypothesized that p21 deficiency would enhance the HoxB4-mediated increase in self-renewal capacity. Indeed the simultaneous modulation of p21 and HoxB4 produced a ten-fold increase in the number of repopulating HSCs after five days of ex vivo culture (Miyake et al. 2006). Another transcription factor, growth factor independence 1 (Gfi1), acts as a negative regulator of HSC self-renewal by limiting proliferation potential (Hock et al. 2004, Zeng H 2004). Gfi1 deficiency increases HSC proliferation and decreases repopulating efficiency during serial transplantation. In fact, Gfi1 is thought to be an upstream regulator of p21, whose expression is decreased in Gfi1 null cells. Factors that govern HSC self-renewal and cellular fate decisions must be tightly controlled throughout the life span of an organism to prevent premature stem cell exhaustion and maintain the integrity of the HSC population.

Recently, epigenetic modifiers of gene expression have emerged as important regulators of stem cells and the ageing process. In addition to direct sequence-specific genetic variations, epigenetic changes to chromatin structure and covalent modification of histones play important roles in regulating gene transcription throughout the life span of an organism. Expression of genes mediating epigenetic processes changes with age (Rossi et al. 2005), suggesting that epigenetic modifications may contribute to HSC ageing. In HSCs a particular class of epigenetic modifiers termed the polycomb group (PcG) proteins are important regulators of self-renewal and cell fate determination. PcG proteins are transcriptional repressors that silence expression of transcripts promoting differentiation in human and mouse embryonic cells (Boyer et al. 2006, Kamminga et al. 2006). One PcG protein in particular, Bmi-1, is required for efficient self-renewing divisions of adult HSCs (Lessard et al. 1998, Park et al. 2003). In transplantation experiments, fetal liver cells deficient in Bmi-1 fail to contribute to long-term hematopoiesis, providing another example of cell intrinsic impairment of self-renewal. Bmi-1 modulates HSC self-renewal by regulating genes important for cellular fate and survival decisions

(Park et al. 2004) and acts as an inhibitor of the tumor suppressor locus that encodes p16^{INK4a} and p19^{ARF} which, as mentioned previously, are key components of the primary senescence checkpoint in human and rodent primary cells . Bmi-1 over-expression in mouse embryonic fibroblasts downregulates p16^{INK4a} and p19^{ARF}, delays replicative senescence, and promotes immortalization (Goodman and Hodgson 1962, McCredie et al. 1971). It is not surprising that Bmi-1 protein levels decline during replicative senescence of human fibroblasts (Lanzkron et al. 1999), whereas levels of p16^{INK4a} and p19^{ARF} accumulate. The increased cellular surveillance and tumor suppressor activity imparted by p16^{INK4a} and p19^{ARF} account, at least in part, for age-related decline in stem cell function (Srour EF et al. 2001). Bmi-1 seems to maintain the stem cell pool by preventing premature senescence through the regulation of key genes implicated in ageing of the HSC compartment.

2.3 Maintaining the HSC Population: Stem Cells Make Life or Death Decisions

In addition to factors that influence HSC self-renewal decisions, factors that regulate cellular death and senescence also play an important role in maintaining stem cell homeostasis and may thereby account for differential ageing patterns observed in various inbred mouse strains. Cellular ageing is associated with the accumulation of DNA damage and genomic instability, and it has been hypothesized that proteins regulating DNA-damage response pathways are key regulators of ageing (Dumble et al. 2007). For example, p53 is a tumor suppressor protein that is activated in response to cellular stresses, including DNA damage and oncogenic activation, to induce cell cycle arrest or apoptosis or both. Mice expressing a hyperactive p53 protein display a heightened tumor resistance and, surprisingly, significantly reduced longevity compared to wild-type mice (Tyner et al. 2002). The mutant mice display a generalized organ atrophy and reduced stress tolerance at middle age that progresses with old age. Investigation of the hematopoietic system revealed that aged p53 mutant mice also display a reduced ability to generate white blood cells after ablation of hematopoietic progenitors with 5-flurouracil (Dumble et al. 2007). It was hypothesized that diminished HSC function drove the decrease in regenerative capacity. Additional experiments in this study showed that C57BL/6 mice expressing hyperactive p53 (p53$^{+/m}$) exhibited no age-associated increases in stem cell number (consistent with wild-type C57BL/6 mice) and in fact showed a reduced number of HSCs compared to age-matched p53$^{+/+}$ and p53$^{+/-}$ mice. Although the mechanism by which p53 reduces HSC proliferation and function is not yet known, a clear role for p53 activity has been established in the HSC ageing process. One plausible scenario is that altered p53 dosage affects the self-renewal capacity of HSCs, which, over time, results in functional differences in the hematopoietic compartments between mutant and wild-type animals (Dumble et al. 2007). The role of altered p53 dosage is now being studied in other stem cell compartments to determine whether the early ageing phenotypes observed in hyperactive p53 mutant mice are a result of a global decline in stem cell function with age.

3 Cell-Intrinsic Mechanisms of Stem Cell Dysfunction

3.1 Signaling Pathways Implicated in Cell-Intrinsic HSC Ageing

The fundamental idea that somatic cells have a limited proliferative capacity predicts that HSCs will show decreased functional ability over time, manifesting as stem cell malfunction. Mouse fetal liver HSCs have 1.6 to 3.0 times higher functional ability than young adult HSCs (Chen et al. 1999). Similarly, young mouse HSCs have 1.6 to 2.0 times higher functional ability compared to old mouse HSCs. Human HSCs from fetal liver, cord blood, and adult bone marrow also show a continuous decline in their ability to proliferate in vitro (Chen et al. 1999). This decline in stem cell frequency and function has been attributed to an increase in apoptosis and senescence of the HSC compartment (Chen et al. 1999, Kamminga and de Haan 2006).

The "free radical theory of ageing" proposed by Harman in 1956 suggests that ageing, as well as age-related degenerative diseases, can be attributed to the deleterious effects that free radicals impose on a cell (Harman 1956; see Passos et al., this volume). Thus, long-lived cells run the risk of accumulating damage from oxidative stress as a result of normal metabolism, exposure to cellular toxins, and cellular stress. Ataxia telangiectasia mutated (ATM) is a protein kinase encoded by the ATM gene (Savitsky et al. 1995), which is mutated in the disease ataxia telangiectasia (AT). ATM functions by regulating telomere length, its primary role, and response to cellular damage (Wang et al. 2005). HSCs in $ATM^{-/-}$ mice, as expected, show signs of telomere dysfunction, increased cellular oxidant levels, and reduced numbers of HSCs. This dysfunction in the HSC compartment results in aplasia and severe anemia in mice as early as six months of age. Administration of the hydrogen peroxide scavenger, N-acetyl-cysteine, to $ATM^{-/-}$ mice restores HSC self-renewal and rescues stem cell aplasia. This reversal in stem cell dysfunction is therefore attributed to mechanisms independent of telomere maintenance (Molofsky et al. 2006). Further investigation of ATM showed that it is a stem cell–specific sensor of the levels of reactive oxygen species (ROS). Increased ROS leads to the activation of the p38 MAPK pathway, which normally takes responsibility for general damage control by responding to a variety of stimuli that signal cellular damage (see also Davis and Kipling, this volume). It is constitutively active in the $ATM^{-/-}$ HSCs, causing an increase in HSC proliferation without self-renewal of the stem cell pool. To mimic the replicative stress that HSCs undergo during ageing, serial transplant experiments were performed in both wild-type and $ATM^{-/-}$ mice. With each transplant, the $ATM^{-/-}$ HSCs showed an elevation of ROS and an increase in p38 MAPK pathway activation compared to wild-type. These data are consistent with the hypothesis that stem cells are depleted during the normal ageing process due to an accumulation of ROS, leading to an increase in p38 MAPK signaling, which in turn results in an increase in cell cycling without the critical function of self-renewal (Ito et al. 2006).

Another potential mechanism of stem cell dysfunction with age is an alteration in the damage control pathway of $p16^{INK4a}$–retinoblastoma (Rb, see Sharpless, this volume). $p16^{INK4a}$ is a cyclin-dependent kinase inhibitor that enforces G1 cell cycle arrest by activating the Rb tumor suppressor, thus inducing irreversible cellular senescence. The Rb pathway is important in HSCs, pancreatic islets, and neural stem cells, making this common mechanism an attractive therapeutic target (Janzen et al. 2006, Krishnamurthy et al. 2006, Molofsky et al. 2006). An increase of $p16^{INK4a}$ with age has been documented in many tissues (Krishnamurthy et al. 2004). At a young age, $p16^{INK4a}$-/- mice do not show variation in HSC frequency or ability to reconstitute an irradiated host compared to wild-type, but by 14 to 24 months, the $p16^{INK4a}$-/- mice have higher HSC frequency, increased repopulating ability, and superior engraftment in serial transplants compared to wild-type. Interestingly, when HSCs from young $p16^{INK4a}$-/- mice are serially transplanted, they undergo premature exhaustion relative to wild-type. Thus, the $p16^{INK4a}$-/- mouse exhibits a true age-specific phenotype in the stem cell population and, as might be expected, these mice are also cancer-prone (Janzen et al. 2006). These experiments demonstrate the importance of the balance between tissue regeneration and tumorigenesis. The ability of $p16^{INK4a}$ to limit stem cell number, by either senescence or apoptosis, is important in its function as a tumor suppressor. These data highlight the importance $p16^{INK4a}$ in the detection or elimination of damaged cells that are subject to malignant transformation.

3.2 HSC-Intrinsic Mechanisms of Functional Decline Lead to Changes in Stem Cell Number with Age

We have cited several studies in which knockout (KO) mice were used to demonstrate a role for certain molecules in the regulation of the hematopoietic system. Unfortunately, embryonic lethality and the unpredictable and nonphysiological downstream effects of gene KO studies prohibit the use of KO or conditional KO strategies to investigate many genes that may have unique roles in adult HSCs. Inbred strains of laboratory mice provide excellent models to study HSC properties because strain-specific phenotypic differences can be directly attributed to underlying genetic differences in the stem cells of mice that are otherwise healthy and viable. For example, the genetic regulation of the age-related decline in stem cell function has been demonstrated extensively by comparing divergent phenotypes among various mouse strains. Stem cells in the relatively long-lived C57BL/6 (B6) mouse strain increase in number during ageing, whereas stem cells in DBA/2 (D2) and BALB/cBy (BALB) strains decrease over time and display overt signs of ageing (Morrison et al. 1996, Chen et al. 1999, de Haan 1999, Kamminga 2005, Yuan et al. 2005). When equal numbers of bone marrow cells from old and young B6 mice are mixed with heterogeneous competitor cells and injected into lethally irradiated recipients, the old cells have a competitive advantage compared to young cells (Harrison et al. 1989, Morrison et al. 1996). In stark contrast to these results, old stem cell populations in the relatively short-lived D2 and BALB mice display a significant competitive disadvantage compared to young cells (Harrison et al. 1989, Van Zant et al. 1990, Chen et al. 2000). Strain-dependent differences in B6 and D2 stem cell ageing were demonstrated in a series of experiments in allophenic (cellular chimeras) mice. Eight–cell embryos from B6 and D2 mice were combined ex vivo to form chimeric embryos that were implanted into the uteri of pseudopregnant females (Van Zant et al. 1990). All tissues in the resulting chimeric mice consisted of a mixture of B6 and D2 cells. During ageing, D2 HSC function progressively declined, evidenced by a shift to the B6 genotype in all blood cell lineages by two years of age. Importantly, the chimeric mice also showed complete tolerance to B6 and D2 cells, thus ruling out all possibility that the D2 genotype is selectively eliminated during ageing (Mintz and Silvers 1967). The chimeras allow simultaneous assessment of two stem cell populations in a common environment, demonstrating the cell intrinsic nature of stem cell ageing properties in DBA and B6 mice. D2 hematopoiesis in the chimeras ceased around the time of the strain's natural life span, which supports the existence of a link between viability of HSC populations and organismal longevity.

Following this important observation in B6/DBA allophenic mice, several other studies followed in which hematopoiesis was characterized more thoroughly in both B6 and DBA strains (de Haan et al. 1997, de Haan and Van Zant 1997, de Haan and Van Zant 1999, Geiger et al. 2001). In B6 mice, the frequency of cells that bear cell surface markers associated with long-term self-renewal increase three- to fivefold as the animals age. It is not surprising that, in whole bone marrow transplant studies, old B6 cells perform better than young ones. Marrow cells derived from old B6 mice simply have a higher proportion of primitive stem cells compared to young mice. However, despite the fact that old B6 cells have a competitive advantage in transplant settings, they possess several qualitative defects. (Nonetheless, in most physiological situations their numerical advantage clearly compensates for qualitative defects.) A true assessment of a stem cell's regenerative capacity is its ability to be serially transplanted through multiple mouse recipients. Serial bone marrow transfer imparts severe proliferative demands on the HSC compartment and leads to stem cell exhaustion. In this setting, stem cell exhaustion is thought to lead to replicative senescence of the transplanted cells, resembling physiological HSC ageing (Bell and Van Zant 2004). Like most inbred strains, old B6 cells cannot be serially transplanted indefinitely and undergo a finite number of population doublings (Siminovitch et al. 1964, Ogden and Mickliem 1976m Harrison 1979, Harrison and Astle 1982, Ross et al. 1982, Kamminga 2005).

Recently, Kamminga et al. (2005) demonstrated that the maximum number of population doublings in primitive cells from B6 mice is approximately 30, and only 20 in D2 Although the maximum expansion potential was greater in B6 mice, serial transplantation resulted in severe functional impairment in both strains, evidenced by decreased clonogenic activity, reduced competitive repopulating activity, and impaired peripheral blood cell recovery. This study again emphasizes that intrinsic genetic regulation influences stem cell function in B6 and DBA mice during ageing.

Similar to the patterns of change in stem cell number in DBA/2 mice with age, the number of primitive hematopoietic cells (CD34+) that can be obtained from human bone marrow aspirates decreases with age. Regression analysis on data collected from almost 50 human marrow samples showed that 24% of the variation in CD34+ cell numbers can be attributed to age (Fig. 6.2). We suspect that comparing genetic and phenotypic differences between the HSCs of B6 and DBA mice will reveal loci involved in regulation of stem cell number in humans. Furthermore, it is conceivable that these loci will prove to be targets of therapeutic intervention that can be manipulated to limit malignancy or stimulate self-renewal or proliferation in bone marrow failure syndromes.

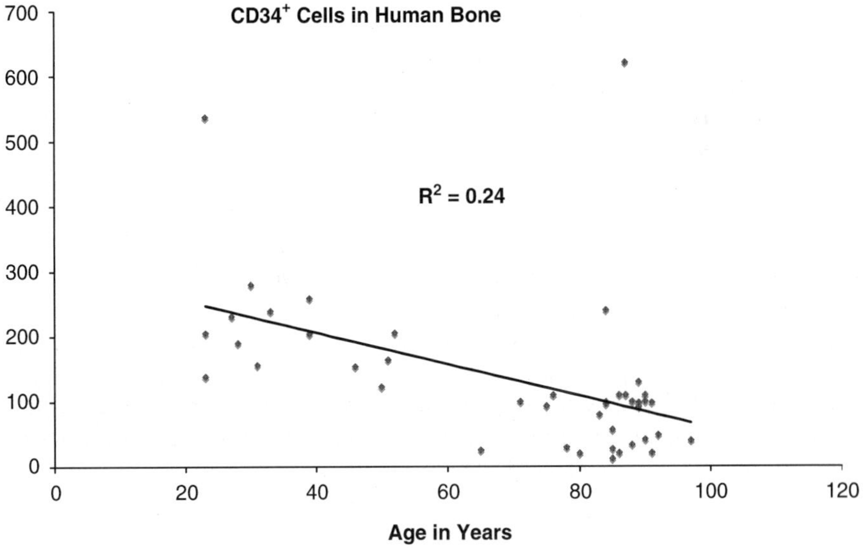

Fig. 6.2 Whole bone marrow aspirates were obtained from 43 adult humans. Mononucleated cells were collected by Ficoll-gradient centrifugation and stained for the presence of CD34. Flow cytometric analysis was used to identify the percentage of CD34+ cells in each sample. Linear regression analysis revealed that approximately 24% of the variation in the number of CD34+ cells in human bone marrow can be attributed to age (*See Color Plate*)

4 Hematopoietic Stem Cell Niches – Extrinsic Regulation of HSCs

The stem cell niche was initially proposed in 1978 as a protective physiological microenvironment that maintains the stem cell in quiescence for self-renewal but can activate the stem cell to proliferate and differentiate to maintain and repair tissue (Schofield 1978). The fact that bone marrow is located within bone cavities suggests that the two tissues interact, providing a protective environment for the HSCs (Li and Li 2006). Recently it has been proposed that more than one niche may exist: the endosteal bone marrow niche, which primarily supports HSCs in steady-state hematopoiesis and maintains HSC quiescence, and a vascular bone marrow niche, which supports HSCs at the fenestrated endothelium of bone marrow sinusoids and allows self-renewing HSCs to access the vasculature. The vascular niche may function as a secondary niche, which requires HSCs to migrate from the primary endosteal niche (Wilson and Trump 2006). Despite the fact that most adult HSCs reside in the bone marrow, administration of ^{89}Sr eliminates, by local irradiation, endosteal niches and results in splenic hematopoiesis (Klassen 1972). This suggests that, although HSCs require a supportive microenvironment, they are not passive residents and can perhaps modify extramedullary niches to enhance support of hematopoiesis outside the bone marrow. The capacity for influencing or remodeling the niche, as we will discuss in a later portion of the chapter, is also an important property of metastatic cancer cells.

4.1 The Vascular Niche

The vascular niche is composed of at least two cell types: HSCs and endothelial cells. During embryogenesis both cell types follow similar, but distinct, patterns of development (Ueno and Weissman 2006). In the adult, vascular endothelial cells contribute to the maintenance of HSCs in vitro (Li et al. 2004) and are required for in vivo hematopoiesis (Avecilla et al. 2004). Recently, HSCs identified by the cell surface marker phenotype $CD150^+$ $CD244^-$ $CD48^-$ were shown to localize in close proximity to sinusoidal endothelium or osteoblasts in the spleen and bone marrow. The spleen contained HSCs associated with the sinusoidal endothelium, with 62% located in the parafollicular regions. In the bone marrow, HSCs localized to the sinusoidal endothelium accounted for $0.0067 \pm 0.0016\%$ of total marrow cells. It was estimated that 60% of the HSCs were in contact with the endothelium, and because only 10% of the bone marrow contacts this area, HSCs are sixfold more likely to contact sinusoids (Kiel et al. 2005).

4.2 The Endosteal Niche

The endosteum, the proposed HSC niche in the long bones, is an active site of bone remodeling with a high concentration of extracellular calcium ions (Adams et al. 2006). Two groups recently proposed this site as an ideal location for a stem cell niche.

Zhang et al. (2003) utilized a bone morphogenic protein (BMP) receptor 1A conditional KO transgenic mouse to demonstrate that the downregulation of BMP signaling results in ectopic trabecular bone formation in the long bone and an increase in HSC number. Calvi et al. (2003) investigated the endosteal niche using a transgenic mouse with constitutively expressed parathyroid hormone and parathyroid hormone-related receptors (PPRs) with a collagen-α1 promoter in osteoblastic cells. These PPR-stimulated osteoblasts showed an increase in proliferation that correlated with an increase in HSC number. The location of this niche was further verified by a study that demonstrated that adhesion of HSCs to osteoblasts on the endosteal surface was essential to maintain HSCs in a quiescent state (Arai et al. 2004). The relationship between osteoblasts and HSCs is not surprising given that during embryogenesis, the hematopoietic system is derived from the mesodermal dorsal aorta and the bone precursors originate from the surrounding mesenchyme, establishing a partnership of the two tissues early in development (Taichman 2005).

4.3 Why Two Stem Cell Niches?

It is plausible that the vascular niche is the location for HSC differentiation and mobilization and that the osteoblastic niche provides a quiescent environment ideal for HSC maintenance. The vascular niche could be the alternative niche for mobilized stem cells, promoting proliferation and further differentiation or maturation and entry into the circulatory system. This description is attractive, considering the sinusoids in the vascular niche consist of a single layer of endothelial cells, designed for blood cell penetration. This would be an ideal location for HSCs to initiate a rapid response to hematopoietic stress or cytokine stimulation. Additionally, at least in mice, an extramedullary vascular niche also exists in the spleen or is established upon ablation of the endosteal microenvironment (Klassen 1972).

4.4 Molecular Interactions within the Niche

Adhesion is perhaps the most obvious interaction between the stem cell and the niche. HSC retention in the vascular niche is mediated in part by CXCL12 (SDF-1), which is also important in homing and mobilization. Several cell types in the niche express CXCL12, including osteoblasts and vascular endothelial cells. The binding of CXCL12 to its receptor, CXCR4, located on HSCs, is a crucial interaction in the retention and maintenance of adult HSCs (Wilson and Trump 2006). CXCL12 is regulated by fibroblast growth factor 2 (FGF2) that binds one of the fibroblast growth factor receptors (FGFR1 IIIc) located on stromal cells of the niche. The binding of FGF2 to FGFRs causes an increase in the degradation of CXCL12 mRNA, reducing its secretion into the extracellular space of the niche. FGF2 can thereby negatively regulate the retention of HSCs in the niche and influence the maintenance of hematopoiesis (Nakayama et al. 2007). Many other adhesion molecules are important in the HSC-niche interaction (Fig. 6.3). VLA4 on HSCs binds vascular cell-adhesion molecule 1 (VCAM1) on sinusoidal endothelial cells. CD44 on HSC membranes binds hyaluronic acid that is present in the niche microenvironment (Wilson and Trump 2006). Another important regulatory molecule in the niche is bone morphogenic protein receptor 1A (BMPR1A), which, as mentioned previously, is normally expressed on osteoblasts lining the endosteal niche (Wilson and Trump 2006). Mice that lack BMPR1A in the stroma display an increase in the number of osteoblasts and repopulating HSCs. Direct contact between HSCs and stromal osteoblasts is also mediated by interactions with N-cadherin, specifically through a group of specialized spindle-shaped N-cadherin-expressing osteoblasts termed "SNOs" (Zhang et al. 2003). In addition to SNOs, HSCs may also express N-cadherin when stimulated by angiopoietin-1 (Ang1). Ang1, which is secreted by osteoblasts, stimulates Tie2 on HSCs, ultimately influencing cell cycle regulation via the phosphoinositide 3-kinase-Akt-p21 pathway. This interaction blocks cell division and mediates adhesion of the HSC to the niche by increasing

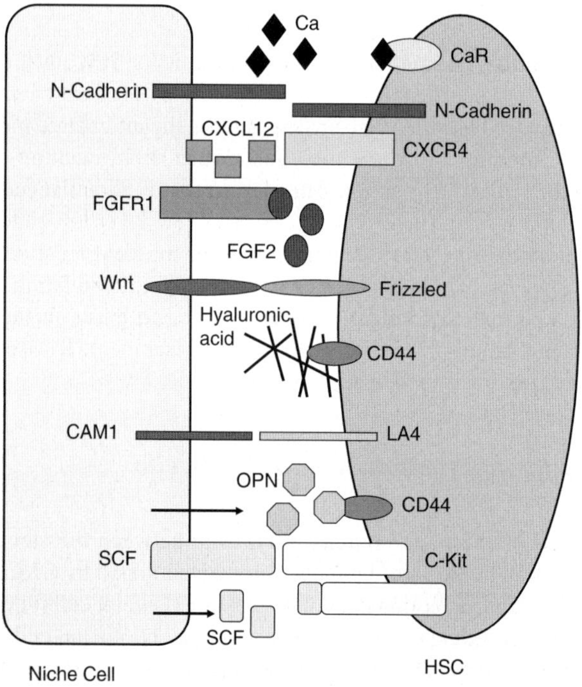

Fig. 6.3 HSCs engage in a number of cell-cell and cell-matrix interactions in the niche. Various cells of the niche (such as osteoblasts and stromal cells) have been shown to express integrins and signaling receptors on the cell surface, and secrete factors such as OPN, SCF, and CXCL12 that have binding partners on HSCs. HSCs also interact with extracellular matrix molecules such as hyaluronic acid via the CD44 molecule on the cell surface. Each of these molecular interactions has been shown to influence HSC functions such as adhesion, proliferation and quiescence (*See Color Plate*)

HSC N-cadherin expression, thus contributing to HSC quiescence and attachment to the niche. These pathways intimately link HSC adhesion to the niche with cell cycle regulation (Arai et al. 2004). Osteoblasts can also influence the number of HSCs in the bone marrow through the secretion of osteopontin (OPN) into the bone matrix. OPN is a glycoprotein that binds to CD44 on HSC membranes (Sugiyama et al. 2006). OPN-deficient mice exhibit a twofold increase in HSCs, suggesting that secretion of OPN by osteoblasts is a mechanism by which the endosteal niche negatively regulates HSC number (Nilsson et al. 2005, Stier et al. 2005).

Another regulator of stem cell frequency is the Wnt pathway, which is activated by the binding of the Wnt protein (from the niche) to its cell-surface receptor complex, frizzled (on the stem cell membrane). Absence of the Wnt-frizzled interaction results in degradation of β-catenin by the proteasome. When Wnt is bound to frizzled this degradation is blocked, and β-catenin accumulates and translocates to the nucleus, resulting in the activation of several target genes. (For a more comprehensive review of this signaling pathway, see Clevers 2006. In the hematopoietic system, activation of the Wnt signaling pathway results in HSC proliferation and inhibition of differentiation, effectively increasing the stem cell pool (Reya and Clevers 2005, Staal and Clevers 2005).

The Notch signaling pathway is a regulator of cell fate in a wide range of developmental processes and cellular systems. Notch is a cell surface receptor located on HSCs that binds the transmembrane protein ligand, Jagged1, located on cells of the endosteal niche. This binding results in the cleavage of the intracellular domain of the notch receptor, which then translocates to the nucleus and activates notch target genes (Kojika and Griffin 2001, Ross and Li 2006). This signaling cascade in HSCs mediates cell fate determination and self-renewal in the endosteal niche. Osteoblasts produce high levels of Jagged1 to increase the frequency of HSCs in the niche via the Notch signaling pathway (Calvi et al. 2003). Integration of the Wnt and Notch signaling pathways contributes to tight regulation of the frequency of HSCs. Notch/Wnt double-reporter mice were used to determine that most cells in the stem cell niche use both pathways at the same time and stimulation of the Wnt pathway causes a transcriptional increase in Notch targets (Duncan et al. 2005).

Another important interaction between the niche and stem cells involves stem cell factor (SCF), which is encoded by the steel (*Sl*) locus and is present in both a membrane-bound (on osteoblasts) and secreted form. SCF binds and activates KIT, which is located on the surface of HSCs. Membrane-bound SCF is a more potent stimulator of KIT, and it supports the retention of HSCs in the microenvironment via activation of VLA4 and VLA5 integrins on the HSC to promote adhesion. Membrane-bound SCF is also essential to osteoblast proliferation and the maintenance of long-term HSC activity (Wilson and Trump 2006). Calcium gradients were also recently described as an important factor in the interaction between the niche and resident HSCs (Adams et al. 2006). Cells bearing a calcium receptor (CaR), including HSCs, respond to this local gradient. Interestingly, $CaR^{-/-}$ mice display hypocellularity of the bone marrow with a disproportionately small number of HSCs, but no defect in differentiation. Moreover, paratrabecular hematopoietic clusters seen in wild-type mice are absent in $CaR^{-/-}$ mice, and the primitive HSCs are found in extramedullary locations. In competitive transplant studies (Adams et al. 2006) HSCs from these mice exhibited a severe competitive disadvantage, indicating that the defect is cell-autonomous and may be attributable to altered homing capacity or diminished retention in the niche. Additional experiments in the same study showed that $CaR^{-/-}$ cells could not adhere to collagen I (the most prominent extracellular matrix protein at the endosteal surface of bone), supporting a potential role for CaR in the retention and lodgment of HSCs in the proper microenvironment. These observations indicate that the local concentration of calcium ions at the endosteal niche is important in the regulation of the HSC population (Adams et al. 2006).

5 Evaluating Stem Cell–Microenvironment Interactions: Transplantation and Mobilization

When Till and McCulloch demonstrated that transplanting mouse bone marrow cells into lethally irradiated recipient mice could rescue them from death and lead to the formation self-renewing hematopoietic colonies on the spleen, they demonstrated the cell-based nature of transplant-mediated rescue and forever incorporated the concept of migratory capacity into our definition of a functional HSC (McCulloch and Till 1960, Till and McCulloch 1961, Becker et al. 1963, Siminovitch et al. 1963). Since that time in the 1960s, the vast majority of experiments designed to investigate HSC function have used some form of transplantation assay to prove the nature and identity of the cells under investigation. In light of the important role that transplantation and mobilization studies have played in the development of our understanding of HSC function, we will briefly review the current understanding of these processes. For more extensive information, see the following reviews: (Lapidot and Petit 2002, Askenasy et al. 2003, Papayannopoulou 2003, Papayannopoulou 2004, Lapidot et al. 2005, Cancelas et al. 2006, Chute 2006, Yaniv et al. 2006).

We have alluded to the migratory capacity of HSCs in our discussion of the results of various transplantation experiments designed to validate the role of specific regulatory molecules. The interaction between stem cells and the microenvironment is fundamental in HSC functions that involve migration. From their genesis early in ontogeny, HSCs begin a journey that leads them through several locations in the developing embryo and eventually inside the bones to the marrow cavity, where they sustain hematopoiesis throughout adult life (Christensen et al. 2004, Ueno and Weissman 2006). Although these cells establish primary residence in the bone marrow of the adult, they have become well known for their ability to migrate. Under physiological conditions, HSCs can leave the bone marrow or mobilize into the peripheral blood and successfully re-enter the bone marrow microenvironment (Wright et al. 2001, Abkowitz JL et al. 2003). HSCs have also been reported to mobilize in response to injury or stress in nonhematopoietic tissue (Kollet et al. 2003). The migratory capacity of HSCs, however, is most apparent in the extreme conditions of clinical transplantation and mobilization.

5.1 Transplantation

When HSCs are removed from their native environment for the purpose of transplantation, only a small percentage of them find their way to a "home" in the recipient's bone marrow and begin contributing to hematopoiesis. Limiting dilution transplant experiments in adult mice using unfractionated bone marrow has demonstrated that despite the robust engraftment accomplished by bone marrow transplant, ultimately only about 5% to12% of functional adult HSCs transplanted intravenously find their way back to the bone marrow within 24 hours posttransplant (Szilvassy et al. 2003). Given this relative inefficiency with which stem cells contribute to engraftment posttransplant, it appears that the unique ability of stem cells to reconstitute long-term hematopoiesis is not due to a special capacity to navigate the vasculature to find the bone marrow, but an ability to adhere to the bone marrow vasculature, extravasate into the bone marrow cavity, and lodge in an hematopoiesis-supporting niche.

Years of transplantation studies and fluorescent-labeling experiments (Papayannopoulou 2003, Lapidot et al. 2005) have demonstrated that transplanted hematopoietic cells are distributed to and deposited in organs throughout the recipient over the first 1 to 3 hours posttransplant, but in less than 48 hours, the transplanted cells are detected almost exclusively in the bone marrow and spleen (Cui et al. 1999, Lanzkron et al. 1999, Srour EF et al. 2001, Szilvassy SJ et al. 2001). Furthermore, although hematopoiesis is initiated in both the spleen and bone marrow, stable engraftment depends on localization of primitive HSC within specialized bone marrow niches. Recent imaging studies point to preferential seeding of quiescent cells to the endosteal regions of the marrow cavity, whereas a combination of quiescent and cycling cells seed the spleen and contribute to radioprotection and short-term engraftment (Nibley and Nibley 1998, Lanzkron et al. 1999, Na Nakorn et al. 2002, Nilsson et al. 2002, Papayannopoulou 2003, Nilsson and Simmons 2004).

5.2 Molecular Mechanisms of HSC Homing

Since it was first implicated as the major HSC-attracting chemokine (Wright et al. 2002), the CXCL12 (SDF-1) ligand and corresponding CXCR4 receptor have emerged as having the most pivotal role in HSC homing and engraftment yet to be realized. This ligand-receptor pair appears to have some influence at virtually every stage of homing and engraftment (Papayannopoulou 2003, Lapidot et al. 2005, Dar et al. 2006). Briefly, it appears that a chemo-attractive gradient of CXCL12 from the peripheral blood to the bone marrow favors localization of HSCs to the bone marrow (Kim and Broxmeyer 1998), adherence to the bone marrow endothelium, extravasation, and lodgment within bone marrow niches. In addition, a host of molecules thought to have a unique role in HSC migration has recently been shown to influence migration, at least in part, by modulating some aspect of the CXCL12/CXCR4 signaling pathway (Peled A et al. 2000, Papayannopoulou et al. 2001, Nilsson et al. 2002, Avigdor A et al. 2004). Upon successful arrival of a stem cell into the extravascular bone marrow microenvironment, the task of finding a niche remains. In addition to the previously mentioned role of CXCL12 in attracting HSCs to the bone marrow, a gradient of CXCL12 might promote migration to or between niches (Dar et al. 2006, Sugiyama et al. 2006). Other molecules have emerged as regulators of HSC trafficking on the basis of their interaction with CXCL12 or CXCR4, for example, the sphingosine 1-phosphate receptor agonist FTY720 (Kimura T et al. 2004) and the well-known CXCR4 receptor antagonist AMD3100 (Vartanian 2000). Other novel modulators of the CXCL12/CXCR4 axis include extracellular UTP (Rossi L et al. 2007b) and the third complement component (C3) (Ratajczak et al. 2006).

Unlike the diverse functions of CXCL12 and CXCR4, E- and P- selectins appear to act more specifically to facilitate HSC homing by promoting the rolling of HSCs on endothelial cells of the bone marrow vasculature (Frenette et al. 1998, Mazo et al. 1998). The integrins very late activation antigen-4/vascular cell adhesion molecule-1(VLA-4/VCAM-1) and lymphocyte function associated antigen-1 (LFA-1) are thought to have similar functions, as well as the ability to promote the actual transendothelial migration/extravasation of HSCs (Peled A et al. 2000; Papayannopoulou et al. 2001).

5.3 *Mechanisms of HSC Mobilization*

Although the migratory capacity of stem cells has long been realized in the setting of transplantation, and stem cells have been identified in the peripheral blood (Till and McCulloch 1961, Goodman and Hodgson 1962), it was not realized until the 1980s that high-dose chemotherapy could be used to mobilize clinically useful numbers of HSCs into the peripheral blood for transplantation (Juttner et al. 1985. Schwartzberg LS et al. 1992, Buckner 1999). Like homing and engraftment, mobilization involves interactions between the stem cell and the bone marrow niche as well as the vasculature. Mobilization is achieved in the clinical setting most often by administration of granulocyte-colony stimulating factor (G-CSF). This method has replaced primary bone marrow harvesting as the method of choice to obtain HSCs for transplantation (Fu and Liesveld 2000, Thomas et al. 2002).

While successful engraftment requires the establishment of strong connections between stem cells and the niche, degradation of portions of the niche is required for stem cell mobilization. For example, a marked increase in neutrophil elastase and cathepsin-G proteases that can degrade VCAM-1 is observed in the bone marrow following administration of G-CSF or cyclophosphamide (Levesque et al. 2002). Similarly, IL-8-induced mobilization is also associated with neutrophil-mediated increases in matrix metalloproteinase-9 (Pruijt et al. 2002).

6 Functional Changes with Age that Involve both HSCs and the Microenvironment

6.1 Effects of Ageing on the Niche

Initially, we cited instances where ageing appears to be associated with changes intrinsic to the stem cell compartment. In fact, multiple lines of evidence, including patterns of stem cell number with age and transcriptional profiling, indicate that in inbred strains of mice, regulation of certain properties of the HSC pool are cell-intrinsic and can be observed in more than one microenvironmental setting (Van Zant et al. 1990, Rossi et al. 2005, Rossi et al. 2007a). However, the stem cell and microenvironment are intimately associated in ways that influence stem cell function, and it is unlikely that the stem cell microenvironment can escape age-related changes such as accumulation of DNA and cellular damage. Given that stem cells are highly dependent on their ability to interact with the niche and the paucity of data on the direct effects of ageing on cells of the HSC niche, we are led to ask whether or not any significant functional consequences can be attributed to an altered interaction between stem and stromal cells with advancing age. Homing, mobilization, and in vitro cell culture have offered insight into this question and suggest that the both the stem cell and the microenvironment are affected by the process of ageing in a way that compromises hematopoiesis (Fig. 6.4).

6.2 Age-Dependent Lineage Skewing of HSCs

It is well documented that a decline in immune function accompanies the natural ageing process in both mice and humans (Morrison et al. 1996, Marley et al. 1999, Liang et al. 2005). This finding has far-reaching implications given the success of stem cell transplantation using stem cells from older individuals and the etiology of age-associated hematopoietic conditions. It has been hypothesized that age-related hematological defects have their origin in cell intrinsic changes in the functional capacity of hematopoietic stem and progenitor cells (Rossi et al. 2005). Indeed, whole genome microarray analysis on purified young and old HSCs revealed almost 1,000 genes to be age-related with a high degree of statistical significance (Rossi et al. 2005). This study found that the expression of numerous genes specifying lymphoid fate and function was downregulated with age, whereas genes directing myeloid specification and function were upregulated with age. These findings are consistent with functional studies in mice, in which transplantation of old HSCs into young recipients revealed a defect in B-cell engraftment and a differentiation pattern skewed toward the myeloid lineage (Kim et al. 2003, Rossi et al. 2005). Recent work by Liang et al. (2005), however, questions a uniquely cell-intrinsic mechanism of compromised transplant success when old donor cells are transplanted into young

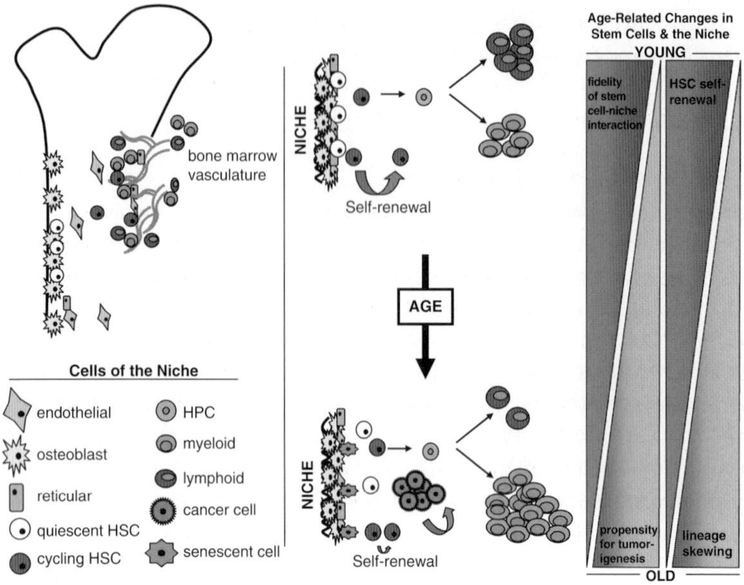

Fig. 6.4 HSCs reside in at least two known niches in the bone marrow, the vascular and endosteal niches. Within these specialized microenvironments, stem cells interact with a variety of cell types such as endothelial, osteoblast, and reticular cells. The fidelity of these interactions influences the balance of quiescence verses cell division and cell fate decisions in more lineage committed progenitors. The niche also plays an important role in the repression or facilitation of tumorigenesis. While a healthy microenvironment can help limit the potential for tumorigenesis, the cells of a tumor microenvironement and premetastatic niche support cancer cell proliferation and metastasis. The fidelity of stem cell – niche interactions and the capacity for HSC self-renewal diminish with age. Propensity for tumorigenesis increases with age and an imbalance in lineage distribution develops, favoring myelopoiesis at the expense of lymphopoiesis (*See Color Plate*)

or old recipients. This study demonstrated that both the stem cells and the microenvironment are affected by the process of ageing in a way that reduces the 24-hour homing efficiency of primitive HSCs. When either (1) old bone marrow cells are transplanted into a young microenvironment or (2) young cells are transplanted into an old microenvironment, HSC homing is significantly lower than when young HSCs are transplanted into young recipients. Even more surprisingly, the combination of old donor HSCs and an old microenvironment did not produce an additive effect in terms of diminished homing relative to age-mismatched transplants (unpublished data). This suggests that the fidelity of HSC-microenvironment interactions, in addition to functional changes unique to either the HSCs or cells of the niche, may be an important aspect of age-associated functional decline in the hematopoietic system. It also suggests that deficits due to aged HSCs and aged stroma are caused by a common mechanism. An exclusively cell-intrinsic origin for impaired B-lymphopoiesis with age is also questionable given the disparity between this interpretation and reports suggesting a causative role for aged stroma in this phenomenon (Stephan et al. 1998, Labrie et al. 2004). In the study by Rossi and colleagues (2005), impaired B-lymphopoiesis was observed when young bone marrow cells were transplanted into an old microenvironment only at early time points posttransplant and B-lymphopoiesis recovered to normal levels by 12 and 26 weeks posttransplant. This may be attributable to the ability of the young stem cells, over time, to re-establish certain cell-cell interactions that support B-lymphopoiesis. This interpretation is consistent with the earlier finding that purified stromal cell populations from older mice were unable to support interleukin-7 (IL-7)–specific B-lymphoid cell lines and that impaired release of IL-7 was not due to an inability to produce IL-7 but the inability to release it upon contact with hematopoietic cells (Stephan et al. 1998).

Analysis of progenitor cell compartments in old mice shows a significant increase in granulocyte-macrophage progenitors and a significant decrease in flk2 expressing common lymphoid progenitors (Rossi et al. 2007a). The lymphoid defects that accompany ageing extend to both B and T cell lineages and, although extensive evidence points to diminished lymphopoietic potential of aged bone marrow, age-associated defects that accumulate in mature B and T cells likely contribute to the decreased immune function observed in aged individuals (Linton and Dorshkind 2004). For example, aged B cell progenitors have defects that compromise their ability to differentiate efficiently and the proliferative potential of certain B cell subsets isolated from the marrow of old mice is reduced (Min et al. 2006). It has further been documented that pro-B cells obtained from old mice do not efficiently respond to lymphopoietic cytokines such as IL-7 (Stephan et al. 1997, Miller and Allman 2003). In regard to age-related changes that occur in the T cell compartment, in both mice and humans, thymic involution results in reduced T cell production and export to secondary lymphoid organs (Linton and Dorshkind 2004). The frequency and absolute number of early T-lineage progenitor (ETP) cells are also reduced in the thymuses of old mice (Min et al. 2005). Decreased precursor input from the bone marrow as well as a decline in the ability of ETPs to proliferate likely contributes to their reduced number with age. In addition to decreased progenitor cell numbers, the responses of mature $CD4^+$ and $CD8^+$ T cells from aged donors are slower and of reduced amplitude compared to responses of younger individuals (Sadighi and Miller 2005). Thus, while the production of myeloid cells is preserved in old mice, the frequency and proliferative capacity of B- and T-cell precursors is reduced during the ageing process. Together, the changes in these populations result in profound diminution of immune function in older individuals (Figs. 6.1 and 6.4).

In addition to compromising immune system function, the lymphoid/myeloid lineage skewing observed with ageing likely contributes to the development of hematological disease in the elderly. In the whole genome microarrays mentioned previously, numerous genes known to be involved in myeloid leukomogenesis were overexpressed in old HSCs compared to young cells (Rossi et al. 2005). This is not surprising since leukemias of myeloid origin present more frequently in aged individuals, while pediatric leukemias tend to be of lymphoid origin. Proto-oncogenes such as Fgfr1, Aml1, Pml, and Eto are also overexpressed in old HSCs, and such genes are often associated with cytogenetic rearrangements that result in the formation of oncogenic fusion proteins (Rossi et al. 2005).

6.3 Ageing, Migration, and Nonhematopoietic Tissue Recruitment of HSCs

Another recent study evaluated changes in stem cell function and stem cell–microenvironment interactions with age by measuring the mobilization efficiency of hematopoietic stem and progenitor cells from young and old mice (Xing et al. 2006). In this study, mobilization efficiency was increased in aged C57BL/6 mice, as might be expected because the stem cell pool in this strain expands with age. Surprisingly, however, additional experiments in the same study demonstrated that young and old stem cells migrate across endothelial cells with equal efficiency, but the older stem cells are much less efficient at adhering to stromal support cells in vitro. Given that the findings from 24-hour homing studies in young and old mice (Liang et al. 2005) were performed in the same mouse strain, it is reasonable to propose that diminished homing and enhanced mobilization with age in this strain can be explained by a reduction in the strength of the association between stem cells and the cells of the bone marrow microenvironment rather than an altered capacity for transendothelial migration in either direction. This further supports the notion that ageing, whether acting upon the stem cell or the microenvironment, or both, alters the integrity of cell-cell and cell-ECM interactions in a way that compromises hematopoiesis.

In addition to homing and mobilization, altered stem cell function associated with the microenvironment can be observed with physiological ageing. For example, a recent publication by Thijssen (2006) indicated that there is a change in physiological mobilization and the regulation of physiologic mobilization with age. This study demonstrated that fewer CD34+ cells were found in peripheral circulation in aged men compared to younger men and that the increased mobilization accompanying acute exercise in young male subjects was present but attenuated in older men. This suggests that, in addition to alterations in the capacity for stem cell egress from the bone marrow (decline from young to old), there is also a change in the function of the regulatory mechanism governing HSC egress (decreased mobilization capacity). The disparity between reduced mobilization with age in this study and enhanced mobilization with age described in the previous study is likely attributed to (1) the murine study being conducted in B6 mice that exhibit unique patterns of stem cell characteristics when compared to other inbred mouse strains or humans and, more importantly, (2) the human study assessing physiological migration, whereas the murine study involved G-CSF-induced mobilization. Regardless of these differences, both studies demonstrate altered cell trafficking with age and suggest a change in stem cell-microenvironment interactions with age. There is an ongoing debate in the scientific community regarding the capacity of bone marrow hematopoietic or "embryonic-like" (Kucia et al. 2006) cells to participate in the regeneration or repair of nonhematopoietic tissues (Dawn and Bolli 2005). Several studies suggest that HSCs participate in repair by transdifferentiation or cell fusion with nonhematopoietic tissues (Eglitis and Mezey 1997, Ferrari et al. 1998, Orlic et al. 2001, Alvarez-Dolado et al. 2003, Kollet O et al. 2003, Wang et al. 2003).

These findings, however, have been challenged by other studies (Wagers et al. 2002, Balsam et al. 2004, Murry et al. 2004), and mechanisms for HSC-mediated repair have not been clearly identified. The controversial yet compelling evidence that HSCs can contribute to nonhematopoietic tissue repair suggests that the availability of HSCs in circulation might be related to the potential for tissue repair throughout the entire organism. It is conceivable that a reduction in circulating stem cells or a diminished ability of resident stem cells to respond to mobilizing stimuli could limit the extent of tissue repair with age. Unfortunately, data on changes in physiological migration of HSCs with age are incomplete at best. It is, perhaps, in the setting of malignancy that we have learned the most about the mutual role of the stem cell and microenvironment in ageing.

7 Stem Cells and Cancer

We have noted that adult stem cell populations exhibit functional decline with age. Ageing is also associated with a dramatic increase in the incidence of cancer, and several recent studies support a stem cell origin for cancer (Campisi 2003, Campisi 2005, Kaplan et al. 2006, Li and Neaves 2006, Rizo et al. 2006, Li et al. 2007). Together, these findings offer strong support to the stem cell theory of ageing by connecting functional decline in the stem cell population to malignancy resulting in limited organismal longevity. Interestingly, from the cell surface, stem cells from a variety of tissues resemble their malignant counterparts. Hematopoietic, mammary, neural, lung, skin, and prostate stem cells can be identified by some of the same cell surface markers used to identify cancer stem cells in these tissues (Li et al. 2007). In addition to cell intrinsic similarities between normal stem cell and cancer cells, an enormous body of literature describes intricate relationships between transformed cells and their supportive niches in the primary tumor and sites of metastasis. Similarities between stem cell-microenvironment and cancer cell-microenvironment interactions have been extensively reviewed by Rizo et al. (2006), Kaplan et al. (2006), Li et al. (2007) and others. Although it is beyond the scope of this publication to adequately summarize all of the relationships that are common to physiological and transformed stem cells and their microenvironment, several key points are briefly summarized below.

7.1 HSCs, Cancer Stem Cells, and their Niches

It has been proposed that cancer stem cells seem to have overpowered the quiescence cues of a restrictive microenvironment and can even escape to sites of metastasis; therefore, these cells are less dependent on a support system such as that offered by the nonmalignant stem cell niche (Li 2006a). Cancer cell metastasis, however, is not random, and sites of metastasis are predictable. This demonstrates that, although cancer cells have indeed escaped certain control mechanisms, they do engage in specific interactions with noncancerous cells of other tissues. In a 2005 review, Campisi summarized several experiments that support the concept that senescence of stromal cells in the stem cell microenvironment either promote malignant transformation of stem cells in the niche or render the microenvironment incapable of keeping damaged or transformed cells in check (Fig. 6.4). Furthermore, recent studies demonstrate that specifically targeting endothelial cells of the niche can inhibit cancerous cells of both hematopoietic and neural origin (Roccaro et al. 2006, Calabrese et al. 2007).

Among the most interesting connections between the stem cell niche and cancer is the recent finding that mesenchymal stem cells, which originate in the bone marrow, support hematopoiesis in physiological conditions, and suppress cancer cell growth in vitro, can actually facilitate tumorigenesis of both hematopoietic and nonhematopoietic cancer cells in vivo (Houghton et al. 2004, Ramasamy et al. 2007). These mesenchymal stem cells may promote tumorigenesis by facilitating the development of cancer stem cell niches (Li et al. 2007, Ramasamy et al. 2007). In a recent review by Kaplan et al. (2006), several ligand-receptor pairs involved in maintaining normal hematopoiesis were shown to be common to cellular interactions in both primary tumors and the premetastatic niche, including VLA-4/VCAM-1, G-CSF/GCSF-R, CXCL12/CXCR4, and Tie-2/Angiopontin to name a few.

7.2 HSCs and Metastatic Cancer Stem Cells use Similar Pathways for Migration

We previously considered some of the pathways involved in migration of HSCs either into or out of the bone marrow microenvironment. Among the numerous similarities between normal HSCs and cancer cells is the ability to utilize several of the same molecular mechanisms for migration (Li and Neaves 2006). For example, the CXCL12/CXCR4 axis, which plays a pivotal role in developmental and adult hematopoiesis and recruitment of HSCs to sites of injury (Kollet et al. 2003), is also involved in the metastasis of breast cancer (Fernandis et al. 2004, Dar et al. 2006, Li et al. 2007). Similarly, matrix metalloproteinases involved in HSC mobilization, migration, (Levesque et al. 2002) and recruitment of HSCs to sites of injury (Kollet et al. 2003) also contribute to cancer cell invasion (John and Tuszynski 2001, Li et al. 2007).

The realization of extensive overlap between stem and cancer cell characteristics has generated a great deal of interest in finding ways to distinguish between the two cell populations. Cancer treatment regimens would be dramatically improved by the ability to preserve the stem cell population while eliminating, with great specificity, cancer cells and cancer stem cells. Although relatively little success has been achieved in these efforts, some recent publications point to a distinction in the role of PTEN and CD44 in normal HSCs versus leukemic cells and therefore may open avenues for highly specific anticancer therapies that are less damaging to normal adult stem cell populations than are traditional chemotherapeutic or radiation regimens (Krause et al. 2006, Yilmaz et al. 2006).

8 Conclusion

In summary, the process of ageing affects the stem cell, the cells of the microenvironment, and the integrity of stem cell-microenvironment interactions. Potential mechanisms for cell-intrinsic changes include the initiation of altered gene expression programs in the stem cell due to accumulated damage, epigenetic modifications, or both. The aged HSC microenvironment may be compromised by accumulation of damage or senescent cells in the stem cell niche. The combined changes cause aged HSCs to interact with the aged microenvironment differently than young HSCs in a young microenvironment. It is easy to see that the resulting changes in stem cell number with age, skewed lineage representation in the peripheral blood, diminished immune response, diminished homing efficiency, reduced stromal cell adherence, and increased propensity for tumorigenesis contradict theories of an "immortal stem cell" in the adult. The effects of age on adult stem cells result in a loss of tissue homeostasis and probably lead to broad functional declines in adult tissues with advancing age, making functional stem cells one of the limiting factors in healthy ageing.

Acknowledgements We gratefully acknowledge the excellent editorial assistance of Paula Thomason.

References

Abkowitz JL, Robinson AE, Kale S, Long MW, JC (2003) Mobilization of hematopoietic stem cells during homeostasis and after cytokine exposure. Blood 102:1249–53
Abramovich C, Humphries R (2005) Hox regulation of normal and leukemic hematopoietic stem cells. Curr Opin Hematol 12:210–16
Abramovich C, Pineault N, Ohta H, Humphries RK (2005) Hox genes: from leukemia to hematopoietic stem cell expansion. Ann NY Acad Sci 1044:109–16
Adams GB et al. (2006) Stem cell engraftment at the endosteal niche is specified by the calcium-sensing receptor. Nature 439:599–603
Alvarez-Dolado M et al. (2003) Fusion of bone-marrow-derived cells with Purkinje neurons, cardiomyocytes and hepatocytes. Nature 425:968–73
Antonchuk J, Sauvageau G, Humphries RK (2002) HOXB4-induced expansion of adult hematopoietic stem cells ex vivo. Cell 109:39–45
Arai F et al. (2004) Tie2/Angiopoietin-1 Signaling regulates hematopoietic stem cell quiescence in the bone marrow niche. Cell 118:149–61
Askenasy N, Stein J, Yaniv I, Farkas D (2003) The topologic and chronologic patterns of hematopoietic cell seeding in host femoral bone marrow after transplantation. Bio Blood Marrow Tr 9:496–504

Avecilla ST et al. (2004) Chemokine-mediated interaction of hematopoietic progenitors with the bone marrow vascular niche is required for thrombopoiesis. Nature Med 10:64–71
Avigdor A et al. (2004) CD44 and hyaluronic acid cooperate with SDF-1 in the trafficking of human CD34+ stem/progenitor cells to bone marrow. Blood 103:2981–89
Balcombe NR, Sinclair A (2001) Ageing: definitions, mechanisms and the magnitude of the problem. Best Pract Res Cl Ga 15:835–49
Balducci L, Ershler WB (2005) Cancer and ageing: a nexus at several levels. Nature 5:655–62
Balsam L, Wagers A, Christensen J, Kofidis T, Weissman I, Robbins R (2004) Haematopoietic stem cells adopt mature haematopoietic fates in ischaemic myocardium. Nature 428:668–73
Bauer H, Tempfer H, Bernroider G, Bauer H (2006) Neuronal stem cells in adults. Exp Gerontol 4:111–16
Becker AJ, McCulloch EA, Till JE (1963) Cytological demonstration of the clonal nature of spleen colonies derived from transplanted mouse marrow cells. Nature 197:452–4
Bell DR, Van Zant G (2004) Stem cells, ageing, and cancer: inevitabilities and outcomes. Oncogene 23:7290–96
Boyer LA et al. (2006) Polycomb complexes repress developmental regulators in murine embryonic stem cells. Nature 441:349–53
Buckner CD (1999) Autologous bone marrow transplants to hematopoietic stem cell support with peripheral blood stem cells: a historical perspective. J Hematother 8:233–36
Calabrese C et al. (2007) A perivascular niche for brain tumor stem cells. Cancer Cell 11:69–82
Calvi LM et al. (2003) Osteoblastic cells regulate the haematopoietic stem cell niche. Nature 425:841–46
Campisi J (2003) Cancer and ageing: rival demons? Nature 3:339–49
Campisi J (2005) Senescent cells, tumor suppression, and organismal ageing: good citizens, bad neighbors. Cell 120:513–22
Cancelas JA, Jansen M, Williams DA (2006) The role of chemokine activation of Rac GTPases in hematopoietic stem cell marrow homing, retention, and peripheral mobilization. Exp Hematol 34:976–85
Chen J, Astle CM, Harrison DE (1999) Development and ageing of primitive hematopoietic stem cells in BALB/cBy mice. Exp Hematol 27:928–35
Chen J, Astle CM, Harrison DE (2000) Genetic regulation of primitive hematopoietic stem cell senescence. Exp Hematol 28:442–50
Cheng T, Rodrigues N, Dombkowski DM, Stier S, Scadden DT (2000a) Stem cell repopulation efficiency but not pool size is governed by p27(kip1). Nature Med 6:1235–40
Cheng T et al. (2000b) Hematopoietic stem cell quiescence maintained by $p21^{cip1/waf1}$. Science 287:1804–08
Christensen JL, Wright DE, Wagers AJ, Weissman IL (2004) Circulation and chemotaxis of fetal hematopoietic stem cells. PLoS Biol 2:0368–77
Chute JP (2006) Stem cell homing. Curr Opin Hematol 13:399–406
Clevers H (2006) Wnt/beta-catenin signaling in development and disease. Cell 127:469–80
Cui J et al. (1999) Bone marrow cell trafficking following intravenous administration. Br J Haematol 107:895–902
Dar A, Kollet O, Lapidot T (2006) Mutual, reciprocal SDF-1/CXCR4 interactions between hematopoietic and bone marrow stromal cells regulate human stem cell migration and development in NOD/SCID chimeric mice. Exp Hematol 34:967–75
Dawn B, Bolli R (2005) Adult bone marrow-derived cells: Regenerative potential, plasticity, and tissue commitment. Basic Res Cardiol 100:494–503
de Haan G, Nijhof W, Van Zant G (1997) Mouse strain-dependent changes in frequency and proliferation of hemaotpoietic stem cells during ageing: Correlation between life span and cycling activity. Blood 89
de Haan G, Van Zant G (1997) Intrinsic and extrinsic control of hemopoietic stem cell numbers: Mapping of a stem cell gene. J Exp Med 186:529–36
de Haan G, Van Zant G (1999) Genetic analysis of hemopoietic cell cycling in mice suggests its involvement in organismal life span. FASEB J 13:707–13

de Haan G, Van Zant, Gary (1999) Dynamic changes in mouse hematopoietic stem cell numbers during ageing. Blood 93:3294–3301
Dubrulle J, McGrew MJ, Pourquie O (2001) FGF signaling controls somite boundary position and regulates segmentation clock control of spatiotemporal Hox gene activation. Cell 106:219–32
Dumble M et al. (2007) The impact of altered p53 dosage on hematopoietic stem cell dynamics during ageing. Blood 109:1736–42
Duncan AW et al. (2005) Integration of Notch and Wnt signaling in hematopoietic stem cell maintenance. Nat Immunol 6:314–22
Edwards B et al. (2002) Annual report to the nation on the status of cancer, 1973–1999, featuring implications of age and ageing on U.S. cancer burden. Cancer 94:2766–92
Eglitis M, Mezey E (1997) Hematopoietic cells differentiate into both microglia and macroglia in the brains of adult mice. Proc. Natl. Acad. Sci. USA 94:4080–05
Ezoe S, Matsumura I, Sato Y, Tanaka H, Kanakura Y (2004) Cell cycle regulation in hematopoietic stem/progenitor cells. Cell Cycle 3:314–18
Fernandis AZ, Prasad A, Band H, Klosel R, Ganju RK (2004) Regulation of CXCR4-mediated chemotaxis and chemoinvasion of breast cancer cells. Oncogene 23:157–67
Ferrari G et al. (1998) Muscle regeneration by bone marrow-derived myogenic progenitors. Science 279:1528–30
Frenette PS, Subbarao S, Mazo I, von Andrian UH, Wagner DD (1998) Endothelial selectins and vascular cell adhesion molecule-1 promote hematopoietic progenitor homing to bone marrow. Proc Natl Acad Sci USA 95:14423–28
Fu S, Liesveld J (2000) Mobilization of hematopoietic stem cells. Blood Rev 14:205–18
Fulle S et al. (2005) Age-dependent imbalance of the antioxidative system in human satellite cells. Exp Gerontol 40:189–97
Geiger H, True JM, de Haan G, Van Zant G (2001) Age- and stage-specific regulation patterns in the hematopoietic stem cell hierarchy. Blood 98:2966–72
Goodman JW, Hodgson GS (1962) Evidence for stem cells in the peripheral blood of mice. Blood 19:702–14
Harman D (1956) Ageing: a theory based on free radical and radiation chemistry. J Gerontol 11:298–300
Harrison DE (1979) Mouse erythropoietic stem cell lines function normally 100 months: loss related to number of transplantations. Mech Ageing Dev 9:427–33
Harrison DE, Astle CM (1982) Loss of stem cell repopulating ability upon transplantation; Effects of donor age, cell number, and transplantation procedure. J Exp Med 156:1767–79
Harrison DE, Astle CM, Stone M (1989) Numbers and functions of transplantable primitive immunohematopoietic stem cells. Effects of age. J Immunol 142:3833
Hock H et al. (2004) Gfi-1 restricts proliferation and preserves functional integrity of haematopoietic stem cells. Nature 431:1002–07
Houghton J et al. (2004) Gastric cancer originating from bone marrow-derived cells. Science 306:1568–71
Ito K et al. (2006) Reactive oxygen species act through p38 MAPK to limit the life span of hematopoietic stem cells. Nature Med 12:446–51
Janzen V et al. (2006) Stem-cell ageing modified by the cyclin-dependent kinase inhibitor p16INK4a. Nature 443:421–26
John A, Tuszynski G (2001) The role of matrix metalloproteinases in tumor angiogenesis and tumor metastasis. Pathol Oncol Res 7:14–23
Juttner CA, To LB, Haylock DN, Brandford A, Kimber RJ (1985) Circulating autologous stem cells collected in very early remission from acute non-lymphoblastic leukaemia produce prompt but incomplete haemopoietic reconstitution after high dose melphalan or supralethal chemoradiotherapy. Br J Haematol 61:739–45
Kamminga LM. vOR, Ausema A, Noach EJ, Weersing E, Dontje B., Vellenga E., de Haan, Gerald (2005) Impaired hematopoietic stem cell functioning after serial transplantation and during normal ageing. Stem Cells 23:82–92

Kamminga LM et al. (2006) The Polycomb group gene Ezh2 prevents hematopoietic stem cell exhaustion. Blood 2006:5
Kamminga LM, de Haan G (2006) Cellular memory and hematopoietic stem cell ageing. Stem Cells 24:1143–9
Kaplan RN, Psaila B, Lyden D (2006) Bone marrow cells in the 'pre-metastatic niche': within bone and beyond. Cancer Metastasis Rev 25:521–29
Keller J (2006) Age-related neuropathology, cognitive decline, and Alzheimer's disease. Ageing Res Rev 5:1–13
Kiel MJ, Yilmaz OH, Iwashita T, Yilmaz OH, Terhorst C, Morrison SJ (2005) SLAM family receptors distinguish hematopoietic stem and progenitor cells and reveal endothelial niches for stem cells. Cell 121:1109–21
Kim CH, Broxmeyer HE (1998) In vitro behavior of hematopoietic progenitor cells under the influence of chemoattractants: stromal cell-derived factor-1, steel factor, and the bone marrow environment. Blood 91:100–10
Kim M, Moon HB, Spangrude GJ (2003) Major age-related changes of mouse hematopoietic stem/progenitor cells. Ann NY Acad Sci 996:195–208
Kimura T et al. (2004) The sphingosine 1-phosphate receptor agonist FTY720 supports CXCR4-dependent migration and bone marrow homing of human CD34+ progenitor cells. Blood 103:4478–86
Klassen LW (1972) Experimental medullary aplasia. J Lab Clin Med 80:8–17
Kojika S, Griffin JD (2001) Notch receptors and hematopoiesis. Exp Hematol 29:1041–52
Kollet O et al. (2003) HGF, SDF-1, and MMP-9 are involved in stress-induced human CD34+ stem cell recruitment to the liver. J Clin Invest 112:160–09
Kondo M et al. (2003) Biology of hematopoietic stem cells and progenitors: Implications for clinical application. Ann Rev Immunol 21:759–806
Krause DJ, Lazarides K, von Andrian UH, Van Etten RA (2006) Requirement for CD44 in homing and engraftment of BCR-ABL-expressing leukemic stem cells. Nature Med 12:1175–80
Krishnamurthy J et al. (2006) p16INK4a induces an age-dependent decline in islet regenerative potential. Nature 443:453–57
Krishnamurthy J et al. (2004) Ink4a/Arf expression is a biomarker of ageing. J Clin Invest 114:12997–11307
Krosl J, Austin P, Beslu N, Kroon E, Humphries RK, Sauvageau G (2003) In vitro expansion of hematopoietic stem cells by recombinant TAT-HOXB4 protein. Nature Med 9:1428–32
Kucia M et al. (2006) A population of very small embryonic-like (VSEL) CXCR4(+)SSEA-1(+) Oct-4+ stem cells identified in adult bone marrow. Leukemia 20:857–69
Labrie JE, Sah AP, Allman DM, Cancro MP, Gerstein RM (2004) Bone marrow microenvironmental changes underlie reduced RAG-mediated recombination and B cell generation in aged mice. J Exp Med 200:411–23
Lanzkron SM, Collector MI, Sharkis SJ (1999) Homing of long-term and short-term engrafting cells in vivo. Ann NY Acad Sci 872:48–54
Lapidot T, Dar A, Kollet O (2005) How do stem cells find their way home? Biol Blood Marrow Tr 106:1901–10
Lapidot T, Petit I (2002) Current understanding of stem cell mobilization: The roles of chemokines, proteolytic enzymes, adhesion molecules, cytokines, and stromal cells. Exp Hematol 30:973–81
Lawrence HJ, Sauvageau G, Humphries RK, Largman C (1996) The role of HOX homeobox genes in normal and leukemic hematopoiesis. Stem Cells 14:281–91
Lessard J, Baban S, Sauvageau G (1998) Stage-specific expression of polycomb group genes in human bone marrow cells. Blood 91:1216–24
Levesque JP, Hendy J, Takamatsu Y, Williams B, Winker IG, Simmons PJ (2002) Mobilization by either cyclophosphamide or granulocyte colony-stimulating factor transforms the bone marrow into a highly proteolytic environment. Exp Hematol 30:440–49
Lewis EB (1978) A gene complex controlling segmentation in Drosophila. Nature 276:565–70
Li F, Tiede B, Massague J, Kang Y (2007) Beyond tumorigenesis: cancer stem cells in metastasis. Cell Res 17:3–14

Li L, Neaves WB (2006) Normal stem cells and cancer stem cells: the niche matters. Cancer Res 66:4553–57

Li W, Johnson SA, Shelley WC, Yoder MC (2004) Hematopoietic stem cell repopulating ability can be maintained in vitro by some primary endothelial cells. Exp Hematol 32:1226–37

Li Z, Li L (2006) Understanding hematopoietic stem-cell microenvironments. Trends Biochem Sci 31:589–95

Liang Y, Van Zant G, Szilvassy SJ (2005) Effects of ageing on the homing and engraftment of murine hematopoietic stem and progenitor cells. Blood 106:1479–87

Linton PJ, Dorshkind K (2004) Age-related changes in lymphocyte development and function. Nature Immunol 5:133–39

Marley SB, Lewis JL, Davidson RJ, Roberts IA, Dokal I, Goldman JM (1999) Evidence for a continuous decline in haemopoietic cell function from birth: an application to evaluating bone marrow failure in children. Br J Haematol 106:162–66

Mazo I, J-C G-R, Frenette PS, Hynes RO, Wagner DD, von Andrian UH (1998) Hematopoietic progenitor cell rolling in bone marrow microvessels: parallel contributions by endothelial selectins and vascular cell adhesion molecule 1. J Exp Med 188:465–74

McCredie KB, Hersh EM, Freireich EJ (1971) Cells capable of colony formation in the peripheral blood of man. Science 171:293–94

McCulloch EA, Till JE (1960) The radiation sensitivity of normal mouse bone marrow cells, determined by quantitative marrow transplantation into irradiated mice. Radiat Res 13:115–125

Miller JP, Allman D (2003) The decline in B lymphopoiesis in aged mice reflects loss of very early B-lineage precursors. J Immunol 171:2326–30

Min H, Montecino-Rodriquez E, Dorshkind K (2005) Effects of ageing on early B- and T- cell development. Immunol Rev 205:7–17

Min H, Montecino-Rodriquez E, Dorshkind K (2006) Effects of ageing on the common lymphoid progenitor to pro-B cell transition. J Immunol 176:1007–12

Mintz B, Silvers WK (1967) "Intrinsic" immunological tolerance in allophenic mice. Science 158:1484–86

Miyake N, Brun AC, Magnusson M, Miyake K, Scadden DT, Karlsson S (2006) HOXB4-induced self-renewal of hematopoietic stem cells is significantly enhanced by p21 deficiency. Stem Cells 24:653–61

Molofsky AV et al. (2006) Increasing p16INK4a expression decreases forebrain progenitors and neurogenesis during ageing. Nature 443:448–52

Morrison SJ, Wandycz AM, Akashi K, Globerson A, Weissman IL (1996) The ageing of hematopoietic stem cells. Nature Med 2:1011–16

Murry C et al. (2004) Haematopoietic stem cells do not transdifferentiate into cardiac myocytes in myocardial infarcts. Nature 428:664–68

Na Nakorn T, Traver D, Weissman IL, Akashi K (2002) Myeloerythroid-restricted progenitors are sufficient to confer radioprotection and provide the majority of day 8 CFU-S. J Clin Invest 29:1326–35

Nakayama T, Mutsuga N, Tosato G (2007) FGF2 posttranscriptionally down-regulates expression of SDF1 in bone marrow stromal cells through FGFR1 IIIc. Blood 109:1363–72

Nibley WE, GJ S (1998) Primitive stem cells alone mediate rapid marrow recovery and multilineage engraftment after transplantation. Bone Marrow Tr 21:345–54

Nilsson SK, Haylock D, Johnston H, Occhiodoro T, Brown T, Simmons PJ (2002) Hyaluronan is synthesized by primitive hemopoietic cells, participates in their lodgement at the endosteum following transplantation and is involved in the regulation of their proliferation and differentiation in vitro. Blood 101:856–62

Nilsson SK et al. (2005) Osteopontin, a key component of the hematopoietic stem cell niche and regulator of primitive hematopoietic progenitor cells. Blood 106:1232–39

Nilsson SK, Simmons PJ (2004) Transplantable stem cells: home to specific niches. Curr Opin Hematol 11:102–06

Ogden DA, Mickliem HS (1976) The fate of serially transplanted bone marrow cell populations from young and old donors. Transplantation 22:287–93

Orlic D et al. (2001) Bone marrow cells regenerate infarcted myocardium. Nature 410:701–05
Papayannopoulou T (2003) Bone marrow homing: the players, the playfield, and their evolving roles. Curr Opin Hematol 10:214–19
Papayannopoulou T (2004) Current mechanistic scenarios in hematopoietic stem/progenitor cell mobilization. Blood 103:1580–85
Papayannopoulou T, Priestley GV, Nakamoto B, Zafiropoulos V, Scott LM (2001) Molecular pathways in bone marrow homing: dominant role of alpha-4 Beta-1 over Beta-2-integrings and selectins. Blood 98:2403–11
Park IK, Morrison SJ, Clarke MF (2004) Bmi-1, stem cells, and senescence regulation. J Clin Invest 113:175–79
Park IK et al. (2003) Bmi-1 is required for maintenance of adult self-renewing haematopoietic stem cells. Nature 423:302–05
Peled A et al. (2000) The chemokine SDF-1 activates the integrins LFA-1, VLA-4, and VLA-5 on immature human CD34(+) cells: role in transendothelial/stromal migration and engraftment of NOD/SCID mice. Blood 95:3289–96
Penninx BWJ, Guralnik JM, Onder G, Ferrucci L, Wallace RB, Pahor M (2003) Anemia and decline in physical performance among older persons. Am J Med 115:104–10
Pinto A, Filippi RD, Frigeri F, Corazzelli G, Normanno N (2003) Ageing and the hemopoietic system. Crit Rev Oncol Hematol 48S:S3–S12
Pruijt JF et al. (2002) Neutrophils are indispensable for hematopoietic stem cell mobilization induced by interleukin-8 in mice. Proc Natl Acad Sci USA 99:6228–33
Ramasamy R, Lam E, Soeiro I, Tisato V, Bonnet D, Dazzi F (2007) Mesencymal stem cells inhibit proliferation and apoptosis of tumor cells: impact on *in vivo* tumor growth. Leukemia 21:304–10
Ratajczak MZ, Reca R, Wysoczynksi M, Yan J, Ratajczak J (2006) Modulation of the SDF-1-CXCR4 axis by the third complement component (C3)-Implications for trafficking of CXCR4+ stem cells. Exp Hematol 34:986–95
Reya T, Clevers H (2005) Wnt signalling in stem cells and cancer. Nature 434:843–50
Rizo A, Vellenga E, de Haan G, Schuringa JJ (2006) Signaling pathways in self-renewing hematopoietic and leukemic stem cells: do all stem cells need a niche? Hum Mol Genet 15:210–19
Roccaro AM et al. (2006) Bortezomib mediates antiangiogenesis in multiple myeloma via direct and indirect effects on endothelial cells. Cancer Res 66:184–91
Ross EA, Anderson N, Micklem HS (1982) Serial depletion and regeneration of the murine hematopoietic system. Implications for hematopoietic organization and the study of cellular ageing. J Exp Med 155:432–44
Ross J, Li L (2006) Recent advances in understanding extrinsic control of hematopoietic stem cell fate. Curr Opin Hematol 13:237–42
Rossi DJ, Bryder D, Weissman IL (2007) Hematopoietic stem cell ageing: Mechanism and consequence. Exp Gerontol in press
Rossi DJ et al. (2005) Cell intrinsic alterations underlie hematopoietic stem cell ageing. Proc Natl Acad Sci USA 102:9194–99
Rossi L et al. (2007) The extracellular nucleotide UTP is a potent inducer of hematopoietic stem cell migration. Blood 109:533–42
Sadighi A, AA, Miller R (2005) Signal transduction in the ageing immune system. Curr Opin Immunol 17:486–91
Sauvageau G et al. (1994) Differential expression of homeobox genes in functionally distinct CD34+ subpopulations of human bone marrow cells. Proc Natl Acad Sci USA 91:12223–27
Sauvageau G et al. (1995) Overexpression of HOXB4 in hematopoietic cells causes the selective expansion of more primitive populations in vitro and in vivo. Genes Dev 9:1753–65
Savitsky K et al. (1995) A single ataxia telangiectasia gene with a product similar to PI-3 kinase. Science 268:1749–53
Schofield R (1978) The relationship between the spleen colony-forming cell and the haemopoietic stem cell. Blood Cells 4:7–25

Schwartzberg LS et al. (1992) Peripheral blood stem cell mobilization by chemotherapy with and without recombinant human granulocyte colony-stimulating factor. J Hematother 1:317–27

Siminovitch L, McCulloch EA, Till JE (1963) The distribution of colony-forming cells among spleen colonies. J Cell Compar Physiol 62:327–36

Siminovitch L, Till JE, McCulloch EA (1964) Decline in colony-forming ability of marrow cells subjected to serial transplantation into irradiated mice. J Cell Physiol 64:23–31

Srour EF et al. (2001) Homing, cell cycle kinetics and fate of transplanted hematopoietic stem cells. Leukemia 15:1681–84

Staal FJ, Clevers H (2005) WNT signalling and haematopoiesis: a WNT-WNT situation. Nat Rev Immunol 5:21–30

Stephan RP, Lill-Elghanian DA, Witte PL (1997) Development of B cells in aged mice: decline in the ability of pro-B cells to respond to IL-7 but not to other growth factors. J Immunol 158:1598–1609

Stephan RP, Reilly CR, Witte PL (1998) Impaired ability of bone marrow stromal cells to support B-lymphopoiesis with age. Blood 91:75–88

Stier S et al. (2005) Osteopontin is a hematopoietic stem cell niche component that negatively regulates stem cell pool size. JEM 201:1781–91

Sugiyama T, Kohara H, Noda M, Nagasawa T (2006) Maintenance of the hematopoietic stem cell pool by CXCL12-CXCR4 chemokine signaling in bone marrow stromal cell niches. Immunity 25:977–88

Szilvassy SJ, Meyerrose TE, Ragland PL, B G (2001) Differential homing and engraftment properties of hematopoietic progenitor cells from murine bone marrow, mobilized peripheral blood, and fetal liver. Blood 98:2108–15

Szilvassy SJ, Ragland PL, Miller CL, Eaves CJ (2003) The marrow homing efficiency of murine hematopoietic stem cells remains constant during ontogeny. Exp Hematol 31:331–38

Taichman RS (2005) Blood and bone: two tissues whose fates are intertwined to create the hematopoietic stem-cell niche. Blood 105:2631–39

Theise N (2006) Gastrointestinal stem cells. III. Emergent themes of liver stem cell biology: niche, quiescence, self-renewal, and plasticity. Am J Gastrointest Liver Physiol 290: G189–G193

Thijssen DH et al. (2006) Haematopoietic stem cells and endothelial progenitor cells in healthy men: effect of ageing and training. Ageing Cell:495–503

Thomas J, Liu F, Link DC (2002) Mechanisms of mobilization of hematopoietic progenitors with granulocyte colony-stimulating factor. Curr Opin Hematol 9:183–89

Till JE, McCulloch EA (1961) A direct measurment of the radiation sensitivity of normal mouse bone marrow cells. Radiat Res 14:213–22

Tyner SD et al. (2002) p53 mutant mice that display early ageing-associated phenotypes. Nature 415:45–53

Ueno H, Weissman IL (2006) Clonal analysis of mouse development reveals a polyclonal origin for yolk sac blood islands. Dev Cell 11:519–33

Van Zant G, Holland B, Eldridge P, Chen J-J (1990) Genotype-restricted growth and ageing patterns in hematopoietic stem cell populations of allophenic mice. J Exp Med 171:1547–65

Vartanian JP (2000) AMD-3100 (AnorMed). IDrugs 3:811–16

Wagers AJ, Sherwood R, Christensen J, Weissman I (2002) Little evidence for developmental plasticity of adult hematopoietic stem cells. Science 297:2256–59

Walkley CR, Fero ML, Chien WM, Purton LE, McArthur GA (2005) Negative cell-cycle regulators cooperatively control self-renewal and differentiation of haematopoietic stem cells. Nat Cell Biol 7:172–78

Wang JC, Warner JK, Erdmann N, Lansdorp PM, Harrington L, Dick JE (2005) Dissociation of telomerase activity and telomere length maintenance in primitive human hematopoietic cells. Proc Natl Acad Sci USA 102:14398–403

Wang X et al. (2003) Cell fusion is the principal source of bone-marrow-derived hepatocytes. Nature 422:897–901

Wilson A, Trump A (2006) Bone-marrow haematopoietic-stem-cell niches. Nat Rev Immunol 6:93–106
Wright DE, Bowman EP, Wagers AJ, Butcher EC, Weissman IL (2002) Hematopoietic stem cells are uniquely selective in their migratory response to chemokines. J Exp Med 195:1145–54
Wright DE, Wagers AJ, Gulati AP, Johnson FL, Weissman IL (2001) Physiological migration of hematopoietic stem and progenitor cells. Science 294
Xing Z et al. (2006) Increased hematopoietic stem cell mobilization in aged mice. Blood 108:2190–97
Yaniv I, Stein J, Farkas D, Askenasy N (2006) The tale of early hematopoietic cell seeding in the bone marrow niche. Stem Cells Dev 15:4–16
Yilmaz OH et al. (2006) Pten dependence distinguishes haematopoietic stem cells from leukaemia-initiating cells. Nature 441
Yu H, Yuan Y, Shen H, Cheng T (2006) Hematopoietic stem cell exhaustion impacted by p18 INK4C and p21 Cip1/Waf1 in opposite manners. Blood 107:1200–06
Yuan R, Astle C, Chen J, Harrison DE (2005) Genetic regulation of hematopoietic stem cell exhaustion during development and growth. Exp Hematol 33:243–50
Yuan Y, Shen H, Franklin DS, Scadden DT, Cheng T (2004) *In vivo* self-renewing divisions of haematopoietic stem cells are increased in the absence of the early G1-phase inhibitor, $p18^{INK4C}$. Nature Cell Biol 6:436–42
Zeng H YR, Kosan C, Klein-Hitpass L, Moroy T. (2004) Transcription factor Gfi1 regulates self-renewal and engraftment of hematopoietic stem cells. EMBO J 23:4116–25
Zhang J et al. (2003) Identification of the haematopoietic stem cell niche and control of the niche size. Nature 425:836–41

Chapter 7
Senescence Signatures of Human Hematopoietic Stem Cells

Stefan Zimmermann and Uwe M Martens(✉)

Abstract Hematopoietic stem cells (HSCs) provide the basis for the enormous regenerative capacity of the hematopoietic system. Despite the need for a lifelong production of blood cells, replicative exhaustion seems to affect the HSC pool. Erosion of telomeres, protective structures at the ends of eukaryotic chromosomes, limits the replicative life span of most somatic cells. The expression of telomerase, the enzyme synthesizing telomeres de novo, is restricted to highly proliferative cells like germ line cells and tumor cells. Adult stem cells such as HSCs still exhibit telomerase activity to a certain level. Here, the telomerase activity does not prevent telomere shortening, suggesting a restriction of the proliferative capacity of HSCs. In this review, we discuss the relevance of telomere maintenance, potential functions of telomerase, and possible clinical applications of telomere manipulation in the hematopoietic stem cell compartment. Furthermore, we discuss new insights affecting the aging of the hematopoietic stem cell pool besides replicative exhaustion.

1 Introduction

The production of billions of new blood cells every day in a human being depends on the undisturbed function of hematopoietic stem cells (HSCs), which represent only about 0.01% of the entire bone marrow cells (Lansdorp et al. 1997). The maintenance of the complex hematopoietic system requires a well-balanced self-renewal and differentiation of HSCs during the entire life span. However, the proliferative life span defining the self-renewal capacity of HSCs appears not to be unrestricted (Lansdorp 1995). A fundamental determinant for the proliferation potential is found in the telomeres, specialized nucleoprotein complexes protecting the ends of eukaryotic chromosomes (reviewed in Blackburn 2001). Telomeres consist of repeated units of G-rich sequences (TTAGGG in humans) that pass in a single-stranded 3′ overhang and that contribute together with telomere-binding proteins to a higher-order

Uwe M Martens
Medical University Centre Freiburg, Department of Hematology/Oncology, D-79106, Freiburg, Germany.
e-mail: uwe.martens@uniklinik-freiburg.de

K.L. Rudolph (ed.), *Telomeres and Telomerase in Ageing, Disease, and Cancer.*

terminal loop (t-loop) structure (Blackburn 2001, Griffith et al. 1999). The number of telomeric repeats varies between different species and within an organism; in humans the telomere length is in the range from 2 to 15 kb (Martens et al. 1998). About 50–100 base pairs (bp) of telomeric DNA are lost with each cell division due to the end replication problem (Watson 1972), nucleolytic processing of the 5′ strand (Lingner et al. 1995), and oxidative damage (von Zglinicki et al. 1995). Maintenance of functional telomeres is crucial for continued proliferation, given that telomere erosion is associated with replicative senescence. Most eukaryotic cells depend on the enzyme telomerase, a reverse transcriptase, for the de novo synthesis of telomeres (reviewed in Cech 2004). Telomerase is a ribonucleoprotein complex, which uses its RNA component (TERC) as a template for the production of telomeric repeats via the catalytic subunit TERT (Greider and Blackburn 1989). Telomerase activity in normal human cells is generally undetectable leading to successive telomere shortening with each cell division, which ultimately limits their proliferative capacity in vitro and most likely in vivo (Harley et al. 1990). Immortal cells like germ line cells (Allsopp et al. 1992), embryonic stem cells (Thomson et al. 1998), and 90% of all tumor cells in humans (Kim et al. 1994) maintain their telomeres by activated telomerase, thus providing an unrestricted life span. Overexpression of hTERT, the catalytic subunit of telomerase, can restore telomerase activity in various human cells (e.g., fibroblasts, hepatocytes, mesenchymal stem cells), which leads to maintenance of telomeres and immortal proliferation (Bodnar et al. 1998, Vaziri and Benchimol 1998, Zimmermann et al. 2004).

The telomere biology of adult stem cells found in highly proliferative tissues such as blood appears to be more complex, because telomeres shorten despite the presence of telomerase activity (see Artandi, this volume).

2 Telomerase Activity in Human HSCs

Stem cells from adult tissues, including skin (Harle-Bachor and Boukamp 1996, Yasumoto et al. 1996), gut (Kolquist et al. 1998), and blood (Broccoli et al. 1995, Chiu et al. 1996, Engelhardt et al. 1997, Morrison et al. 1996) in principle harbor low levels of telomerase activity. Among them, blood progenitor and stem cells show graduations of these basal telomerase levels (Chiu et al. 1996, Engelhardt et al. 1997, Hiyama et al. 1995, Yui et al. 1998). Thus, higher telomerase activity can be detected in bone marrow (BM) $CD34^{+}CD38^{+}$ committed progenitor cell fraction than in the $CD34^{+}CD38^{-/low}$ fraction containing primitive hematopoietic progenitor cells (Hiyama et al. 1995). In turn, BM $CD34^{+}CD38^{-}$ cells show lower telomerase activity compared to corresponding cells isolated from fetal liver (Yui et al. 1998). Initial basal telomerase levels in HSCs can be transiently upregulated in the course of in vitro cultures upon stimulation with cytokines including interleukin 3 (IL-3) (Engelhardt et al. 1997, Yui et al. 1998, Zimmermann et al. 2004). There is arguably a link between the cell cycle of HSCs and telomerase activity demonstrating an elevated activity in actively cycling cells and a downregulation upon differentiation and re-entry into G_0, where most resting HSCs can be found

(Engelhardt et al. 1997, Zhu et al. 1996). Since repopulating HSCs represent only a minor fraction of the $CD34^+CD38^-$ cells (Bhatia et al. 1997) and candidate human HSCs considered as nonobese diabetic severe combined immunodeficient (NOD/SCID) mice BM repopulation cells (SRCs) contain heterogeneous subpopulations with distinct engraftment and differentiation capacities (Glimm et al. 2001), the extent of telomerase activity in the rare SRCs is not known. A recent study using an hTERT-reporting adenoviral vector in cord blood (CB) $CD34^+$ cells demonstrates an upregulated hTERT expression in proliferating short-term SRCs, which remains relatively high in committed colony-forming progenitor cells, but is downregulated in mature myeloid cells (Jaras et al. 2006). Nevertheless, a decreased self-renewal capacity of SRCs is found in this study as a consequence of hTERT upregulation.

3 Telomere Dynamics in Human HSCs

Irrespective of the actual degree of telomerase activity in HSCs, there is clear evidence that proliferation in this compartment is accompanied by extensive telomere shortening during cell culture in vitro (Engelhardt et al. 1997, Zimmermann et al. 2004) and aging in vivo (Vaziri et al. 1994). This is reflected by about four-kilo base pairs (kbp) shorter telomeres in $CD34^+CD38^{low}$ candidate HSCs from adult bone marrow compared to their counterparts from fetal liver (Vaziri et al. 1994). A study of highly purified subpopulations of BM cells representing different HSC candidate populations demonstrated significantly longer telomeres in $CD34^+CD38^-$ and, to some extent, in side population (SP) cells compared to $CD34^+CD38^+$ cells, which positively correlate with the known proliferation potential of these populations (Van Ziffle et al. 2003). Calculations on telomere length dynamics of HSCs expanded in vitro for three to four weeks indicate an overall loss of 35–75 bp of telomeric repeats per population doubling (PD) in the HSC pool (Engelhardt et al. 1997, Vaziri et al. 1994), which appears slightly smaller in relation to other somatic cells that lose between 50–100 bp / PD (Allsopp et al. 1992, Harley et al. 1990, Vaziri et al. 1993). Even though the in vitro studies are inconsistent in terms of heterogeneous starting cell populations such as $CD34^+CD38^+$, $CD34^+CD38^{-/low}$, $AC133^+$, and lineage-depleted (lin^-) fractions from different tissues (BM, CB, FL, and peripheral blood [PB]) and different expansion protocols using variable media and additives (serum, stroma, and different cytokine combinations), telomerase seems not to countervail against telomere erosion in these cells (Engelhardt et al. 1997, Vaziri et al. 1994, Zimmermann et al. 2004). However, it has to be considered that studying the actual telomere dynamics in HSCs is complicated due to a lack of definite HSC markers. Even $CD34^+CD38^-$ purified cells represent a heterogeneous population, within which the real HSCs are representing only a minor fraction (Fig. 7.1). In general, there seems to be an ontogeny-related decline in the hematopoietic capacity of HSCs manifesting in a reduced proliferative and NOD/SCID engrafting potential of stem cells from adult mobilized PB compared to CB $CD34^+$ cells (Lansdorp et al. 1994, Tanavde et al. 2002, Traycoff et al. 1995).

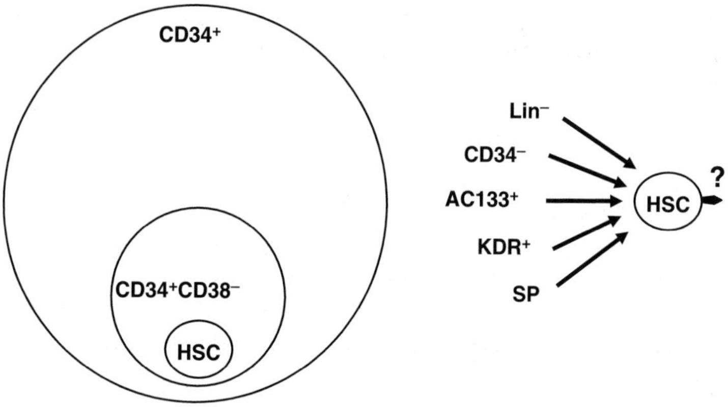

Fig. 7.1 Scheme of HSC isolation markers. HSCs are typically enriched in the CD34+ or its CD34+CD38− compartment. However, these subfractions still constitute heterogeneous cell populations, only a small percentage of which only contain real HSCs. Other HSC purification strategies include depletion of mature hematopoietic cells (lin−) or even CD34 (CD34−), and the use of novel markers such as vascular growth factor receptor 2 (KDR) and AC133, and Hoechst 33342 dye efflux ability of side population (SP) cells, in part overlapping the previous fractions (for a review see Bonnet 2002). The question mark points to the need for a unique marker to enrich pure HSC populations in order to study telomere dynamics in these cells, which may be diluted in more heterogeneous fractions

Repopulation of the hematopoietic system in the course of bone marrow or hematopoietic stem cell transplantations (HSCT) displays an interesting model to study telomere dynamics of HSCs in vivo. Earlier reports indicated a significantly reduced telomere length in recipients after allogeneic stem cell transplantation compared to their respective donors (Akiyama et al. 2000, Akiyama et al. 1998, Lee et al. 1999, Notaro et al. 1997, Wynn et al. 1998). Subsequent data suggest a less severe influence of such transplantations on the telomeres, mainly depending on the size of the transplanted stem cell pool and the age of the donor (Brummendorf et al. 2001b, Roelofs et al. 2003, Rufer et al. 2001a, Thornley et al. 2002). The impact of the used cell source on telomere shortening appears also negligible, resulting in the same degree of shortening for BM and peripheral blood stem cell (PBSC) transplants (Robertson et al. 2001). Nevertheless, most recently it has been shown that longer telomeres characterizing CB HSCs may be indicative of a replicative advantage in comparison to PBSCs in the course of allogeneic stem cell transplantions (Pipes et al. 2006).

4 Telomere Dynamics in Mature Hematopoietic Cells

Certain mature hematopoietic cells like T lymphocytes also have low levels of telomerase activity rising transiently in response to antigen stimulation similar to HSCs (Liu et al. 1999, Weng et al. 1998, Weng et al. 1996). Since extensive cell division and clonal expansion are critical for effective immune function, telomere dynamics are of particular interest in these cells. Again, limited telomerase levels are not sufficient to prevent telomere shortening and replicative senescence. However, overexpression of hTERT is able to reconstitute a constant high level of telomerase activity and extend the life span of T cells (Hooijberg et al. 2000, Rufer et al. 2001b). Although the rate of telomere shortening is lower in hTERT-transduced T cells, long-term studies indicate here a progressive telomere loss eventually resulting in even shorter telomeres at senescence than in controls (Roth et al. 2005). Interestingly, elimination of endogenous hTERT in human T lymphocytes using a dominant-negative mutant (DN-hTERT) leads to a decreased life span and cytogenetic abnormalities, which indicates the major influence of hTERT on the longevity of these cells–without preventing overall telomere shortening but maybe playing a role in the repair of sporadic telomere attrition (Roth et al. 2003).

Telomere shortening featured by mature hematopoietic cells in vivo follows a cubic function over time, which is marked by a significant drop within the first year of life and a slower, steady decline thereafter (Rufer et al. 1999) (reviewed in (Ohyashiki et al. 2002)). A recent longitudinal study of telomere length from newborn baboons confirmed the rapid decline in telomere length in granulocytes and lymphocytes in the first year after birth. After around one year the telomere length appeared to stabilize in all cell types, suggesting that HSCs switch to a different functional mode characterized by a decreased turnover rate after an initial phase of rapid expansion (Baerlocher et al. 2007). Individual replicative histories of lymphocytes are represented by heterogeneous telomere length distributions in different subpopulations, namely a shorter mean telomere length in T cells compared with B cells within which memory cells have longer telomeres than naive cells (Baerlocher and Lansdorp 2003, Martens et al. 2002, Rufer et al. 1999). A later study indicates that B cells are capable of inducing telomerase after stimulation, resulting in telomere maintenance during differentiation from naive to memory B cells (Son et al. 2003). Since granulocytes have a short life span and do not replicate, their age-related telomere loss is much less pronounced than in lymphocytes. Therefore, the rather homogeneous telomere length of granulocytes seems to be a good surrogate marker for HSC proliferation kinetics under the assumption that the HSCs exhibit a constant telomere shortening during replication and differentiation from the HSC to granulocytes (Rufer et al. 1999) (reviewed in Verfaillie et al. 2002). A stochastic simulation method based upon granulocyte telomere lengths was used to estimate HSC replication rates in vivo and found human HSCs replicating only once per 45 weeks (Shepherd et al. 2004), which is substantially slower than the estimates for murine HSCs (1 per 2.5 weeks) (Abkowitz et al. 2000). Such infrequent cell divisions in human HSCs would put their above mentioned calculated telomere loss into perspective. Understanding these HSC dynamics in vivo could be important to assess the consequences of allogeneic stem cell transplantation discussed above. Besides an overall trend in age-related telomere shortening, the telomere length of hematopoietic cells at any given age shows significant differences (Frenck et al. 1998, Rufer et al. 1998), which are primarily determined genetically as twin studies revealed (Rufer et al. 1999, Slagboom et al. 1994).

5 Telomere Manipulation in Adult Stem Cells

As seen above for T cells, modulation of telomerase expression can be a valuable tool for studying its function in various cell types. Along with other groups we could demonstrate that hTERT overexpression in hematopoietic progenitors such as $CD34^+$ CB cells results in a significant elevation of telomerase activity that cannot prevent overall telomere shortening and is incapable of increasing the replicative capacity of these cells (Akimov et al. 2005, Elwood et al. 2004, Wang et al. 2005, Zimmermann et al. 2004). Interestingly, only cotransduction of hTERT with the human papillomavirus type 16 (HPV16) E6 and E7 oncogenes in $CD34^+$ CB cells resulted in continuous proliferation, giving rise to permanent cell lines with a myelo-erythroid/mast cell progenitor phenotype. These cell lines exhibited stabilization of telomere length and minimal chromosomal aberrations in contrast to CB cell cultures expressing only HPV16 E6/E7, which went through a crisis period resulting in highly aneuploid cell lines (Akimov et al. 2005). An eventual stabilization of telomere length without changes in telomerase activity was observed during establishment of leukemic cell lines from normal cord blood cells, indicating that additional genetic or epigenetic alterations are required for telomere maintenance in immortalized human hematopoietic cells (Wang et al. 2005). Overall, evidence is accumulating that hTERT could function independently of its enzymatic activity as a prosurvival factor (Armstrong et al. 2005, Cao et al. 2002, Elwood et al. 2004, Folini et al. 2005, Gorbunova and Seluanov 2003, Massard et al. 2006, Yamada et al. 2003). In this context, recent data demonstrated that overexpression of hTERT in a cytokine-dependent, human hematopoietic progenitor cell line, as well as in normal CB $CD34^+$ cells, resulted in protection from apoptosis in the absence of cytokine stimulation but did not favor unlimited replicative potential (Li et al. 2006).

Extensive CB $CD34^+$ cell expansion for up to 18 weeks in long-term cultures with sustained telomerase activity and a minimal telomere loss could be demonstrated only on the basis of refined culture conditions, including special cytokine cocktails and stroma support (Gammaitoni et al. 2004). Similarly, the application of an optimized serum-free and cytokine-limited defined medium in CB $AC133^+$ cell expansion provided a very early uncommitted HSC proliferation and was accompanied by high levels of telomerase activity to maintain telomere length (Yao et al. 2006). No matter which approach is used, in vitro expansion of HSCs for a prolonged period without telomere shortening has striking clinical implications for allogeneic transplantation, considering the limited HSC numbers in individual CB samples and other HSC sources (Jaroscak et al. 2003, Moore 2000).

Other stem cells which are prevalently found in the bone marrow are mesenchymal stem cells (hMSCs). These cells can differentiate along variable lineages, including bone, cartilage, adipose and muscle cells (Jiang et al. 2002). Unlike HSCs and other adult stem cells, hMSCs lack telomerase expression, which limits their replicative capacity in tissue culture (Simonsen et al. 2002, Zimmermann et al. 2003). Ectopic telomerase expression is able to expand the replicative capacity of hMSCs while the cells are retaining their functional characteristics (Shi et al. 2002, Simonsen et al. 2002). Nevertheless, premalignant changes were observed in the course of long-term cultures of hTERT overexpressing hMSCs (Serakinci et al. 2004), evoking some concerns about potential clinical applications (Keith 2004). The fact that stem cells and cancer cells share several common features, such as the same factors regulating self-renewal of normal HSCs and leukemic cells (Lessard and Sauvageau 2003), has led to the "cancer stem cell" hypothesis (reviewed in (Marx 2003)). If there is a stem cell origin for certain cancers, these stem cells would already have telomerase and it would not need to be reactivated, although the enzyme activity may increase in later stages of carcinogenesis (reviewed in (Armanios and Greider 2005). In reverse, limited telomerase activity accompanied by a finite life span in adult stem cells may promote aging but may prevent cancer. On the other hand there is evidence that telomere shortening can induce chromosomal instability and cancer initiation (see Rudolph, this volume).

6 Consequences of Telomerase Inhibition in HSCs

When telomeres shorten until they are significantly eroded, a DNA damage cascade is activated, and cells usually undergo replicative senescence or apoptosis (Harley et al. 1994, Maser and DePinho 2002). Because the majority of tumor cells is reliant on telomerase for telomere stabilization, inhibition of this enzyme represents an attractive concept for cancer therapy (reviewed in Zimmermann 2007). Proof of the underlying principle for such strategies was given by studies in which telomerase was abolished by overexpression of DN-hTERT in immortal cancer cell lines, resulting in telomere erosion and induction of senescence or apoptosis (Hahn et al. 1999, Zhang et al. 1999). In addition to genetic approaches, different synthetic telomerase inhibitors are tested at present with promising results for specific killing of tumor cells (Asai et al. 2003, Damm et al. 2001, El-Daly et al. 2005, Herbert et al. 1999) (reviewed in Zimmermann 2007), some of which progress significantly toward clinical application (Dikmen et al. 2005, Djojosubroto et al. 2005, Gellert et al. 2006). The main portion of normal human cells considered as telomerase-negative should not be affected by telomerase inhibition, but potential side effects by such treatment modalities concern telomerase-positive stem cells such as HSCs, which most likely turn out less dramatically in light of the low estimated HSC replication rates in vivo. As seen above, the exact role of telomerase in these cells is not completely understood, yet there is clear evidence that telomerase activity is required for health and viability within the entire life span (Mitchell et al. 1999, Vulliamy et al. 2001). A particular model to study the role of telomerase in vivo is the human disease Dyskeratosis congenita (DKC), which is characterized by anemia, immune deficiency, skin and nail lesions, chromosomal instability, and cancer (reviewed in Collins and Mitchell 2002; see also Du et al., this volume). Two genes encoding proteins of the telomerase complex were found to be mutated, resulting in a partial telomerase inhibition; abnormally short telomeres, and a dramatically limited proliferative capacity of hematopoietic and epithelial tissues in DKC patients (Mitchell et al. 1999). Autosomal-dominant forms of the disease are caused by mutations in the core RNA component of telomerase hTERC (Goldman et al. 2005, Vulliamy et al. 2001) (reviewed in Vulliamy and Dokal 2003). The unusual appearance of two distinct mutations in both hTERC alleles in a DKC patient was reported recently (Ly et al. 2005b), whereby haploinsufficiency of hTERC is most likely resulting in an anticipation of the autosomal-dominant DKC disease forms correlated with an increase in telomere shortening in successive generations of families with this disease (Vulliamy et al. 2004). The hematological abnormalities developed in most DKC patients clarify that highly proliferative cell lineages like the hematopoietic progenitor cells would suffer most in the course of telomerase inhibition.

Another human disease possibly associated with telomerase is the bone marrow failure syndrome aplastic anemia (AA), which is characterized by pancytopenia related to reduced or absent immature hematopoietic cells (Maciejewski et al. 1994, Marsh et al. 1990, Scopes et al. 1994). The higher turnover of HSCs is here expressed in significantly shorter telomeres in mature blood cells like peripheral lymphocytes and granulocytes than those in age-adjusted healthy controls (Ball et al. 1998, Brummendorf et al. 2001a, Lee et al. 2001). As in DKC, mutations in telomerase hTERC (Vulliamy et al. 2002) may be responsible for the compromised telomerase activity resulting in the observed phenotypes. Most recently, evidence for a direct link between hTERC sequence variants found in AA patients and abolished telomerase activity was described (Ly et al. 2005a). Analogously, mutations in hTERT were recently identified among AA patients, again associated with short telomeres and low telomerase enzymatic activity (Yamaguchi et al. 2005). Bone marrow failure of variable severity due to DKC may be present in otherwise phenotypic normal adults and may masquerade as AA, whereas common mutations in hTERC are linking the two diseases (Dokal and Vulliamy 2003, Fogarty et al. 2003, Marrone et al. 2004). Interestingly, hTERC haploinsufficiency in autosomal-dominant DKC cases, which is associated with a modest reduction of telomerase activation of around 50%, is sufficient to induce the severe phenotypes of the disease described above. Overall, this indicates the need of a tight control of telomerase levels throughout life in human HSCs (reviewed in Collins and Mitchell 2002).

Recent studies suggest that even normal human cells like primary fibroblasts harbor some previously undetected telomerase activity (Masutomi et al. 2003). These basal telomerase levels seem to be important for the maintenance of the 3′ overhang, cell proliferation, and cellular life span without any consequences on the overall telomere loss (Masutomi et al. 2003). It would be interesting to demonstrate a similar function of telomerase in the stem cell compartment. We found that telomerase ablation in CB $AC133^{+}$ cells by DN-hTERT overexpression is accompanied by a reduction in clonogenic growth without changing the mean telomere length, supporting the hypothesis above that telomerase even in HSCs may have additional functions beyond simple telomere lengthening (Zimmermann et al. 2004). In addition, we observed that high concentrations of the small molecule telomerase inhibitor BIBR1532 specifically kill tumor cells of the hematopoietic system and do not harm proliferation and clonogenic growth of normal $CD34^+$ cells, reflecting potential differences between genetic and pharmacological approaches of telomerase inhibition (El-Daly et al. 2005).

7 Lessons from Mice

Mice have by far longer telomeres than humans, which complicates a direct comparison of telomere-dependent replicative capacities in cells from both species. Nevertheless, valuable insights are given from telomerase-deficient mice, in which both alleles of mTERC, the murine RNA component of telomerase, are deleted. The first four to five generations of such $mTERC^{-/-}$ mice have unobtrusive phenotypes until telomeres get critically short. Then in following generations, similar to DKC patients, cytogenetic aberrations, a comprised bone marrow, immune deficiency, tumor formation, and an overall reduced life span appear in these telomerase-null mice (Blasco et al. 1997, Hande et al. 1999, Herrera et al. 1999, Rudolph et al. 1999). Wild-type mice HSCs already have a finite replicative life span, based upon their limitation to four to seven serial transplantations into irradiated mice (Harrison and Astle 1982, Harrison et al. 1978, Siminovitch et al. 1964). As in human HSCs, there seems to be telomere shortening despite telomerase activity in murine HSCs, resulting in more than 40% loss of total telomeric DNA after four rounds of serial transplantation (Allsopp et al. 2001, Allsopp et al. 2003b). Similar to the human situation, it is not yet clear how telomere shortening in HSCs is influencing the more differentiated offspring. However, it has been shown that stimulated T cells isolated from HSC transplant recipients could rejuvenate their telomeres by activation of telomerase (Allsopp et al. 2002). HSCs from telomerase-deficient mice, including the $mTERC^{-/-}$ mice above and $mTERT^{-/-}$ mice (Liu et al. 2000), can only be serially transplanted for two rounds due to an even accelerated telomere loss compared to wild-type controls (Allsopp et al. 2003a). This indicates a role for telomerase in HSCs in at least limiting the rate of telomere shortening during cell divisions to allow extended proliferation, the prerequisite of (hematopoietic) stem cell function throughout life.

In contrast to wild-type, transgenic mice overexpressing mTERT exhibit fourfold elevated levels of telomerase activity and stable telomeres during serial transplantations (Allsopp et al. 2003b). However, the transplantation capacity could not be increased for mTERT overexpressing HSCs, which indicates telomere-independent barriers for the transplantation of mice HSC, possibly reflecting a form of premature senescence or possible dilution and loss of real stem cells during serial transplantation (Allsopp et al. 2003b, Wright and Shay 2002). In a similar transgenic mice strain overexpressing mTERT, the proliferation rate of hematopoietic cells is not elevated, although these mice are also characterized by increased telomerase activity and maintained telomere length in hematopoietic and many other tissues compared to nontransgenic control mice (Artandi et al. 2002). Interestingly, robust TERT expression is accompanied by an increased susceptibility of breast cancer in aging females of these mice, which possibly indicates a direct oncogenic role of the enzyme (Artandi et al. 2002). Most recently, it was shown, that p21 deletion improved stem cell function without rescuing telomere function in mTERC knockout mice (Choudhury et al. 2007) (see also Gutierrez and Ju, this volume).

Finally, it seems that short telomeres themselves and not telomerase per se are limiting tissue renewal capacity. This raises questions about the importance of telomerase dosage effects on telomere length and disease phenotypes as seen above for DKC (Hao et al. 2005).

8 Other Factors Determining Self-Renewal and Aging in HSCs

The self-renewal process is crucial for maintaining HSC pool size throughout life, but the factors guiding this were widely unknown. Comprehensive transcriptional profiling studies elucidated the molecular background behind many stem cell attributes, including their self-renewal capacity (Ivanova et al. 2002, Ramalho-Santos et al. 2002). Since then, several genetic regulatory programs have been identified to play important roles in self-renewal decisions of HSCs. Among them are HOXB4 (Sauvageau et al. 1995), Notch1 (Carlesso et al. 1999), Bmi-1 (Lessard and Sauvageau 2003, Park et al. 2003), and the β-catenin pathway (Reya et al. 2003). Later it was shown that some of these diverse intracellular pathways might be integrated by NF-Ya, the regulatory and DNA-binding subunit of the trimeric transcription factor NF-Y. Overexpression of NF-Ya in primitive hematopoietic cells activates the transcription of not only multiple HOX4 paralogs, Notch-1, LEF-1, but also telomerase RNA. HSCs overexpressing NF-Ya are biased toward primitive hematopoiesis in vitro and show strikingly increased in vivo repopulating abilities making NF-Ya a potent cellular regulator of HSC self-renewal (Zhu et al. 2005). In addition it was demonstrated that NF-Y binding to the CCAAT region of the hTERC promoter is decisive for promoter activity (Zhao et al. 2000). Essential attributes of stemness defined by Ramalho-Santos involve also high resistance to stress with upregulated DNA repair and detoxifier systems, interaction with the extracellular matrix, engagement in the cell cycle, and a remodeled chromatin (Ramalho-Santos et al. 2002), which all or in part could be affected in aging stem cells (see also Waterstrat et al., this volume). Thus, studies on mice deficient for the integral nucleotide excision repair protein ERCC1 indicate that accumulation of DNA damage leads to exhaustion of HSC activity (Prasher et al. 2005). In addition, the dependence of the HSC self-renewal capacity on inhibition of oxidative stress was shown in studies with mice deficient for the "ataxia telangiectasia mutated" (*Atm*) gene resulting in progressive BM failure (Ito et al. 2004). In general, detoxification of reactive oxygen species (ROS) is crucial to prevent protein carbonylation, which is associated with age-related human diseases such as Parkinson's disease, Alzheimer's disease, and cancer (reviewed in Dalle-Donne et al. 2003, Nystrom 2005). It is yet to be shown if detoxification systems are disturbed in aging stem cells, thus eventually leading to an accumulation of carbonylated proteins.

The $p16^{INK4a}$ protein inhibits cell cycle progression and induces cellular senescence, and its expression increases with age in many tissues, in line with the accumulation of senescent cells (reviewed in Collins and Sedivy 2003, Campisi 2005). A recent study found that $p16^{INK4a}$ levels accumulate with age in mouse BM HSCs and modulate specific age-associated HSC functions, whereas knocking out $p16^{INK4a}$ diminishes HSC repopulating defects and apoptosis associated with an improved stress tolerance and survival of cells (Janzen et al. 2006).

In addition, novel data also suggest that the chromatin structure and its modifications regarded as cellular memory play an important role during HSC aging (reviewed in Kamminga and de Haan 2006). Possibly heterochromatin, which is a transcriptionally silent part of the genome, influences HSC aging and differentiation, as reflected by a higher number of distinct transcripts in HSCs than in committed progenitors (Terskikh et al. 2003). Although extensive formulation of heterochromatin is now considered as a hallmark of cellular senescence (Zhang et al. 2005), the role of such epigenetic changes for the aging of stem cells has to be investigated.

9 Conclusions and Outlook

Hematopoietic stem cells maintain hematopoiesis through tightly regulated layers of self-renewal, differentiation, and cell death during the entire life span of vertebrates. Nevertheless, programmed telomere shortening as a hallmark of replicative senescence occurs in HSCs, which is not completely understood, since conventional correlations between telomere maintenance and telomerase activity appear to be only partially valid in these cells. Importantly, telomerase activity and telomere dynamics seem to play a different role in embryonic stem cells compared to adult stem cells. Although stem cell proliferation generally allows for tissue repair and regeneration, it also bears the risk of cancer, which is impaired by tumor suppressor mechanisms, either eliminating potential cancer cells by apoptosis or limiting their proliferation by senescence. Apparently, factors ensuring stem cell proliferation as a prerequisite of relative longevity must be well balanced with mechanisms that prevent cancer (reviewed in Beausejour and Campisi 2006). In this context, coexistence of telomerase expression – maybe here rather as an antiapoptotic factor – and telomere shortening might appear less contradictory in the hematopoietic stem cell compartment.

The paucity of markers for cellular senescence makes it difficult to evaluate the relevance of this phenomenon for in vivo aging – not only in stem cells. Most recently the combinatory use of three biomarkers (telomere dysfunction, activation of the ATM DNA-damage response, and heterochromatinization of the nuclear genome) in aging baboons provided evidence that senescent cells exist in vivo and can account for more than 15% of skin fibroblasts in aged animals (Herbig et al. 2006). Whether the stem cell compartment is affected to the same extent remains elusive. Hence, the identification of further senescence markers could facilitate a

Dokal I, Vulliamy T (2003) Dyskeratosis congenita: its link to telomerase and aplastic anaemia. Blood Rev 17: 217–25

El-Daly H, Kull M, Zimmermann S, Pantic M, Waller CF, Martens UM (2005) Selective cytotoxicity and telomere damage in leukemia cells using the telomerase inhibitor BIBR1532. Blood 105: 1742–49

Elwood NJ, Jiang XR, Chiu CP, Lebkowski JS, Smith CA (2004) Enhanced long-term survival, but no increase in replicative capacity, following retroviral transduction of human cord blood CD34+ cells with human telomerase reverse transcriptase. Haematologica 89: 377–78

Engelhardt M, Kumar R, Albanell J, Pettengell R, Han W, Moore MA (1997) Telomerase regulation, cell cycle, and telomere stability in primitive hematopoietic cells. Blood 90: 182–93

Fogarty PF, Yamaguchi H, Wiestner A, Baerlocher GM, Sloand E, Zeng WS, Read EJ, Lansdorp PM, Young NS (2003) Late presentation of dyskeratosis congenita as apparently acquired aplastic anaemia due to mutations in telomerase RNA. Lancet 362: 1628–30

Folini M, Brambilla C, Villa R, Gandellini P, Vignati S, Paduano F, Daidone MG, Zaffaroni N (2005) Antisense oligonucleotide-mediated inhibition of hTERT, but not hTERC, induces rapid cell growth decline and apoptosis in the absence of telomere shortening in human prostate cancer cells. Eur J Cancer 41: 624–34

Frenck RW, Jr., Blackburn EH, Shannon KM (1998) The rate of telomere sequence loss in human leukocytes varies with age. Proc Natl Acad Sci U S A 95: 5607–10

Gammaitoni L, Weisel KC, Gunetti M, Wu KD, Bruno S, Pinelli S, Bonati A, Aglietta M, Moore MA, Piacibello W (2004) Elevated telomerase activity and minimal telomere loss in cord blood long-term cultures with extensive stem cell replication. Blood 103: 4440–48

Gellert GC, Dikmen ZG, Wright WE, Gryaznov S, Shay JW (2006) Effects of a novel telomerase inhibitor, GRN163L, in human breast cancer. Breast Cancer Res Treat 96: 73–81

Glimm H, Eisterer W, Lee K, Cashman J, Holyoake TL, Nicolini F, Shultz LD, von Kalle C, Eaves CJ (2001) Previously undetected human hematopoietic cell populations with short-term repopulating activity selectively engraft NOD/SCID-beta2 microglobulin-null mice. J Clin Invest 107: 199–206

Goldman F, Bouarich R, Kulkarni S, Freeman S, Du HY, Harrington L, Mason PJ, Londono-Vallejo A, Bessler M (2005) The effect of TERC haploinsufficiency on the inheritance of telomere length. Proc Natl Acad Sci USA 102: 17119–24

Gorbunova V, Seluanov A (2003) Telomerase as a growth-promoting factor. Cell Cycle 2: 534–37

Greider CW, Blackburn EH (1989) A telomeric sequence in the RNA of Tetrahymena telomerase required for telomere repeat synthesis. Nature 337: 331–37

Griffith JD, Comeau L, Rosenfield S, Stansel RM, Bianchi A, Moss H, de Lange T (1999) Mammalian telomeres end in a large duplex loop. Cell 97: 503–14

Hahn WC, Stewart SA, Brooks MW, York SG, Eaton E, Kurachi A, Beijersbergen RL, Knoll JH, Meyerson M, Weinberg RA (1999) Inhibition of telomerase limits the growth of human cancer cells. Nat Med 5: 1164–70

Hande MP, Samper E, Lansdorp P, Blasco MA (1999) Telomere length dynamics and chromosomal instability in cells derived from telomerase null mice. J Cell Biol 144: 589–601

Hao LY, Armanios M, Strong MA, Karim B, Feldser DM, Huso D, Greider CW (2005) Short telomeres, even in the presence of telomerase, limit tissue renewal capacity. Cell 123: 1121–31

Harle-Bachor C, Boukamp P (1996) Telomerase activity in the regenerative basal layer of the epidermis inhuman skin and in immortal and carcinoma-derived skin keratinocytes. Proc Natl Acad Sci USA 93: 6476–81

Harley CB, Futcher AB, Greider CW (1990) Telomeres shorten during ageing of human fibroblasts. Nature 345: 458–60

Harley CB, Kim NW, Prowse KR, Weinrich SL, Hirsch KS, West MD, Bacchetti S, Hirte HW, Counter CM, Greider CW, et al. (1994) Telomerase, cell immortality, and cancer. Cold Spring Harb Symp Quant Biol 59: 307–15

Harrison D, Astle C (1982) Loss of stem cell repopulating ability upon transplantation. Effects of donor age, cell number, and transplantation procedure. J Exp Med 156: 1767–79

Harrison DE, Astle CM, Delaittre JA (1978) Loss of proliferative capacity in immunohemopoietic stem cells caused by serial transplantation rather than aging. J Exp Med 147: 1526–31

Herbert B, Pitts AE, Baker SI, Hamilton SE, Wright WE, Shay JW, Corey DR (1999) Inhibition of human telomerase in immortal human cells leads to progressive telomere shortening and cell death. Proc Natl Acad Sci USA 96: 14276–81

Herbig U, Ferreira M, Condel L, Carey D, Sedivy JM (2006) Cellular senescence in aging primates. Science 311: 1257

Herrera E, Samper E, Blasco MA (1999) Telomere shortening in mTR–/– embryos is associated with failure to close the neural tube. Embo J 18: 1172–81

Hiyama K, Hirai Y, Kyoizumi S, Akiyama M, Hiyama E, Piatyszek MA, Shay JW, Ishioka S, Yamakido M (1995) Activation of telomerase in human lymphocytes and hematopoietic progenitor cells. J Immunol 155: 3711–15

Hooijberg E, Ruizendaal JJ, Snijders PJ, Kueter EW, Walboomers JM, Spits H (2000) Immortalization of human CD8+ T cell clones by ectopic expression of telomerase reverse transcriptase. J Immunol 165: 4239–45

Ito K, Hirao A, Arai F, Matsuoka S, Takubo K, Hamaguchi I, Nomiyama K, Hosokawa K, Sakurada K, Nakagata N, Ikeda Y, Mak TW, Suda T (2004) Regulation of oxidative stress by ATM is required for self-renewal of haematopoietic stem cells. Nature 431: 997–1002

Ivanova NB, Dimos JT, Schaniel C, Hackney JA, Moore KA, Lemischka IR (2002) A Stem Cell Molecular Signature. Science 298: 601–04

Janzen V, Forkert R, Fleming HE, Saito Y, Waring MT, Dombkowski DM, Cheng T, DePinho RA, Sharpless NE, Scadden DT (2006) Stem-cell ageing modified by the cyclin-dependent kinase inhibitor p16INK4a. Nature 443: 421–26

Jaras M, Edqvist A, Rebetz J, Salford LG, Widegren B, Fan X (2006) Human short-term repopulating cells have enhanced telomerase reverse transcriptase expression. Blood 108: 1084–91

Jaroscak J, Goltry K, Smith A, Waters-Pick B, Martin PL, Driscoll TA, Howrey R, Chao N, Douville J, Burhop S, Fu P, Kurtzberg J (2003) Augmentation of umbilical cord blood (UCB) transplantation with ex vivo-expanded UCB cells: results of a phase 1 trial using the AastromReplicell System. Blood 101: 5061–67

Jiang Y, Jahagirdar BN, Reinhardt RL, Schwartz RE, Keene CD, Ortiz-Gonzalez XR, Reyes M, Lenvik T, Lund T, Blackstad M, Du J, Aldrich S, Lisberg A, Low WC, Largaespada DA, Verfaillie CM (2002) Pluripotency of mesenchymal stem cells derived from adult marrow. Nature 418: 41–49

Kamminga LM, de Haan G (2006) Cellular memory and hematopoietic stem cell aging. Stem Cells 24: 1143–49

Keith WN (2004) From stem cells to cancer: balancing immortality and neoplasia. Oncogene 23: 5092–94

Kim NW, Piatyszek MA, Prowse KR, Harley CB, West MD, Ho PL, Coviello GM, Wright WE, Weinrich SL, Shay JW (1994) Specific association of human telomerase activity with immortal cells and cancer. Science 266: 2011–15

Kolquist KA, Ellisen LW, Counter CM, Meyerson M, Tan LK, Weinberg RA, Haber DA, Gerald WL (1998) Expression of TERT in early premalignant lesions and a subset of cells in normal tissues. Nat Genet 19: 182–86

Lansdorp PM (1995) Telomere length and proliferation potential of hematopoietic stem cells. J Cell Sci 108: 1–6

Lansdorp PM, Dragowska W, Thomas TE, Little MT, Mayani H (1994) Age-related decline in proliferative potential of purified stem cell candidates. Blood Cells 20: 376–80; discussion 380–81

Lansdorp PM, Poon S, Chavez E, Dragowska V, Zijlmans M, Bryan T, Reddel R, Egholm M, Bacchetti S, Martens U (1997) Telomeres in the haemopoietic system. Ciba Found Symp 211: 209–18; discussion 219–22

Lee J, Kook H, Chung I, Kim H, Park M, Kim C, Nah J, Hwang T (1999) Telomere length changes in patients undergoing hematopoietic stem cell transplantation. Bone Marrow Tr 24: 411–15

Lee JJ, Kook H, Chung IJ, Na JA, Park MR, Hwang TJ, Kwak JY, Sohn SK, Kim HJ (2001) Telomere length changes in patients with aplastic anaemia. Br J Haematol 112: 1025–30

Lessard J, Sauvageau G (2003) Bmi-1 determines the proliferative capacity of normal and leukaemic stem cells. Nature 423: 255–60
Li S, Ferguson MJ, Hawkins CJ, Smith C, Elwood NJ (2006) Human telomerase reverse transcriptase protects hematopoietic progenitor TF-1 cells from death and quiescence induced by cytokine withdrawal. Leukemia 20: 1270–78
Lingner J, Cooper JP, Cech TR (1995) Telomerase and DNA end replication: no longer a lagging strand problem? Science 269: 1533–34
Liu K, Schoonmaker MM, Levine BL, June CH, Hodes RJ, Weng NP (1999) Constitutive and regulated expression of telomerase reverse transcriptase (hTERT) in human lymphocytes. Proc Natl Acad Sci USA 96: 5147–52
Liu Y, Snow BE, Hande MP, Yeung D, Erdmann NJ, Wakeham A, Itie A, Siderovski DP, Lansdorp PM, Robinson MO, Harrington L (2000) The telomerase reverse transcriptase is limiting and necessary for telomerase function in vivo. Curr Biol 10: 1459–62
Ly H, Calado RT, Allard P, Baerlocher GM, Lansdorp PM, Young NS, Parslow TG (2005a) Functional characterization of telomerase RNA variants found in patients with hematologic disorders. Blood 105: 2332–39
Ly H, Schertzer M, Jastaniah W, Davis J, Yong SL, Ouyang Q, Blackburn EH, Parslow TG, Lansdorp PM (2005b) Identification and functional characterization of 2 variant alleles of the telomerase RNA template gene (TERC) in a patient with dyskeratosis congenita. Blood 106: 1246–52
Maciejewski JP, Anderson S, Katevas P, Young NS (1994) Phenotypic and functional analysis of bone marrow progenitor cell compartment in bone marrow failure. Br J Haematol 87: 227–34
Marrone A, Stevens D, Vulliamy T, Dokal I, Mason PJ (2004) Heterozygous telomerase RNA mutations found in dyskeratosis congenita and aplastic anemia reduce telomerase activity via haploinsufficiency. Blood 104: 3936–42
Marsh JC, Chang J, Testa NG, Hows JM, Dexter TM (1990) The hematopoietic defect in aplastic anemia assessed by long-term marrow culture. Blood 76: 1748–57
Martens UM, Brass V, Sedlacek L, Pantic M, Exner C, Guo Y, Engelhardt M, Lansdorp PM, Waller CF, Lange W (2002) Telomere maintenance in human B lymphocytes. Br J Haematol 119: 810–18
Martens UM, Zijlmans JM, Poon SS, Dragowska W, Yui J, Chavez EA, Ward RK, Lansdorp PM (1998) Short telomeres on human chromosome 17p. Nat Genet 18: 76–80
Marx J (2003) Cancer research. Mutant stem cells may seed cancer. Science 301: 1308–10
Maser RS, DePinho RA (2002) Connecting chromosomes, crisis, and cancer. Science 297: 565–69
Massard C, Zermati Y, Pauleau AL, Larochette N, Metivier D, Sabatier L, Kroemer G, Soria JC (2006) hTERT: a novel endogenous inhibitor of the mitochondrial cell death pathway. Oncogene 25: 4505–14
Masutomi K, Yu EY, Khurts S, Ben-Porath I, Currier JL, Metz GB, Brooks MW, Kaneko S, Murakami S, DeCaprio JA, Weinberg RA, Stewart SA, Hahn WC (2003) Telomerase maintains telomere structure in normal human cells. Cell 114: 241–53
Mitchell JR, Wood E, Collins K (1999) A telomerase component is defective in the human disease dyskeratosis congenita. Nature 402: 551–55
Moore MA (2000) Umbilical cord blood: an expandable resource. J Clin Invest 105: 855–56
Morrison SJ, Prowse KR, Ho P, Weissman IL (1996) Telomerase activity in hematopoietic cells is associated with self-renewal potential. Immunity 5: 207–16
Notaro R, Cimmino A, Tabarini D, Rotoli B, Luzzatto L (1997) In vivo telomere dynamics of human hematopoietic stem cells. Proc Natl Acad Sci USA 94: 13782–85
Nystrom T (2005) Role of oxidative carbonylation in protein quality control and senescence. Embo J 24: 1311–17
Ohyashiki JH, Sashida G, Tauchi T, Ohyashiki K (2002) Telomeres and telomerase in hematologic neoplasia. Oncogene 21: 680–87
Park IK, Qian D, Kiel M, Becker MW, Pihalja M, Weissman IL, Morrison SJ, Clarke MF (2003) Bmi-1 is required for maintenance of adult self-renewing haematopoietic stem cells. Nature 423: 302–05

Pipes BL, Tsang T, Peng SX, Fiederlein R, Graham M, Harris DT (2006) Telomere length changes after umbilical cord blood transplant. Transfusion 46: 1038–43

Prasher JM, Lalai AS, Heijmans-Antonissen C, Ploemacher RE, Hoeijmakers JH, Touw IP, Niedernhofer LJ (2005) Reduced hematopoietic reserves in DNA interstrand crosslink repair-deficient Ercc1–/– mice. Embo J 24: 861–71

Ramalho-Santos M, Yoon S, Matsuzaki Y, Mulligan RC, Melton DA (2002) "Stemness": Transcriptional profiling of embryonic and adult stem cells. Science 298: 587–600

Reya T, Duncan AW, Ailles L, Domen J, Scherer DC, Willert K, Hintz L, Nusse R, Weissman IL (2003) A role for Wnt signalling in self-renewal of haematopoietic stem cells. Nature 423: 409–14

Robertson JD, Testa NG, Russell NH, Jackson G, Parker AN, Milligan DW, Stainer C, Chakrabarti S, Dougal M, Chopra R (2001) Accelerated telomere shortening following allogeneic transplantation is independent of the cell source and occurs within the first year post transplant. Bone Marrow Tr 27: 1283–86

Roelofs H, de Pauw ES, Zwinderman AH, Opdam SM, Willemze R, Tanke HJ, Fibbe WE (2003) Homeostasis of telomere length rather than telomere shortening after allogeneic peripheral blood stem cell transplantation. Blood 101: 358–62

Roth A, Baerlocher GM, Schertzer M, Chavez E, Duhrsen U, Lansdorp PM (2005) Telomere loss, senescence, and genetic instability in CD4+ T lymphocytes overexpressing hTERT. Blood 106: 43–50

Roth A, Yssel H, Pene J, Chavez EA, Schertzer M, Lansdorp PM, Spits H, Luiten RM (2003) Telomerase levels control the life span of human T lymphocytes. Blood 102: 849–57

Rudolph KL, Chang S, Lee HW, Blasco M, Gottlieb GJ, Greider C, DePinho RA (1999) Longevity, stress response, and cancer in aging telomerase-deficient mice. Cell 96: 701–12

Rufer N, Brummendorf TH, Chapuis B, Helg C, Lansdorp PM, Roosnek E (2001a) Accelerated telomere shortening in hematological lineages is limited to the first year following stem cell transplantation. Blood 97: 575–77

Rufer N, Brummendorf TH, Kolvraa S, Bischoff C, Christensen K, Wadsworth L, Schulzer M, Lansdorp PM (1999) Telomere fluorescence measurements in granulocytes and T lymphocyte subsets point to a high turnover of hematopoietic stem cells and memory T cells in early childhood. J Exp Med 190: 157–67

Rufer N, Dragowska W, Thornbury G, Roosnek E, Lansdorp PM (1998) Telomere length dynamics in human lymphocyte subpopulations measured by flow cytometry. Nat Biotechnol 16: 743–7

Rufer N, Migliaccio M, Antonchuk J, Humphries RK, Roosnek E, Lansdorp PM (2001b) Transfer of the human telomerase reverse transcriptase (TERT) gene into T lymphocytes results in extension of replicative potential. Blood 98: 597–603

Sauvageau G, Thorsteinsdottir U, Eaves CJ, Lawrence HJ, Largman C, Lansdorp PM, Humphries RK (1995) Overexpression of HOXB4 in hematopoietic cells causes the selective expansion of more primitive populations in vitro and in vivo. Genes Dev 9: 1753–65

Scopes J, Bagnara M, Gordon-Smith EC, Ball SE, Gibson FM (1994) Haemopoietic progenitor cells are reduced in aplastic anaemia. Br J Haematol 86: 427–30

Serakinci N, Guldberg P, Burns JS, Abdallah B, Schrodder H, Jensen T, Kassem M (2004) Adult human mesenchymal stem cell as a target for neoplastic transformation. Oncogene 23: 5095–98

Shepherd BE, Guttorp P, Lansdorp PM, Abkowitz JL (2004) Estimating human hematopoietic stem cell kinetics using granulocyte telomere lengths. Exp Hematol 32: 1040–50

Shi S, Gronthos S, Chen S, Reddi A, Counter CM, Robey PG, Wang CY (2002) Bone formation by human postnatal bone marrow stromal stem cells is enhanced by telomerase expression. Nat Biotechnol 20: 587–91

Siminovitch L, Till JE, McCulloch EA (1964) Decline in colony-forming ability of marrow cells subjected to serial transplantation into irradiated mice. J Cell Physiol 64: 23–31

Simonsen JL, Rosada C, Serakinci N, Justesen J, Stenderup K, Rattan SI, Jensen TG, Kassem M (2002) Telomerase expression extends the proliferative life-span and maintains the osteogenic potential of human bone marrow stromal cells. Nat Biotechnol 20: 592–96

Slagboom PE, Droog S, Boomsma DI (1994) Genetic determination of telomere size in humans: a twin study of three age groups. Am J Hum Genet 55: 876–82
Son NH, Joyce B, Hieatt A, Chrest FJ, Yanovski J, Weng NP (2003) Stable telomere length and telomerase expression from naive to memory B-lymphocyte differentiation. Mech Ageing Dev 124: 427–32
Tanavde VM, Malehorn MT, Lumkul R, Gao Z, Wingard J, Garrett ES, Civin CI (2002) Human stem-progenitor cells from neonatal cord blood have greater hematopoietic expansion capacity than those from mobilized adult blood. Exp Hematol 30: 816–23
Terskikh AV, Miyamoto T, Chang C, Diatchenko L, Weissman IL (2003) Gene expression analysis of purified hematopoietic stem cells and committed progenitors. Blood 102: 94–101
Thomson JA, Itskovitz-Eldor J, Shapiro SS, Waknitz MA, Swiergiel JJ, Marshall VS, Jones JM (1998) Embryonic stem cell lines derived from human blastocysts. Science 282: 1145–47
Thornley I, Sutherland R, Wynn R, Nayar R, Sung L, Corpus G, Kiss T, Lipton J, Doyle J, Saunders F, Kamel-Reid S, Freedman M, Messner H (2002) Early hematopoietic reconstitution after clinical stem cell transplantation: evidence for stochastic stem cell behavior and limited acceleration in telomere loss. Blood 99: 2387–96
Traycoff CM, Kosak ST, Grigsby S, Srour EF (1995) Evaluation of ex vivo expansion potential of cord blood and bone marrow hematopoietic progenitor cells using cell tracking and limiting dilution analysis. Blood 85: 2059–68
Van Ziffle JA, Baerlocher GM, Lansdorp PM (2003) Telomere length in subpopulations of human hematopoietic cells. Stem Cells 21: 654–60
Vaziri H, Benchimol S (1998) Reconstitution of telomerase activity in normal human cells leads to elongation of telomeres and extended replicative life span. Curr Biol 8: 279–82
Vaziri H, Dragowska W, Allsopp RC, Thomas TE, Harley CB, Lansdorp PM (1994) Evidence for a mitotic clock in human hematopoietic stem cells: loss of telomeric DNA with age. Proc Natl Acad Sci USA 91: 9857–60
Vaziri H, Schachter F, Uchida I, Wei L, Zhu X, Effros R, Cohen D, Harley CB (1993) Loss of telomeric DNA during aging of normal and trisomy 21 human lymphocytes. Am J Hum Genet 52: 661–67
Verfaillie CM, Pera MF, Lansdorp PM (2002) Stem cells: hype and reality. Hematology Am Soc Hematol Educ Program: 369–91
von Zglinicki T, Saretzki G, Docke W, Lotze C (1995) Mild hyperoxia shortens telomeres and inhibits proliferation of fibroblasts: a model for senescence? Exp Cell Res 220: 186–93
Vulliamy T, Dokal I (2006) Dyskeratosis congenita. Semin Hematol 43: 157–66
Vulliamy T, Marrone A, Dokal I, Mason PJ (2002) Association between aplastic anaemia and mutations in telomerase RNA. Lancet 359: 2168–70
Vulliamy T, Marrone A, Goldman F, Dearlove A, Bessler M, Mason PJ, Dokal I (2001) The RNA component of telomerase is mutated in autosomal dominant dyskeratosis congenita. Nature 413: 432–35.
Vulliamy T, Marrone A, Szydlo R, Walne A, Mason PJ, Dokal I (2004) Disease anticipation is associated with progressive telomere shortening in families with dyskeratosis congenita due to mutations in TERC. Nat Genet 36: 447–49
Wang JC, Warner JK, Erdmann N, Lansdorp PM, Harrington L, Dick JE (2005) Dissociation of telomerase activity and telomere length maintenance in primitive human hematopoietic cells. Proc Natl Acad Sci USA 102: 14398–403
Watson JD (1972) Origin of concatemeric T7 DNA. Nat New Biol 239: 197–201
Weng NP, Hathcock KS, Hodes RJ (1998) Regulation of telomere length and telomerase in T and B cells: a mechanism for maintaining replicative potential. Immunity 9: 151–57
Weng NP, Levine BL, June CH, Hodes RJ (1996) Regulated expression of telomerase activity in human T lymphocyte development and activation. J Exp Med 183: 2471–79
Wright WE, Shay JW (2002) Historical claims and current interpretations of replicative aging. Nat Biotechnol 20: 682–88
Wynn RF, Cross MA, Hatton C, Will AM, Lashford LS, Dexter TM, Testa NG (1998) Accelerated telomere shortening in young recipients of allogeneic bone- marrow transplants [see comments]. Lancet 351: 178–81

Yamada O, Akiyama M, Kawauchi K, Adachi T, Yamada H, Kanda N, Aikawa E (2003) Overexpression of telomerase confers a survival advantage through suppression of TRF1 gene expression while maintaining differentiation characteristics in K562 cells. Cell Transplant 12: 365–77

Yamaguchi H, Calado RT, Ly H, Kajigaya S, Baerlocher GM, Chanock SJ, Lansdorp PM, Young NS (2005) Mutations in TERT, the gene for telomerase reverse transcriptase, in aplastic anemia. N Engl J Med 352: 1413–24

Yao CL, Feng YH, Lin XZ, Chu IM, Hsieh TB, Hwang SM (2006) Characterization of serum-free ex vivo-expanded hematopoietic stem cells derived from human umbilical cord blood CD133(+) cells. Stem Cells Dev 15: 70–78

Yasumoto S, Kunimura C, Kikuchi K, Tahara H, Ohji H, Yamamoto H, Ide T, Utakoji T (1996) Telomerase activity in normal human epithelial cells. Oncogene 13: 433–9

Yui J, Chiu CP, Lansdorp PM (1998) Telomerase activity in candidate stem cells from fetal liver and adult bone marrow. Blood 91: 3255–62

Zhang R, Poustovoitov MV, Ye X, Santos HA, Chen W, Daganzo SM, Erzberger JP, Serebriiskii IG, Canutescu AA, Dunbrack RL, Pehrson JR, Berger JM, Kaufman PD, Adams PD (2005) Formation of MacroH2A-containing senescence-associated heterochromatin foci and senescence driven by ASF1a and HIRA. Dev Cell 8: 19–30

Zhang X, Mar V, Zhou W, Harrington L, Robinson MO (1999) Telomere shortening and apoptosis in telomerase-inhibited human tumor cells. Genes Dev 13: 2388–99

Zhao JQ, Glasspool RM, Hoare SF, Bilsland A, Szatmari I, Keith WN (2000) Activation of telomerase rna gene promoter activity by NF-Y, Sp1, and the retinoblastoma protein and repression by Sp3. Neoplasia 2: 531–39

Zhu J, Zhang Y, Joe GJ, Pompetti R, Emerson SG (2005) NF-Ya activates multiple hematopoietic stem cell (HSC) regulatory genes and promotes HSC self-renewal. Proc Natl Acad Sci USA 102: 11728–33

Zhu X, Kumar R, Mandal M, Sharma N, Sharma HW, Dhingra U, Sokoloski JA, Hsiao R, Narayanan R (1996) Cell cycle-dependent modulation of telomerase activity in tumor cells. Proc Natl Acad Sci USA 93: 6091–95

Zimmermann S, Glaser S, Ketteler R, Waller CF, Klingmüller U, Martens UM (2004) Effects of telomerase modulation in human hematopoietic progenitor cells. Stem Cells 22: 741–49

Zimmermann S, Martens, U.M. (2007) Telomeres and telomerase as targets for cancer therapy. Cell Mol Lif Sci [Epub ahead of print]

Zimmermann S, Voss M, Kaiser S, Kapp U, Waller CF, Martens UM (2003) Lack of telomerase activity in human mesenchymal stem cells. Leukemia 17: 1146–49

Chapter 8
Telomere Shortening Induces Cell Intrinsic Checkpoints and Environmental Alterations Limiting Adult Stem Cell Function

Luis Guachalla Gutierrez and Zhenyu Ju(✉)

Abstract A decline in adult stem cell function is one of the features of ageing, which is associated with impaired tissue maintenance and reduced regenerative capacity. Telomere shortening occurs in the vast majority of human tissues during ageing and is also observed in ageing stem cell compartments. Studies of somatic cells have shown that telomere dysfunction induces DNA damage checkpoints, which lead to cell cycle arrest (senescence) or apoptosis. Whether similar checkpoint responses occur in stem cells in response to telomere shortening and whether this contributes to impaired organ maintenance during ageing remains to be investigated. In addition to the activation of cell intrinsic checkpoints, telomere shortening can also induce cell extrinsic, environmental alterations that limit stem cell function. Understanding of adult stem cell ageing and could ultimately lead to the identification of targets for therapies aiming to improve organ maintenance during ageing. In this chapter we summarize our current knowledge on checkpoint responses and environmental alteration induced by telomere dysfunction.

1 Telomere Shortening and Replicative Ageing

Primary human cells have a limited replicative capacity (Hayflick and Moorhead 1961), correlating with the telomere reserve of the cells (Harley et al. 1990). Studies on overexpression of telomerase have proven that telomere shortening is the cause of replicative senescence, which limits proliferation of human cells to a finite number of cell doublings (Bodnar et al. 1998; see Allsopp, this volume).

Telomeres are nucleoprotein complexes, which are composed of tandem double-strand DNA repeats, a 3′ single-strand overhang and associated telomere binding proteins (Blackburn 1991; see Dillin and Karlseder and see Blasco, this volume). The main function of telomeres is to cap the ends of chromosomes and protect them

Zhenyu Ju
Department of Gastroenterology, Hepatology and Endocrinology, Medical School Hannover, 30625 Hannover, Germany.
e-mail: ju.Zhenyu@mh-hannover.de

Institute of Laboratory Animal Science, Chinese Academy of Medical Sciences (CAMS), Beijing, China

K.L. Rudolph (ed.), *Telomeres and Telomerase in Ageing, Disease, and Cancer*.

from end-to-end fusions, degradation, and genetic instability (Blackburn 2001; see Rudolph, this volume). To counteract telomere shortening that occurs as a consequence of cell division, the enzyme telomerase is required. Telomerase consists of two essential components: (1) an RNA component (*Terc*) and (2) a reverse transcriptase (*Tert*). In humans, telomerase expression is regulated very restrictively. It is active during embryogenesis, but its expression is suppressed postnatally in most somatic tissues. Telomerase activity only remains active in germ cells and some adult stem cell compartments, for example, hematopoietic stem cells and intestinal basal crypts (Broccoli et al. 1995; see Artandi, this volume).

In telomerase-negative cells, telomeres shorten with each round of cell division, which is due to (1) the end replication problem of DNA polymerase (for review see Shay and Wright, 2000a), (2) postreplication processing of telomere ends (Sfeir et al. 2005), and (3) additional factors, including reactive oxygen species (von Zglinicki 2002). When telomeres reach a critically short length they lose capping function and the chromosomal ends become unprotected. Upon telomere uncapping, the open chromosomal ends activate DNA damage responses, inducing senescence or apoptosis (see Allsopp, this volume). In contrast to the growing understanding of these checkpoint responses in somatic cells, little is known about checkpoint responses downstream of telomere dysfunction in stem cells.

A number of studies have shown that telomere shortening occurs in almost all tissues during human ageing (for review Djojosubroto et al. 2003). Telomere shortening also affects stem cell compartment, such as $CD34^+$ hematopoietic stem cells (see Zimmermann and Martens, this volume). In addition, accelerated telomere shortening is seen in chronic diseases such as atherosclerosis and associated cardiovascular disease (Fuster and Andres 2006), liver cirrhosis (Wiemann et al. 2002), and Alzheimer's disease (Panossian et al. 2003). In people over 60 years of age, a correlation between telomere length in peripheral blood cells and increased mortality exists, suggesting that telomere length could represent a prognostic marker of human ageing and age-related diseases (Cawthon et al. 2003). It is difficult to use a single measurement of telomere length for prognostic purposes given the variability of telomere length between individuals (Rando 2006). However, it has been shown that the shortest telomere rather than the mean telomere length is important for organismal ageing (Hemann et al. 2001b). Moreover, it is known that a sentinel of dysfunctional telomeres in a cell is sufficient to induce senescence (Zou et al. 2004). Thus it is conceivable that telomere dysfunction could be a better marker than average telomere length for cellular and organismal ageing, a hypothesis that needs to be explored.

There is also evidence that telomere shortening plays a causal role in human ageing and disease. Mutations in the enzyme telomerase have been linked to impaired telomere maintenance and human disease. The autosomal-dominant form of dyskeratosis congenita (DC) is induced by a mutation in the RNA component of telomerase (*TERC*) gene (Vulliamy et al. 2001). These patients show accelerated telomere shortening and reduced life span (Vulliamy et al. 2004). A different set of mutations that affect hTERT has been identified in a subpopulation of patients with idiopathic pulmonary fibrosis (Armanios et al. 2007, Tsakiri et al. 2007). Telomerase gene mutations affecting *TERT* and *TERC* have also been found in different forms of anemia other than DKC (Vulliamy et al. 2002, Yamaguchi et al. 2005). In most of these diseases, telomerase gene mutations usually affect only one allele of *TERT* or *TERC*. These data indicate that telomere length in humans is limited, since a gene dose reduction in telomerase can have deleterious effects on organ maintenance and life span.

The exclusive expression of telomerase in stem cell compartments indicates that telomerase expression is a prerequisite for maintaining stem cell function during ageing. Nevertheless, telomeres shorten in human stem cells during "normal aging" without mutations in the telomerase enzyme (see Zimmermann and Martens, this volume). Therefore, we can assume that telomere shortening can limit adult stem cell function also during normal ageing.

2 Telomere Shortening and Stem Cell Ageing

A gradual decline in regeneration and organ homeostasis is an important aspect of human ageing, which is associated with impaired function and maintenance of stem cells (Schlessinger and Van Zant 2001; see Waterstrat et al., this volume). The extent to which the ageing of stem cells determines impaired maintenance of organs and tissues during ageing remains to be defined. It is possible that stem cell function is a major determinant of declining organ homeostasis and repair capacity, specifically in tissues with high rates of cell turnover. In line with this assumption, genetic experiments on telomerase knockout mice have shown that telomere dysfunction severely affects the maintenance of high-proliferative organs (Lee et al. 1998, Rudolph et al. 1999, Herrera et al. 1999, Choudhury et al. 2007). Similarly, DKC patients with decreased telomerase function die from bone marrow failure, which points again to high-turnover organs as being most sensitive to telomere shortening (see Du et al., this volume). However, there is also evidence that telomere shortening can negatively influence postmitotic organs. Studies on human ageing have shown correlations between telomere shortening and Alzheimer disease (Panossian et al. 2003) and heart disease (Leri et al. 2001, Leri et al. 2003). Moreover, telomerase gene mutation is the underlying cause in some cases of idiopathic lung fibrosis (see above). There is evidence that adult stem cells reside in all organs, including heart, brain, muscle, and lung. It is possible that the negative influence of telomere shortening on organ function (both in low and high turnover organs) is caused by an inhibition of adult stem cells in response to telomere dysfunction. In agreement with this hypothesis, a reduction in adult neurogenesis occurs in telomerase knockout mice with shortened telomeres (Ferron et al. 2004).

Stem cells are functionally defined by their ability to differentiate and self-renew. Over the life span, adult stem cells have the ability to regenerate the tissue from which they are derived. However, accumulating evidence suggests that stem cells do age (Van Zant and Liang 2003; see Waterstrat et al., this volume). Serial transplantation experiments show that the self-renewal capacity of hematopoietic stem cells (HSCs) is finite (Harrison and Astle 1982).

Telomerase is active in some human stem cell compartments, such as HSCs and intestinal basal crypts which represent the stem and progenitor cell compartment of the gut. Telomerase activity in stem cells seems to be important in maintaining their self-renewal ability (Morrison et al. 1996, Allsopp and Weissman 2002).

In humans, telomere shortening occurs in HSCs during ageing (Vaziri et al. 1994), which suggests that the low level of telomerase activity is not sufficient to maintain telomere length during ageing. In addition, accelerated telomere shortening occurs in response to replicative stress after bone marrow transplantation (Engelhardt et al. 1997, Wynn et al. 1998). Similarly, mouse experiments have also revealed an accelerated telomere shortening in HSCs in response to serial transplantation (Allsopp et al. 2001). HSCs from first-generation telomerase-deficient mice show normal repopulating capacity after transplantation into lethally irradiated recipients. However, these G1 *Terc*$^{-/-}$ HSCs are exhausted after only two rounds of serial transplantation, while wild-type HSCs can be serially transplanted four times (Allsopp et al. 2003a). A possible explanation for these findings is that telomerase is required to slow down telomere shortening in order to maintain the replicative life span of HSCs. Alternatively, telomerase deletion could directly impair the self-renewal of HSCs, leading to a depletion of HSCs under stress conditions.

Research work on telomerase transgenic mice has revealed experimental evidence for the telomere length-independent effects of *TERT* expression on the function and maintenance of stem cells in the hair bulge (see Artandi, this volume). However, over-expression of telomerase does not increase the proliferative capacity of mouse HSCs in serial transplantation experiments, although it stabilized telomere length in HSCs (Allsopp et al. 2003b). Together, these data suggest that telomere-ength-independent barriers can limit the self-renewing capacity of HSCs during serial transplantation. The molecular basis of these checkpoints remains to be determined and could involve the accumulation of reactive oxygen species (Ito et al. 2006; see Passos et al., this volume) or p16 (see Sharpless, this volume), as well as the induction of stress-induced senescence, among other possibilities. In summary, it is likely that telomere-dependent and telomere-independent mechanisms limit stem cell function during ageing. In the following sections we focus on telomere-dependent mechanisms of stem cell ageing.

3 Cell Intrinsic Checkpoints in Response to Dysfunctional Telomeres

The generation of telomerase knockout mice (*Terc*$^{-/-}$) provides a unique experimental model to study the in vivo consequences of telomere dysfunction (Blasco et al. 1997). Late-generation *Terc*$^{-/-}$ mice possess several premature ageing phenotypes that predominantly affect high turnover organs (Chang, this volume), indicating that stem cell function is compromised in these mice. Indeed, experimental data show impaired maintenance and function of hematopoietic, intestinal, and neuronal stem cells in ageing *Terc*$^{-/-}$ mice (Rudolph et al. 1999, Samper et al. 2002, Ferron et al. 2004, Choudhury 2007). It is an open question in the field of stem cell ageing what molecular checkpoints limit the function of stem cells in response to telomere dysfunction and other forms of DNA damage accumulation during ageing. There is experimental evidence that DNA damage responses are different in embryonic stem cells as compared to somatic cells (Hong and Stambrook 2004). It is conceivable that different pathways control DNA damage responses in adult stem cells compared to somatic cells. Thus telomere dysfunction and DNA damage accumulation could have different effects on stem cells compared to somatic cells.

The most upstream response to telomere uncapping in mammalian cells is the formation of DNA-damage foci at dysfunctional telomeres (d'Adda di Fagagna et al. 2003). These DNA damage foci activate ATM/ATR signaling (d'Adda di Fagagna et al. 2003), p53 (Chin et al. 1999), and its downstream target p21, which initiates a G1 cell cycle arrest known as senescence (Herbig et al. 2004). In yeast cells it has been shown that the induction of DNA damage response is exonuclease 1 (Exo1) dependent (Maringele and Lydall 2002). A current model indicates that Exo1 processes dysfunctional telomeres leading to the formation of single-stranded DNA (ssDNA) at dysfunctional telomeres and subsequent generation of DNA damage signals. Whether this model applies to mammalian cells is currently unknown, but there is evidence for ssDNA formation in response to DNA damage in mammalian cells (Zou and Elledge 2003). Considering that the activation of the DNA damage pathways in response to DNA damage accumulation and telomere dysfunction is a causal factor of ageing, it is of utmost importance to delineate these checkpoint responses in vivo and the effect of these checkpoints on stem cell function. The late-generation *Terc*$^{-/-}$ mice provide a unique tool to analyze these questions in vivo. In the following section we summarize experimental data on checkpoint response in telomere dysfunctional mice (Fig. 8.1).

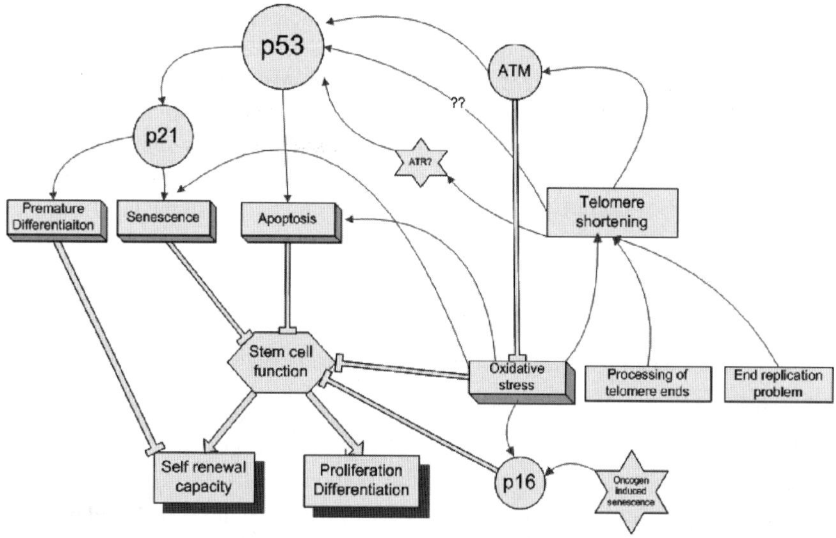

Fig. 8.1 Telomere shortening induces activation of cell cycle checkpoint proteins such as *Atm*, p53, and p21, which probably affects stem cell function by also triggering apoptosis and/or senescence. Although the role of *Atm* in activating p53 is well documented, additional *Atm*-independency p53-activation pathways might be operational in stem cells. *Atm* also functions in stem cell maintenance by regulating reactive oxygen species and p16 levels. Differentiation signaling also impairs stem cell function which can be rescued by deletion of p21 in dysfunctional telomere mice

3.1 *Atm*

The ataxia telangiectasia mutated gene (*Atm*) encodes for a kinase that phosphorylates a number of substrates in response to DNA damage. The same was found to be mutated in the rare autosomal-recessive disease ataxia telangiectasia (AT), a syndrome characterized by accelerated telomere loss, genomic instability, neurodegeneration, immunodeficiency, predisposition to malignancy, clinical radiosensitivity, and incomplete sexual maturation (Shiloh 2003). The phenotype observed in these patients was in part recapitulated in a number of knockout mice generated for *Atm* (Barlow et al. 1996, Elson et al. 1996, Xu et al. 1996).

Atm itself has a role in maintaining length and function of telomere in yeast and mammalian cells (Vaziri et al. 1997, Hande et al 2001, Greenwell et al 1995, Naito 1998). In addition, various studies have shown that *Atm* contributes to DNA damage signals induced by telomere dysfunction (d'Adda di Fagagna 2003). *Atm* deficiency in *Terc*$^{-/-}$ mice leads to an acceleration of telomere shortening associated with a general proliferation defect on stem and progenitor cells levels also affecting neuronal stem cells (Wong et al. 2003). The life span of double mutant mice is significantly shorter than that of late-generation *Terc*$^{-/-}$ mice. However, there was substantial elimination of T cell lymphoma associated with *Atm* deficiency in Terc$^{-/-}$ mice. Reduced life span of *Terc*$^{-/-}$, *Atm*$^{-/-}$ mice correlates with an aggravation of defects in organ maintenance occurring in *Terc*$^{-/-}$ knockout mice. Together, these results indicate that *Atm*-independent pathways can mediate the activation of DNA damage signals in telomere dysfunctional mice. The aggravation of organ phenotypes in *Terc*$^{-/-}$, *Atm*$^{-/-}$ double knockout mice is presumably due to the role of *Atm* in maintaining telomere length and function (see above), adding up to the level of telomere dysfunction in *Terc*$^{-/-}$ mice. In addition, *Atm* appears to play a role in detoxification of reactive oxygen species (ROS). In *Terc*$^{+/+}$ mice, *Atm* deletion induces bone marrow failure characterized by an impaired repopulating capacity of *Atm*$^{-/-}$ HSC and elevated levels of ROS and p16^{Ink4a} in the HSC compartment (Ito 2004). The loss of HSC function in *Atm*$^{-/-}$ mice could not be rescued by overexpression of telomerase (Ito et al. 2006), indicating that telomere-independent functions of *Atm* are required for the maintenance of stem cell function.

3.2 *p53*

p53 is a major tumor suppressor and a key component of DNA damage signaling that induces either cell cycle arrest or apoptosis. p53 also has a significant role in the induction of checkpoints in response to telomere shortening (Shay et al. 1991, Hara et al. 1991, Chin et al. 1999). According to the M1/M2 model of cell mortality, telomere dysfunction induces two checkpoints (mortality stages) limiting the proliferation of primary human cells (Wright and Shay 1992): Low levels of telomere dysfunction induce cell cycle arrest and replicative senescence; high levels of telomere dysfunction induce apoptosis. It has been shown that inhibition of p53 by viral oncogenes or dominant negative mutations can extend the life span of human cells by inhibiting the induction of senescence (Wright and Shay 1992, Bond et al. 1994). There is evidence that p53-independent pathways can limit the proliferation of cells in response to telomere dysfunction by inducing p16 (Bond et al. 2004; see Sharpless, this volume). Abrogation of both Rb- and p53-checkpoint function prevents the induction of senescence cell cycle arrest in response to telomere dysfunction (Wright and Shay 1992). However, cell proliferation beyond the M1 checkpoint results in further telomere shortening, eventually inducing the second mortality stage, M2. M2 is characterized by massive telomere shortening, chromosomal instability, and cell death (see Allsopp; see Rudolph, this volume). In agreement with this cell culture model, low levels of telomere dysfunction in mouse liver induced p53-dependent senescence, whereas high levels induced p53-independent apoptosis (Lechel et al. 2005).

Studies in telomerase-deficient mice have shown cell type specific differences in response to telomere dysfunction (Lee et al. 1998). Late-generation *Terc*$^{-/-}$ mice show an impairment in organ homeostasis correlating with elevated rates of apoptosis in thymic lymphocytes and male germ line (Lee et al. 1998, Hemann et al. 2001a). In contrast, cell cycle arrest was the major checkpoint limiting liver regenerating in response to telomere shortening (Satyanarayana et al. 2003). In the intestine, an impairment in epithelia maintenance correlated with both cell cycle arrest and apoptosis in the basal crypts of intestinal epithelium, which carries the stem and progenitor cells (Choudhury et al. 2007).

The in vivo role of p53 in response to telomere shortening has been studied in mice doubly mutant for *Terc* and p53 (Chin et al. 1999). Deletion of p53 rescued germ cell apoptosis and testis atrophy in late-generation *Terc*$^{-/-}$ mice (Chin et al. 1999). This resulted in increased fertility as observed in the propagation of telomere dysfunctional mice until the eighth generation as compared to the mice with intact p53, who became infertile in the sixth generation. Nevertheless, *Terc*$^{-/-}$germ line atrophy and infertility reappeared in G7/8 mTerc -/-, p53 -/- mice, indicating that p53-independent checkpoints can limit germ cell maintenance in response to severe telomere dysfunction (Chin et al. 1999).

Deletion of p53 did not result in improved survival of telomere dysfunctional mice because of early tumor development in these mice (Artandi et al. 2000). Tumors in doubly mutant mice were characterized by high rates of chromosomal instability. Specifically *Terc*$^{-/-}$ p53$^{+/-}$ mice exhibited a shift in the tumor spectrum toward a higher incidence of epithelial cancers (Artandi et al. 2000). These indicated that telomere dysfunction can cooperate with loss of p53 to induce chromosomal instability and cancer initiation. The tumor spectrum in *Terc*$^{-/-}$, *p53*$^{+/-}$ mice mimicked the high incidence of epithelial cancer in ageing humans, indicating that similar mechanisms might be involved in tumor initiation during human ageing (see Rudolph, this volume). The high incidence of cancer and early death of *Terc*$^{-/-}$, *p53*$^{-/-}$ mice prevented study of the influence of p53 on organ maintenance and stem cell function in response to telomere shortening and ageing. There is experimental evidence that overactivation of p53 induces premature ageing in mice (Donehower 2002). Mice expressing a mutant, overactive form of p53 display premature ageing phenotypes such as reduced longevity, osteoporosis, generalized organ atrophy, and diminished stress tolerance (Tyner et al. 2002). Moreover, it has been shown that hypermorphic levels of p53 in these mice affect hematopoietic stem cell (HSC) number, proliferation, and _ functionality with age, whereas deletion of one copy of p53 increased HSC proliferation during mouse ageing (Dumble et al. 2007). In contrast, telomere dysfunctional mice carrying an extra copy of wild-type p53 gene under its natural promoter (total three copies of p53) did not show an increase in tissue atrophy or a shortening in life span (Garcia-Cao et al. 2006). These results were surprising since telomere dysfunctional mice carrying an extra copy of p53 are expected to have a 1.5-fold increase in p53 gene dosage when compared to telomere dysfunctional mice with only two copies of p53. Notably, the in vivo load of telomere-derived chromosomal damage was significantly decreased in super-p53/telomerase-null mice compared with normal-p53/telomerase-null mice, indicating that the removal of damaged cells may have prevented the acceleration of ageing phenotypes in these mice.

Together these data suggest that although p53 per se plays a role in stem cell function and organ maintenance, telomere-driven ageing can also act through p53-independent pathways. To this end, it remains an open question to delineate p53-dependent as well as p53- independent pathways controlling adult stem cell ageing in the context of dysfunctional telomeres and ageing.

3.3 *p21*

Cdkn1a (p21) represents one of the first identified downstream targets of p53 mediating p53-dependent cell cycle arrest (El-Diery et al. 1993). In primary human fibroblasts p21 induces cell cycle arrest at the senescence stage in response to telomere dysfunction, whereas p21 deletion leads to an increased life span of these cells (Brown et al. 1997). In contrast to the well-established role of p21 in cellular senescence in cell culture, its role for in vivo ageing remains unclear. An increase in p21 expression occurs during human and rodent ageing in muscle (Edwards et al. 2007, Welle et al. 2004) and muscle stem cells (Machida and Booth 2004). p21 expression has also been observed in some age-associated diseases such as atherosclerosis (Matthews et al. 2006) and tubular fibrosis of the kidney (Ding et al. 2001). In contrast, heart myocytes of ageing mouse showed a decrease in p21 expression during ageing (Torella et al. 2004). Similarly, p21 expression is decreased in aging rat liver (Kitano et al. 1996). Based on the senescence marker SA-βGal some studies have described an accumulation of senescent cells in aged skin of baboons (Herbig et al. 2006) as well as in human skin (Dimri et al. 1995). SA-βGal-positive cells were also detected in atherosclerotic plaques (for review see Minamino and Komuro 2007) and liver cirrhosis (Wiemann et al. 2002). Together, these data indicate that p21 expression and senescence could contribute to tissue pathology during ageing and chronic disease in some tissues.

Studies in telomerase-deficient mice revealed a functional role of p21 on in vivo aging in the context of telomere dysfunction. p21 deletion had no influence on ageing and life span of *Terc*$^{+/-}$ mice with functional telomeres but significantly elongated the life span of *Terc*$^{-/-}$ mice with dysfunctional telomeres (Choudhury et al. 2007). Increased life span of the *Terc*$^{-/-}$ mice correlated with improved organ maintenance specifically observed in the intestinal epithelia and lymphocyte compartment. Notably, p21 deletion rescued the depletion of stem and progenitor cells in the gut and the hematopoietic system of ageing *Terc*$^{-/-}$ mice, indicating that p21-dependent checkpoints limit stem cell self-renewal in these organs in response to telomere dysfunction. Moreover, p21 deletion improved the function of HSCs from *Terc*$^{-/-}$ mice when tested in transplantation experiments.

Deletion of p21 rescued cell cycle arrest in intestinal stem and progenitor cells of telomere dysfunctional mice but did not prevent the induction of apoptosis in this compartment. Similarly, p21 deletion did not rescue germ cell apoptosis and testicular atrophy in *Terc*$^{-/-}$ mice. In the hematopoietic system, p21 deletion prevented the depletion of the HSC compartment (lineage negative cells, Sca1^{+}, c-Kit^{+} =LSK cells). This rescue was not associated with increased proliferation or suppression of apoptosis. Instead, there was evidence that p21 deletion prevented the induction of stem cell differentiation reducing the HSC pool size in *Terc*$^{-/-}$ mice. Together, these results provided the first experimental evidence that inhibition of DNA-damage checkpoints can improve the maintenance and function of adult stem cells and organismal survival in the context of telomere dysfunction. Notably, SA-β-Gal staining did not reveal a strong accumulation of senescent cells in ageing *Terc*$^{-/-}$, p21$^{+/+}$ compared to *Terc*$^{-/-}$, p21$^{-/-}$ mice. These data indicate that p21-arrested cells may not persist indefinitely in aged tissues but might be removed by secondary mechanisms possibly involving apoptosis (Choudhury et al. 2007) or immune responses targeting senescent cells (Xue et al. 2007).

In contrast to the findings on p21 deletion in *Terc*$^{-/-}$ mice, deletion of p21 itself can also have negative effects on stem cell maintenance in telomerase wild-type mice with long telomere reserves. It has been reported that deletion of p21 leads to increased rates of HSC turnover (Cheng et al. 2000) and increased neuronal stem cell proliferation after ischemic injury (Qiu et al. 2004) in certain mouse strains. Increased rates of stem cell turnover resulted in depletion of HSCs when the animals were exposed to chemotherapy or under replication stress by repeated bone marrow transplantation (Cheng et al. 2000). Together, these findings suggest that p21 has a dual role in stem cell aging. On one hand, p21 appears to be important to maintain stem cell quiescence thus preventing stem cell exhaustion during ageing and in response to acute stress. On the other hand, p21 upregulation becomes disadvantageous in the context of telomere dysfunction in old mice, leading to depletion of stem and progenitor cell compartments by induction of cell cycle arrest or differentiation (for review see Ju et al. 2007a).

Notably, p21 deletion improved stem cell function, organ homeostasis, and life span of telomere-dysfunctional mice without accelerating cancer formation (Choudhury et al. 2007). Similarly, there was no evidence for an increase in genomic instability in bone marrow cells of *Terc*$^{-/-}$, p21$^{-/-}$ mice compared to

Terc$^{-/-}$, p21$^{+/+}$ mice. These data indicate that p21-independent checkpoints can prevent the accumulation of chromosomal instability and the initiation of cancer in telomere-dysfunctional mice (see Rudolph, this volume).

In conclusion, p21 might be considered as a potential therapeutic target to improve regeneration and organ homeostasis in chronic diseases and aging-related disorders associated with telomere dysfunction and reduced survival.

3.4 *Ink4a/Arf*

The *Ink4a/Arf* locus encodes for two gene products (p16 and p19ARF) that are both upregulated in various tissues during mammalian ageing (Krishnamurthy et al. 2004). Further investigations reinforced that p16^{INK4a} induces an age-dependent decline in the regenerative potential of HSCs (Janzen et al. 2006), neural stem cells (Molofsky et al. 2006), and potential pancreatic stem cells (Krishnamurthy et al. 2006). Moreover, p16 upregualtion is a hallmark feature of senescence in primary human fibroblasts (Alcorta et al. 1996). However, deletion of *Ink4a/Arf* did not attenuate cellular responses and tissue phenotypes associated with telomere dysfunction in ageing *Terc*$^{-/-}$ mice (Khoo et al. 2007). A possible explanation is that *Ink4a/ARF* has a negligible role of p53 activation in response to DNA damage. Thus ARF-independent induction of p53 may have negative effects on organ maintenance independent of ARF/p16 gene status.

3.5 *p18, p15, p27, p57*

A variety of cell cycle inhibitors have been implicated in stem cell function during ageing. However, it is momentarily unclear whether these cell cycle inhibitors contribute to stem cell defects induced by telomere dysfunction. Therefore, we do not discuss these cell cycle inhibitor in greater detail here.

4 Telomere Dysfunction Induces Environmental Alterations Limiting Stem Cell Function and Engraftment

In addition to cell intrinsic checkpoints, environmental alterations can limit the function of adult stem cells during ageing. In mouse models the function of stem cells declines in various compartments, including HSCs and muscle stem cells. The decline in adult stem cell function during mouse ageing has been associated with downregulation of notch signaling in muscle satellite cells (Conboy et al. 2003) and a switch in CeBP/alpha signaling in ageing liver (Iakova et al. 2003). A recent landmark study has shown that the connection of blood flow from young and old mice (parabiosis) improves stem cell function in various tissues of ageing mice, correlating with a reversion of some of the age-associated changes in signaling pathways in aged stem cells (Conboy et al. 2005). This study suggested that there are changes in the macroenvironment (systemic milieu) that negatively influence stem cell function. It is possible that an accumulation of "negative factors" occurs during ageing, impairing adult stem cell function. Alternatively, there might be an age-dependent decline in "positive factors" that are necessary for normal stem cell function. In these experiments, parabiosis could have led to an improvement in stem cell function by diluting "negative factors" and/or by increasing the concentration of "positive factors" in the aged organism (Rando 2006).

In addition to alterations in the macroenvironment, there is also evidence for age-associated alterations in the microenvironment (the stem cell niche) that limit stem cell function during ageing (Fig. 8.2). Ageing associated alterations in stem cell niche function appear to affect maintenance and function of melanocyte stem cells (Nishimura et al. 2005), intestinal stem cells (Kim et al. 2005), spermatogonia (Ryu et al. 2006), muscle stem cells (Shefer et al. 2006), neural stem cells (Alvarez-Buylla and Lim 2004) as well as hematopoietic stem cells (Miller and Van Zant 2006; see Waterstrat et al., this volume). To understand the influence of ageing on the stem cell microenvironment it will be important to define the cell types that compose the stem cell niche in different organs. In the hematopoietic system, bone marrow stromal cells can contribute to the formation of stem cell niches. In addition, it has been proposed that there might be different niches composed of either osteoblasts or endothelial cells (Wilson and Trumpp 2006). A major function of the stem cell niche is to control the balance between HSC self-renewal and differentiation as well as the balance between stem cell quiescence and proliferation (Wilson and Trumpp 2006). As a major cell extrinsic factor, cytokines influence stem cell function and the cross talk between stem cells and their surrounding niches (Adams and Scadden 2006). Thus the macroenvironment (see above) could also influence the microenvironment (niche function).

An emerging field in stem cell research is to delineate molecular factors that are the cause of age-associated alterations in the stem cell environment. Studies on *Terc*$^{-/-}$ mice have provided the first experimental evidence that telomere dysfunction can induce alterations in the environment that limit hematopoietic stem cell function (Ju et al. 2007b). One of the hallmark phenotypes of ageing *Terc*$^{-/-}$ mice is a decrease in B-lymphopoiesis in bone marrow accompanied by an increase in myeloid proliferation. Cross-transplantation experiments of HSCs from *Terc*$^{+/+}$ into *Terc*$^{-/-}$ and vice versa revealed that these alterations in HSC function were mainly due to the environment and not caused by HSC intrinsic defects.

In addition to the effects on the function of HSCs, environmental alterations in telomere dysfunctional mice impaired engraftment of transplanted HSCs into stem cell niches. These results could have implications for new approaches in regenerative medicine aiming to use cell transplantation therapies to improve organ functions (Hornsby 2004, Shay and Wright 2000b). The studies on *Terc*$^{-/-}$ mice indicate that alterations of the environment could limit such approaches by limiting the engraftment and function of transplanted cells.

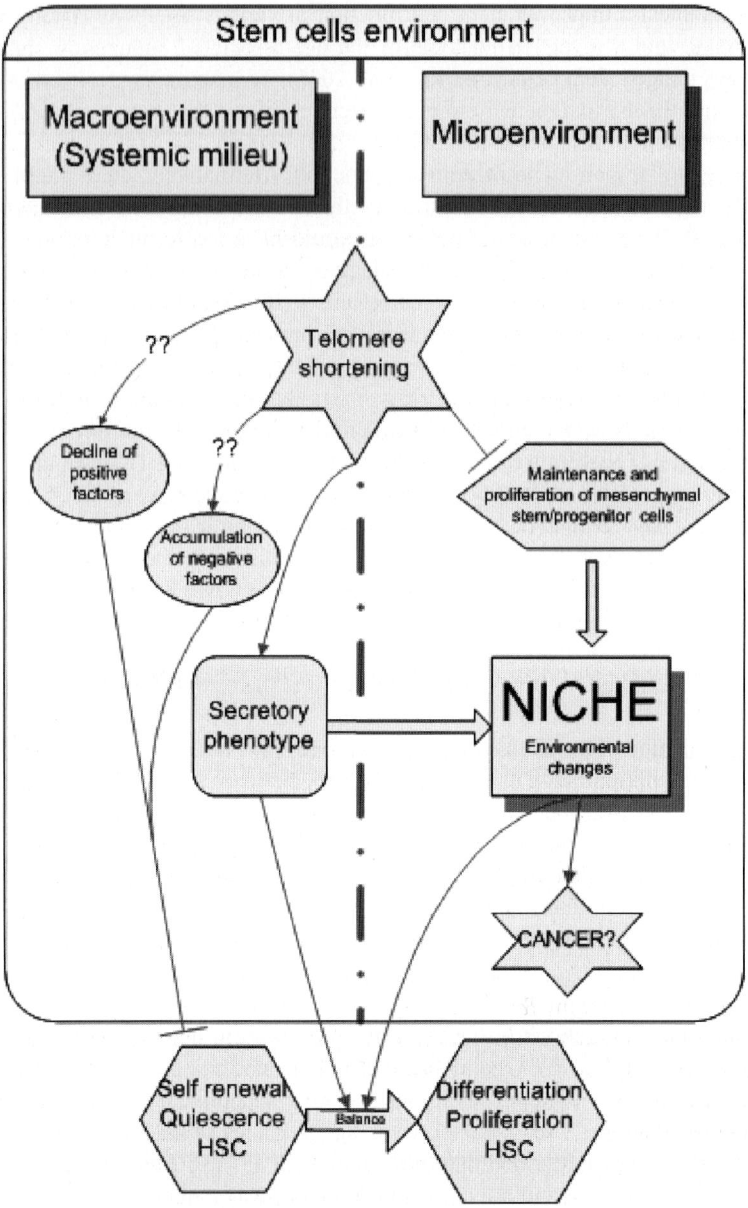

Fig. 8.2 Telomere shortening reduces positive factors, accumulates negative factors, and induces an age-associated secretory phenotype in the macroenvironment; in the microenvironment it compromises the stromal cell compartment and alters cytokine profiles, thus influencing stem cell function in terms of balance between self-renewal and differentiation of HSC. This suggests that changes in the stem cell niche can also be involved in carcinogenesis

Telomere dysfunction & the macroenvironment. It has first been recognized in cell culture experiments that telomere induced senescence induces a secretory phenotype resulting in elevated secretion of pro-inflammatory cytokines and growth factors (Krtolica et al. 2001). In accordance with these data, blood sera of late generation *Terc*$^{-/-}$ show an age-dependent increase in the level of various cytokines and growth factors (Ju et al. 2007b). One of the upregulated factors was G-CSF, which showed an age-dependent increase in *Terc*$^{-/-}$ mice, but not in age-matched *Terc*$^{+/-}$ mice. Bone marrow transplantation experiments revealed that the elevation of G-CSF levels in *Terc*$^{-/-}$ mice was independent of telomere dysfunction in HSCs but dependent on telomere dysfunction in the host environment. Treatment with recombinant human G-CSF mimicked some of the phenotypes of ageing *Terc*$^{-/-}$ mice including the impairment in B-cell development in bone marrow as well as the reduced engraftment of transplanted HSCs. In contrast, application of neutralizing antibodies lowered endogenous G-CSF levels in the plasma of aged *Terc*$^{-/-}$ mice and improved the homing of HSCs into the deep bone marrow stroma of Terc$^{-/-}$ mice (Ju et al. 2007b). Together, these studies demonstrate that telomere dysfunction induces alteration in the macroenvironment that can limit HSC function and engraftment.

4.1 Telomere Dysfunction and the Microenvironment

In telomere-dysfunctional mice, progressive telomere shortening in the bone marrow stromal compartment is associated with decreased proliferative capacity of bone marrow-derived MSCs. In bone marrow of *Terc*$^{-/-}$ mice the total number of stromal cells is decreased. In addition to its effect on the number of stromal cells, telomere dysfunction compromised the capacity of bone marrow stromal cells to maintain functional HSCs. When wild-type HSCs were co-cultured on bone marrow derived stromal cell layers, there was a remarkable decline in HSC function for cells co-cultured on stromal layers derived from *Terc*$^{-/-}$ mice compared to stromal cell layers from *Terc*$^{+/+}$ mice (Ju et al. 2007b).

Together, these studies have shown that telomere dysfunction induces alteration in the micro- and macroenvironments that can limit HSC function and engraftment. It remains to be investigated whether environmental changes in response to telomere dysfunction could contribute to the increased cancer initiation that has been observed in ageing telomere-dysfunctional mice (Rudolph et al. 1999, Rudolph et al. 2001). The senescent cells might stimulate the proliferation of cells that harbor preneoplastic mutations. It is known that the cancer-associated stromal cells facilitate the progression of epithelial cancers by influencing the microenvironment of the malignant phenotype (Olumi et al. 1999, Tlsty and Hein 2001). Moreover, the secretory phenotype of senescent fibroblasts can disrupt the functional and morphological differentiation of epithelial cells, and can contribute to the altered differentiation and proliferation of co-cultured epithelial cells (Parrinello et al. 2005, Campisi, 2005). Indeed, there is mounting evidence that senescent fibroblasts create a local tissue environment that promotes the growth of initiated or preneoplastic epithelial cells both in culture and in vivo (Krtolica et al. 2001, Roninson 2002). Since stem cells are prime targets for carcinogenesis, a possible consequence of cellular senescence may be its impact on stem cells and their niches (Li and Neaves 2006).

5 Outlook

Experimental evidence from studies on animal models supports the concept that stem cell ageing compromises organ homeostasis and regeneration. Recent experimental evidence suggests that aged adult stem cells still possess the ability to participate in tissue maintenance and repair if provided with optimal environmental factors. In addition, there is emerging evidence that telomere shortening induces cell intrinsic checkpoints and environmental alterations that limit stem cell maintenance and function. Therefore, it is of utmost importance to explore the molecular nature of cell intrinsic checkpoints and environmental alterations that limit adult stem cell function during ageing. An understanding of these mechanisms could ultimately lead to the development of new therapies aiming to improve the function of adult stem cells during ageing or the engraftment of transplanted cells in aged tissues.

References

Adams GB, Scadden DT (2006) The hematopoietic stem cell in its place. Nat Immunol 7(4): 333–37

Alcorta DA, Xiong Y, Phelps D, Hannon G, Beach D, Barrett JC (1996) Involvement of the cyclin-dependent kinase inhibitor p16 (INK4a) in replicative senescence of normal human fibroblasts. Proc Natl Acad Sci USA 93(24): 13742–47

Allsopp RC, Cheshier S, Weissman IL (2001) Telomere shortening accompanies increased cell cycle activity during serial transplantation of hematopoietic stem cells. J Exp Med 193(8): 917–24

Allsopp RC, Morin GB, DePinho R, Harley CB, Weissman IL (2003a) Telomerase is required to slow telomere shortening and extend replicative life span of HSCs during serial transplantation. Blood 102(2): 517–20

Allsopp RC, Morin GB, Horner JW, DePinho R, Harley CB, Weissman IL (2003b) Effect of TERT over-expression on the long-term transplantation capacity of hematopoietic stem cells. Nat Med 9(4): 369–71

Allsopp RC, Weissman IL (2002) Replicative senescence of hematopoietic stem cells during serial transplantation: does telomere shortening play a role? Oncogene 21(21): 3270–3

Alvarez-Buylla A, Lim DA (2004) For the long run: maintaining germinal niches in the adult brain. Neuron 41(5): 683–86

Armanios MY, Chen JJ, Cogan JD, Alder JK, Ingersoll RG, Markin C, Lawson WE, Xie M, Vulto I, Phillips JA 3rd, Lansdorp PM, Greider CW, Loyd JE (2007) Telomerase mutations in families with idiopathic pulmonary fibrosis. N Engl J Med. 356(13): 1317–26

Artandi SE, Chang S, Lee SL, Alson S, Gottlieb GJ, Chin L, DePinho RA (2000) Telomere dysfunction promotes non-reciprocal translocations and epithelial cancers in mice. Nature 406(6796): 641–45

Barlow C, Hirotsune S, Paylor R, Liyanage M, Eckhaus M, Collins F, Shiloh Y, Crawley JN, Ried T, Tagle D, Wynshaw-Boris A (1996) Atm-deficient mice: a paradigm of ataxia telangiectasia. Cell 86: 159–71

Blackburn EH (1991) Structure and function of telomeres. Nature 350: 569–73

Blackburn EH (2001) Switching and signaling at the telomere. Cell 106: 661–73

Blasco MA, Lee HW, Hande MP, Samper E, Lansdorp PM, DePinho RA, Greider CW (1997) Telomere shortening and tumor formation by mouse cells lacking telomerase RNA. Cell 91(1): 25–34

Bodnar AG, Ouellette M, Frolkis M, Holt SE, Chiu CP, Morin GB, Harley CB, Shay JW, Lichtsteiner S, Wright WE (1998) Extension of life-span by introduction of telomerase into normal human cells. Science 279: 349–52

Bond J, Jones C, Haughton M, DeMicco C, Kipling D, Wynford-Thomas D (2004) Direct evidence from siRNA-directed "knock down" that p16(INK4a) is required for human fibroblast senescence and for limiting ras-induced epithelial cell proliferation. Exp Cell Res 292(1): 151–56

Bond JA, Wyllie FS, Wynford-Thomas D (1994) Escape from senescence in human diploid fibroblasts induced directly by mutant p53.Oncogene. 9(7): 1885–89

Broccoli D, Young JW, de Lange T (1995) Telomerase activity in normal and malignant hematopoietic cells. Proc Natl Acad Sci USA 92(20): 9082–9086

Brown JP, Wei W, Sedivy JM (1997) Bypass of senescence after disruption of p21CIP1/WAF1 gene in normal diploid human fibroblasts. Science 277(5327): 831–34

Campisi J (2005) Senescent cells, tumor suppression, and organismal aging: good citizens, bad neighbors. Cell 120(4): 513–22

Cawthon RM, Smith KR, O'Brien E, Sivatchenko A, Kerber RA (2003) Association between telomere length in blood and mortality in people aged 60 years or older. Lancet 361: 393–95

Cheng T, Rodrigues N, Shen H, Yang Y, Dombkowski D, Sykes M, Scadden DT (2000) Hematopoietic stem cell quiescence maintained by p21cip1/waf1. Science 287: 1804–08

Chin S, Artand, Q, Shen A, Tam S, Lee G, Gottlieb C, Greider R, DePinho (1999) p53 deficiency rescues the adverse effects of telomere loss and cooperates with telomere dysfunction to accelerate carcinogenesis. Cell 97(4): 527–28

Choudhury AR, Ju Z, Djojosubroto MW, Schienke A, Lechel A, Schaetzlein S, Jiang H, Stepczynska A, Wang C, Buer J, Lee HW, von Zglinicki T, Ganser A, Schirmacher P, Nakauchi H, Rudolph KL (2007). Cdkn1a deletion improves stem cell function and life span of mice with dysfunctional telomeres without accelerating cancer formation. Nat Genet 39: 99–105

Conboy IM, Conboy MJ, Smythe GM, Rando TA (2003) Notch-mediated restoration of regenerative potential to aged muscle. Science 302: 1575–77

Conboy IM, Conboy MJ, Wagers AJ, Girma ER, Weissman IL, Rando TA (2005) Rejuvenation of aged progenitor cells by exposure to a young systemic environment. Nature 433: 760–4

d'Adda di Fagagna F, Reaper PM, Clay-Farrace L, Fiegler H, Carr P, Von Zglinicki T, Saretzki G, Carter NP, Jackson SP (2003) A DNA damage checkpoint response in telomere-initiated senescence. Nature 426: 194–98

Dimri GP, Lee X, Basile G, Acosta M, Scott G, Roskelley C, Medrano EE, Linskens M, Rubelj I, Pereira-Smith O, et al (1995) A biomarker that identifies senescent human cells in culture and in aging skin in vivo. Proc Natl Acad Sci USA 92(20): 9363–67

Ding G, Franki N, Kapasi AA, Reddy K, Gibbons N, Singhal PC (2001) Tubular cell senescence and expression of TGF-beta1 and p21(WAF1/CIP1) in tubulointerstitial fibrosis of aging rats. Exp Mol Pathol 70(1): 43–53

Djojosubroto MW, Choi YS, Lee HW, Rudolph KL (2003) Telomeres and telomerase in aging, regeneration and cancer. Mol Cells 15: 164–75

Donehower LA (2002) Does p53 affect organismal aging? J Cell Physiol 192: 23–33

Dumble M, Moore L, Chambers SM, Geiger H, Van Zant G, Goodell MA, Donehower LA (2007) The impact of altered p53 dosage on hematopoietic stem cell dynamics during aging. Blood 109(4): 1736–42

Edwards MG, Anderson RM, Yuan M, Kendziorski CM, Weindruch R, Prolla TA (2007) Gene expression profiling of aging reveals activation of a p53-mediated transcriptional program. BMC Genomics 8: 80

El-Diery WS, Harper JW, O'Connor PM, Velculescu VE, Canman CE, Jackman J, Pitenpol JA, Burrell M, Hill DE, Wang Y (1993) WAF1, a potential mediator of p53 tumour suppression. Cell 75: 817–25

Elson A, Wang Y, Daugherty CJ, Morton CC, Zhou F, Campos-Torres J. Leder P (1996) Pleiotropic defects in ataxia-telangiectasia protein-deficient mice. Proc Nat Acad Sci 93: 13084–89

Engelhardt M, Kumar R, Albanell J, Pettengell R, Han W, Moore MA (1997) Telomerase regulation, cell cycle, and telomere stability in primitive hematopoietic cells. Blood 90(1): 182–93

Ferron S, Mira H, Franco S, Cano-Jaimez M, Bellmunt E, Ramirez C, Farinas I, Blasco MA (2004) Telomere shortening and chromosomal instability abrogates proliferation of adult but not embryonic neural stem cells. Development 131(16): 4059–70

Fuster JJ, Andres V (2006) Telomere biology and cardiovascular disease. Circ Res 24: 99(11): 1167–80

Garcia-Cao I, Garcia-Cao M, Tomas-Loba A, Martin-Caballero J, Flores JM, Klatt P, Blasco MA, Serrano M (2006) Increased p53 activity does not accelerate telomere-driven ageing. EMBO Res 7(5): 546–52

Greenwell PW, Kronmal SL, Porter SE, Gassbenhuber J, Obermaier B, Petes TD (1995) TEL1, a gene involved in controlling telomere length in S. cerevisiae, is homologous to the human ataxia telangiectasia gene. Cell 82: 823–29

Hande MP, Balajee AS, Tchirkov A, Wynshaw-Boris A, Lansdorp PM (2001) Extra-chromosomal telomeric DNA in cells from Atm–/– mice and patients with ataxia-telangiectasia. Hum Molec Genet 10: 519–28

Hara E, Tsurui H, Shinozaki A, Nakada S, Oda K. (1991) Cooperative effect of antisense-Rb and antisense-p53 oligomers on the extension of life span in human diploid fibroblasts, TIG-1. Biochem Biophys Res Commun 179(1): 528–34

Harley CB, Futcher AB, Greider CW (1990) Telomeres shorten during ageing of human fibroblasts. Nature 345: 458–60

Harrison DE, Astle CM (1982) Loss of stem cell repopulating ability upon transplantation. Effects of donor age, cell number, and transplantation procedure. J Exp Med 156:1767–79

Hayflick L, Moorehead PS (1961). The serial cultivation of human diploid cell strains. Exp Cell Res 25: 585–621

Hemann MT, Rudolph KL, Strong MA, DePinho RA, Chin L, Greider CW (2001a) Telomere dysfunction triggers developmentally regulated germ cell apoptosis. Mol Biol Cell 12: 2023–30

Hemann MT, Strong MA, Hao LY, Greider CW (2001b) The shortest telomere, not average telomere length, is critical for cell viability and chromosome stability. Cell 107(1): 67–77

Herbig U, Jobling WA, Chen B, Chen DJ, Sedivy JM (2004) Telomere shortening triggers senescence of human cells through a pathway involving ATM, p53, and p21CIP1, but not p16INK4a. Molec Cell 14(4): 501–13

Herbig U, Ferreira M, Condel L, Carey D, Sedivy JM (2006) Cellular senescence in aging primates. Science 311(5765): 1257

Herrera E, Samper E, Blasco MA (1999) Telomere shortening in mTR–/– embryos is associated with failure to close the neural tube. EMBO J 18(5): 1172–81

Hong Y, Stambrook PJ (2004) Restoration of an absent G1 arrest and protection from apoptosis in embryonic stem cells after ionizing radiation. Proc Natl Acad Sci USA 101(40): 14443–48

Hornsby PJ (2004) Telomerization of mammalian cells and transplantation of telomerized cells in immunodeficient mice. Methods Mol. Biol. 240: 147–66

Iakova P, Awad SS, Timchenko NA (2003) Aging reduces proliferative capacities of liver by switching pathways of C/EBPalpha growth arrest. Cell 113(4): 495–506

Ito K, Hirao A, Arai F, Matsuoka S, Takubo K, Hamaguchi I, Nomiyama K, Hosokawa K, Sakurada K, Nakagata N, Ikeda Y, Mak, TW, Suda TW (2004) Regulation of oxidative stress by ATM is required for self-renewal of haematopoietic stem cells. Nature 431:997–1002

Ito K, Hirao A, Arai F, Takubo K, Matsuoka S, Miyamoto K, Ohmura M, Naka K, Hosokawa K, Ikeda Y, Suda T (2006) Reactive oxygen species act through p38 MAPK to limit the life span of hematopoietic stem cells Nat Med 12:446–51

Janzen V, Forkert R, Fleming HE, Saito Y, Waring MT, Dombkowski DM, Cheng T, DePinho RA, Sharpless NE, Scadden DT (2006) Stem-cell ageing modified by the cyclin-dependent kinase inhibitor p16(INK4a). Nature 443: 421–26

Ju Z, Choudhury AR, Rudolph KL (2007a) A dual role of p21 in stem cell ageing. Ann NY Acad Sci 1100: 333–44

Ju Z, Jiang H, Jaworski M, Gompf A, Rathinam C, Klein C, Trumpp A, Rudolph KL (2007b) Telomere dysfunction induces environmental defects limiting hematopoietic stem cell function and engraftment. Nat Med 13(6): 742–7

Khoo CM, Carrasco DR, Bosenberg MW, Paik JH, Depinho RA (2007) Ink4a/Arf tumor suppressor does not modulate the degenerative conditions or tumor spectrum of the telomerase-deficient mouse. Proc Natl Acad Sci USA 104(10): 3931–36

Kim JY, Siegmund KD, Tavare S, Shibata D (2005) Age-related human small intestine methylation: evidence for stem cell niches. BMC Med 3: 10

Kitano S, Venable S, Smith JR, Reed TD, Roth GS (1996) Effect of aging on regulation of sdi-1 in rat hepatocytes. Biochem Biophys Res Commun 5; 225(1): 122–27

Krishnamurthy J, Ramsey MR, Ligon KL, Torrice C, Koh A, Bonner-Weir S, Sharpless NE (2006) p16INK4a induces an age-dependent decline in islet regenerative potential. Nature 443: 453–57

Krishnamurthy J, Torrice C, Ramsey MR, Kovalev GI, Al-Regaiey K, Su L, Sharpless NE (2004) Ink4a/Arf expression is a biomarker of aging. J Clin Invest 114(9): 1299–3107

Krtolica A. Parrinello S. Lockett S, Desprez P, Campisi J (2001). Senescent fibroblasts promote epithelial cell growth and tumorigenesis: a link between cancer and aging. Proc Natl Acad Sci USA 98: 12072–77

Lechel A, Satyanarayana A, Ju Z, Plentz RR, Schaetzlein S, Rudolph C, Wilkens L, Wiemann SU, Saretzki G, Malek NP, Manns MP, Buer J, Rudolph KL (2005) The cellular level of telomere dysfunction determines induction of senescence or apoptosis in vivo. EMBO Rep 6(3): 275–81

Lee HW, Blasco MA, Gottlieb GJ, Horner JW, Greider CW, DePinho RA (1998) Essential role of mouse telomerase in highly proliferative organs. Nature 392: 569–74

Leri A, Barlucchi L, Limana F, Deptala A, Darzynkiewicz Z, Hintze TH, Kajstura J, Nadal-Ginard B, Anversa P (2001) Telomerase expression and activity are coupled with myocyte proliferation and preservation of telomeric length in the failing heart. Proc Natl Acad Sci USA 98: 8626–31

Leri A, Franco S, Zacheo A, Barlucchi L, Chimenti S, Limana F, Nadal-Ginard B, Kajstura J, Anversa P, Blasco MA (2003) Ablation of telomerase and telomere loss leads to cardiac dilatation and heart failure associated with p53 upregulation. EMBO J 22: 131–39

Li L, Neaves WB (2006) Normal stem cells and cancer stem cells: the niche matters. Cancer Res 66(9): 4553–57

Machida S, Booth FW (2004) Increased nuclear proteins in muscle satellite cells in aged animals as compared to young growing animals. Exp Gerontol 39(10): 1521–25

Maringele L, Lydall D (2002) EXO1-dependent single-stranded DNA at telomeres activates subsets of DNA damage and spindle checkpoint pathways in budding yeast yku70Delta mutants. Genes Dev 16(15): 1919–33

Matthews C, Gorenne I, Scott S, Figg N, Kirkpatrick P, Ritchie A, Goddard M, Bennett M (2006) Vascular smooth muscle cells undergo telomere based senescence in human atherosclerosis: effects of telomerase and oxidative stress. Cir Res 99:156–64

Miller A, Van Zant G (2006) Advances in hematopoietic stem cell research through mouse genetics. Curr Opin Hematol 13(4):209–15

Minamino T, Komuro I (2007) Vascular cell senescence: contribution to atherosclerosis. Circ Res 100(1): 15–26

Molofsky AV, Slutsky SG, Joseph NM, He S, Pardal R, Krishnamurthy J, Sharpless NE, Morrison SJ (2006) Increasing p16INK4a expression decreases forebrain progenitors and neurogenesis during ageing. Nature 443(7110): 448–52

Morrison SJ, Prowse KR, Ho P, Weissman IL (1996) Telomerase activity in hematopoietic cells is associated with self-renewal potential. Immunity 5(3): 207–16

Naito T, Matsuura A, Ishikawa F (1998) Circular chromosome formation in a fission yeast mutant defective in two ATM homologues. Nature Genet 20: 203–06

Nishimura EK, Granter SR, Fisher DE (2005) Mechanisms of hair graying: incomplete melanocyte stem cell maintenance in the niche. Science 307: 720–24

Olumi AF, Grossfeld GD, Hayward SW, Carroll PR, Tlsty TD, Cunha GR (1999) Carcinoma-associated fibroblasts direct tumor progression of initiated human prostatic epithelium, Cancer Res 59: 5002–11

Panossian LA, Porter VR, Valenzuela HF, Zhu X, Reback E, Masterman D, Cummings JL, Effros RB (2003) Telomere shortening in T cells correlates with Alzheimer's disease status. Neurobiol Aging 24(1): 77–84

Parrinello S, Coppe J, Krtolica A, Campisi J (2005) Stromal-epithelial interactions in aging and cancer: senescent fibroblasts alter epithelial cell differentiation. J Cell Sci 118: 485–96

Qiu J, Takagi Y, Harada J, Rodrigues N, Moskowitz MA, Scadden DT, Cheng T (2004) Regenerative response in ischemic brain restricted by p21cip1/waf1. J Exp Med 199(7): 937–45

Rando TA (2006) Stem cells, ageing and the quest for immortality. Nature 441: 1080–86

Roninson IB (2002) Oncogenic functions of tumour suppressor p21(Waf1/Cip1/Sdi1): association with cell senescence and tumour-promoting activities of stromal fibroblasts. Cancer Lett 179: 1–14

Rudolph KL, Chang S, Lee HW, Blasco M, Gottlieb GJ, Greider C, DePinho RA (1999) Longevity, stress response, and cancer in ageing telomerase-deficient mice. Cell 96: 701–12

Rudolph KL, Millard M, Bosenberg MW, DePinho RA (2001) Telomere dysfunction and evolution of intestinal carcinoma in mice and humans. Nat Genet 28: 155–59

Ryu BY, Orwig KE, Oatley JM, Avarbock MR, Brinster RL (2006) Effects of aging and niche microenvironment on spermatogonial stem cell self-renewal. Stem Cells 24(6): 1505–11

Samper E, Fernandez P, Eguia R, Martin-Rivera L, Bernad A, Blasco MA, Aracil M (2002) Long-term repopulating ability of telomerase-deficient murine hematopoietic stem cells. Blood 99(8): 2767–75

Satyanarayana A, Wiemann SU, Buer J, Lauber J, Dittmar KE, Wustefeld T, Blasco MA, Manns MP, Rudolph KL (2003) Telomere shortening impairs organ regeneration by inhibiting cell cycle re-entry of a subpopulation of cells. EMBO J 22(15): 4003–13

Schlessinger D, Van Zant G (2001) Does functional depletion of stem cells drive aging? Mech Ageing Dev 122: 1537–54

Sfeir AJ, Chai W, Shay JW, Wright WE. Telomere-end processing the terminal nucleotides of human chromosomes. Mol Cell 2005 Apr 1;18(1): 131–38

Shay JW, Pereira-Smith OM and Wright WE (1991) A role for both Rb and p53 in the regulation of human cellular senescence. Exp Cell Res 196: 33–39

Shay JW, Wright WE (2000a) Hayflick, his limit and cellular ageing. Nature Rev Mol Cell Biol 1: 72

Shay JW, Wright WE (2000b) The use of telomerized cells for tissue engineering. Nat Biotechnol 18: 22–23

Shefer G, Van de Mark DP, Richardson JB, Yablonka-Reuveni Z (2006) Satellite-cell pool size does matter: defining the myogenic potency of aging skeletal muscle. Dev Biol 294: 50–66

Shiloh Y (2003) Atm and related protein kinases: safeguarding genome integrity. Nature Rev Cancer 3: 155–68

Tlsty TD, Hein PW (2001) Know thy neighbor: stromal cells can contribute oncogenic signals. Curr Opin Genet Dev 11: 54–59

Tsakiri KD, Cronkhite JT, Kuan PJ, Xing C, Raghu G, Weissler JC, Rosenblatt RL, Shay JW, Garcia CK (2007) Adult-onset pulmonary fibrosis caused by mutations in telomerase. Proc Natl Acad Sci USA 104(18): 7552–57

Tyner SD, Venkatachalam S, Choi J, Jones S, Ghebranious N, Igelmann H, Lu X, Soron G, Cooper B, Brayton C, Hee Park S, Thompson T, Karsenty G, Bradley A, Donehower LA (2002) p53 mutant mice that display early ageing-associated phenotypes. Nature 415: 45–53

Torella D, Rota M, Nurzynska D, Musso E, Monsen A, Shiraishi I, Zias E, Walsh K, Rosenzweig A, Sussman MA, Urbanek K, Nadal-Ginard B, Kajstura J, Anversa P, Leri A (2004) Cardiac stem cell and myocyte aging, heart failure, and insulin-like growth factor-1 overexpression.Circ Res 94(4): 514–24

Van Zant G, Liang Y (2003) The role of stem cells in aging. Exp Hematol 31: 659–72

Vaziri H, Dragowska W, Allsopp RC, Thomas TE, Harley CB, Lansdorp PM (1994) Evidence for a mitotic clock in human hematopoietic stem cells: loss of telomeric DNA with age. Proc Natl Acad Sci USA 91(21): 9857–60

Vaziri H, West MD, Allsopp RC, Davison TS, Wu YS, Arrowsmith CH, Poirier GG, Benchimol S (1997) ATM-dependent telomere loss in aging human diploid fibroblasts and DNA damage lead to the post-translational activation of p53 protein involving poly(ADP-ribose) polymerase. EMBO J 16(19): 6018–33

von Zglinicki T (2002) Oxidative stress shortens telomeres. Trends Biochem Sci 27(7): 339–44

Vulliamy T, Marrone A, Goldman F, Dearlove A, Bessler M, Mason PJ, Dokal I (2001) The RNA component of telomerase is mutated in autosomal dominant dyskeratosis congenita. Nature 413: 432–35

Vulliamy T, Marrone A, Dokal I, Mason PJ (2002) Association between aplastic anaemia and mutations in telomerase RNA. Lancet 359: 2168–70

Vulliamy T, Marrone A, Szydlo R, Walne A, Mason PJ, Dokal I (2004) Disease anticipation is associated with progressive telomere shortening in families with dyskeratosis congenita due to mutations in TERC. Nat Genet 36: 447–49

Welle S, Brooks AI, Delehanty JM, Needler N, Bhatt K, Shah B, Thornton CA (2004) Skeletal muscle gene expression profiles in 20–29 year old and 65–71 year old women. Exp Gerontol 39(3): 369–77

Wiemann SU, Satyanarayana A, Tsahuridu M, Tillmann HL, Zender L, Klempnauer J, Flemming P, Franco S, Blasco MA, Manns MP, Rudolph KL (2002) Hepatocyte telomere shortening and senescence are general markers of human liver cirrhosis. FASEB J 16(9): 935–42

Wilson A, Trumpp A (2006) Bone-marrow haematopoietic-stem-cell niches. Nat Rev Immunol 6: 93–106

Wong KK, Maser RS, Bachoo RM, Menon J, Carrasco DR, Gu Y, Alt FW, DePinho RA (2003) Telomere dysfunction and Atm deficiency compromises organ homeostasis and accelerates ageing. Nature 421: 643–48

Wright WE, Shay JW (1992) The two-stage mechanism controlling cellular senescence and immortalization. Exp Gerontol 27: 383–89

Wynn RF, Cross MA, Hatton C, Will AM, Lashford LS, Dexter TM, Testa NG (1998) Accelerated telomere shortening in young recipients of allogeneic bone-marrow transplants. Lancet 17(351): 178–81

Xu Y, Ashley T, Brainerd EE, Bronson RT, Meyn MS, Baltimore D (1996) Targeted disruption of ATM leads to growth retardation, chromosomal fragmentation during meiosis, immune defects, and thymic lymphoma. Genes Dev 10: 2411–22

Xue W, Zender L, Miething C, Dickins RA, Hernando E, Krizhanovsky V, Cordon-Cardo C, Lowe SW (2007) Senescence and tumour clearance is triggered by p53 restoration in murine liver carcinomas. Nature 445(7128): 656–60

Yamaguchi H, Calado RT, Ly H, Kajigaya S, Baerlocher GM, Chanock SJ, Lansdorp PM, Young NS (2005) Mutations in TERT, the gene for telomerase reverse transcriptase, in aplastic anemia. N Engl J Med 352: 1413–24

Zou Y, Sfeir A, Gryaznov SM, Shay JW, Wright WE (2004) Does a sentinel or a subset of short telomeres determine replicative senescence? Mol Biol Cell 15(8): 3709–18

Zou L, Elledge SJ (2003) Sensing DNA damage through ATRIP recognition of RPA-ssDNA complexes. Science 300(5625): 1542–48

Chapter 9
p16^{INK4a} and Stem Cell Ageing: A Telomere-Independent Process?

Norman E. Sharpless

Abstract Long-lived mammals replace lost or consumed cells at a tremendous pace: an adult human replaces ~1% of their many trillion red blood cells every day through de novo synthesis. Similarly, cell division at various rates is at work throughout life to continuously replace lost cells in many epithelial tissues such as the gut, skin, breast, and lung. Moreover, several tissues (e.g., memory T-lymphocytes, pancreatic β-cells) possess a potential for "facultative growth" in the adult; that is, under certain circumstances (e.g., infection, pregnancy), these usually quiescent cells re-enter the cell cycle en masse to increase the mass of a given tissue through controlled proliferation. These forms of tissue replacement and repair require that many adult mammalian tissues contain reservoirs of stem cells capable of generating terminally differentiated effector cell types. The unique cellular property that enables stem cells to maintain such function throughout life is the ability to produce large numbers of differentiated cell types while also *self-renewing* themselves so that their reserves do not become depleted. In this chapter, I will review some of the evidence to support the notion that certain aspects of mammalian ageing result from an age-dependent decline in the function of self-renewing stem cells, and discuss the relationship of p16^{INK4a} and telomeres with regard to stem cell ageing.

1 Properties of Self-Renewal Relevant to Molecular Ageing

A few characteristics of self-renewal bear mention. First, most proliferation in an adult mammal is not self-renewing. Self-renewal appears to be a specific and unusual property of stem cells and other rare cell types that are capable of limited numbers of self-renewing divisions after periods of dormancy. Strictly speaking, not all self-renewing cells are "stem cells": adult tissue-specific stem cells (e.g., hematopoietic stem cells) are "multipotent" in that they can give rise to differentiated progeny of several

---

Norman E. Sharpless
Department of Medicine and Genetics, The Lineberger Comprehensive Cancer Centre, The University of North Carolina Chapel Hill, North Carolina, United States.
e-mail: NES@med.unc.edu

K.L. Rudolph (ed.), *Telomeres and Telomerase in Ageing, Disease, and Cancer.*

different types. In contrast, there are a few kinds of self-renewing cells such as pancreatic β-cells and memory T-cells that have a restricted potential for differentiation, thus generating progeny similar to the parental cell. This type of self-renewing cell is sometimes termed a "unipotent progenitor." Admittedly, the distinction between tissue-specific stem cells and unipotent progenitors is somewhat arbitrary, with the latter representing a subset of the former. For convenience, the term *stem cell* will be used in reference to both types of adult self-renewing cells in this chapter unless otherwise indicated. In contrast, the term *progenitor* will designate cell types derived from stem cells that are relatively undifferentiated but that are nonetheless incapable of long-term self-renewal. Progenitors in this sense are also sometimes called *transit amplifying cells*.

Many stem cells are believed to produce differentiated cells through a series of increasingly more committed progenitor intermediates. This hierarchical structuring has been characterized in greatest detail in the hematopoietic system, where long-term hematopoietic stem cells (LT-HSC) produce multipotent progenitor subsets that retain full lineage potential yet have little or no capacity for self-renewal (Bryder et al. 2006). These multipotent progenitors in turn give rise to oligopotent progenitors (Akashi et al. 2000, Kondo et al., 1997), which in turn give rise to more lineage-restricted progenitors from which all of the mature blood cells eventually arise (Fig. 9.1). While Figure 9.1 depicts HSC and progeny, similar hierarchies are thought to exist in many other tissues (brain, gut, liver, lung, etc.) containing tissue-specific stem cells. In the hematopoietic system, these stem and progenitor cell subsets can be purified to near-homogeneity by differences in immunophenotype. For example, true murine LT-HSC express low levels of lineage markers (lin-) specific to the differentiated progeny, high levels of the Kit receptor (Kit+), high levels of a cell surface marker called Sca1 (Sca+), and low levels of another antigen, Cd34 (Cd34 lo). As the Lin-Kit+Sca+Cd34 lo LT-HSC divide asymmetrically, they produce another LT-HSC and a more differentiated daughter cell with reduced potential for self-renewal (a short-term HSC or ST-HSC) which harbors increased Cd34 expression (Osawa et al. 1996). Thus, LT-HSC and ST-HSC can be distinguished immunophenotypically by expression of Cd34. Similar multimarker panels can be used to identify the various blood intermediates shown in Figure 9.1.

This hierarchical differentiation scheme appears to have advantages. It allows for an enormous amplification in the numbers of differentiated cells from a single stem cell by combining subsequent steps in differentiation with rapid proliferation (Hodgson and Bradley 1984, Passegue et al. 2005). Also, the multistep differentiation scheme places limited proliferative demand on the self-renewing stem cells themselves so that these cells divide very infrequently (Bradford et al. 1997, Cheshier et al. 1999). In the hematopoietic system, greater than 90% of HSC are in the G_0/G_1 phase of the cell cycle in the unstressed organism (Yamazaki et al. 2006), while downstream progenitors cycle much more rapidly (Passegue et al. 2005). The minimal proliferative pressure on stem cells has the benefit of not subjecting them to the potentially mutagenic hazards of DNA replication and cell division, and may thus contribute to the integrity and longevity required of these cells. Also, as stem cells appear to be a substrate for oncogenic mutations that lead to cancer, this quiescence may be important to tumor suppression in this compartment. As G_0 is a relatively metabolically

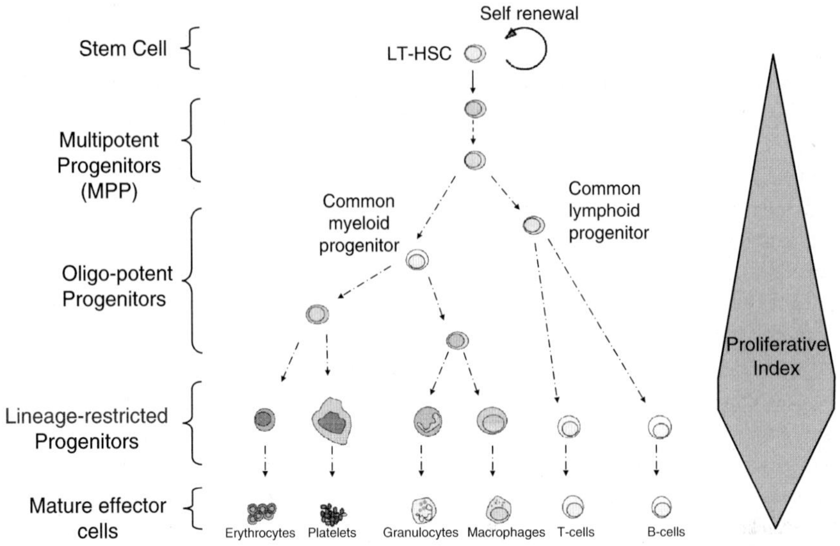

Fig. 9.1 Hematopoietic developmental hierarchy. Self-renewing HSC reside at the top of the hierarchy, giving rise to a number of multipotent progenitors. Multipotent progenitors give rise to oligopotent progenitors, including the common lymphoid progenitor (CLP), which gives rise to mature B lymphocytes, T lymphocytes, and natural killer (NK) cells. The common myeloid progenitor (CMP) gives rise to granulocyte/macrophage progenitors (GMP) that differentiate into monocytes/macrophages and granulocytes, and megakaryocyte/erythrocyte progenitors (MEP) that differentiate into megakaryocytes/platelets, and erythrocytes. Development from the oligopotent progenitors through to mature blood cells proceeds through a number of intermediate progenitors (not shown). The developmental passage of HSC through multipotent progenitors, oligopotent progenitors, and lineage-specific progenitors is generally associated with increases in proliferative index, although this trend is not absolute and has not been resolved for all stages of development (*See Color Plate*)

inactive phase of the cell cycle, it has been suggested that stem cells may be subjected to lower levels of DNA damage-inducing metabolic by-products and reactive oxygen species than more metabolically active differentiated cell types (Rossi et al. 2005).

Moreover, while once believed to be far more restricted, tissue-specific stem cells are now known to be involved in controlling homeostasis in many tissues, including those with lower turnover rates such as the brain (Weissman 2000). While biologists may disagree over the physiological significance of such cells in the adult, their existence in rodents is beyond dispute, and their existence in human tissue is strongly suggested. Given this fact, it is reasonable to postulate that at least some characteristics of ageing once thought to be solely "degenerative" might in fact reflect a diminution of regenerative capacity by the resident stem cells of the affected tissue. It is not obvious which features of ageing reflect solely tissue degeneration versus stem cell functional decline (or some combination), and elucidating the relative contributions of these two mechanisms of ageing on a tissue-by-tissue basis will be a principal focus of future research in mammalian ageing. Moreover, as stem cells reside in specialized niches, where they receive and respond to external cues to proliferate, differentiate, or self-renew (Moore and Lemischka 2006, Schofield 1978), it seems likely that the stem cell microenvironment exerts significant influence on stem cell ageing (see Waterstrat et al. this volume).

Lastly, it is important to note that the capacity for self-renewal appears to come with some danger for the entire organism (Campisi 2003, Krishnamurthy et al. 2004, Reya et al. 2001). Genetic lesions that evade repair in stem cells can be propagated to their self-renewing progeny, and in such a way can accumulate with age. In particular, mutations that provide a growth advantage in a self-renewing compartment provide positive selection for the mutant cell, with malignant transformation being the endpoint of the accumulation of multiple cancer-promoting mutations. In contrast, mutations in proliferating cells that are fated to differentiate or die are thought to have little or no oncogenic potential unless the mutagenic process establishes the capacity for self-renewal on otherwise non-self-renewing cells. As stem cells are exposed to an array of endogenous and exogenous mutagens, oncogenic events in stem cell compartments should occur throughout the mammalian life span. To counter this, stem cells appear to have evolved mechanisms aimed at maintaining genomic integrity beyond that of other proliferating cells, for example, expression of ABC/MDR transporter genes (Rossi et al. 2005, Zhou et al. 2001) to efflux xenobiotic toxins from the cell. When mutations occur despite these error prevention mechanisms, evolutionarily perfected tumor-suppressor mechanisms (e.g., senescence) exist that sense malignant growth and censor would-be malignant stem cell clones. However, the function of these checkpoints in the stem cell compartment needs to be further explored (see Gutierrez and Ju, this volume). This relationship between self-renewing cells and cancer prevention predicts that such tumor suppressor mechanisms may also inadvertently contribute to ageing by contributing to stem cell attrition.

2 Stem Cell Function Declines with Ageing

As extensive tissue replacement does not occur in adults of short-lived invertebrates, such model systems have to date been less useful in understanding the mechanisms of stem cell ageing. A notable exception in this regard, however, have been the elegant studies of germ line mortality genes by Ahmed and colleagues (see, for example, Ahmed and Hodgkin 2000, Meier et al. 2006), which have identified several pathways required for germ line stem cell longevity, including maintenance of telomerase. Nonetheless, the bulk of studies of the determinants of stem cell longevity have been carried out in rodents, with confirmatory analyses in humans where possible. Of these rodent studies, the majority in particular have focused on hematopoietic stem cells in mice, which are the most well-characterized tissue specific stem cells.

2.1 Hematopoietic Stem Cell Ageing

Advancing age is accompanied by a number of pathophysiological changes in the hematopoietic system whose etiology suggests a loss of homeostatic control (see also Waterstrat et al., this volume). The most clinically significant of these are the diminution and decreased competence of the adaptive immune system in the elderly (Linton and Dorshkind 2004), the dramatically increased incidence of myelogenous diseases, including myelodysplasia and myeloid leukemias, that accompanies ageing (Lichtman and Rowe 2004), and the onset of anemia in the elderly (Beghe et al. 2004, Guralnik et al. 2004). Elderly patients are also more likely to suffer adverse side effects and prolonged toxicity to myelosuppressive drugs (e.g., chemotherapy for certain cancers; see, for example, Appelbaum et al. 2006, Brunello et al. 2005, Lenhoff et al. 2006), while increasing donor age in bone marrow transplantation is a predictor of transplant-related mortality (Ash et al. 1991, Buckner et al. 1984, Castro-Malaspina et al. 2002, Kollman et al. 2001, Yakoub-Agha et al. 2006), suggesting the reconstituting ability of HSC from old donors is intrinsically compromised. Finally, the finding of familial aplastic anemia, a profound form of bone marrow failure, in patients with congenital defects in telomerase activity is of particular interest in this regard (see also Du et al., this volume). These provocative findings in humans, however, are greatly strengthened by studies in rodents where HSC can be purified to near homogeneity and transplanted to quantitatively evaluate functional autonomy.

Perhaps counterintuitively, reserves of hematopoietic stem cells have been shown to expand with age in many, but not all, laboratory strains of mice through quantitation by immunophenotyping (Morrison et al. 1996, Rossi et al. 2005, Sudo et al. 2000), side population activity (Pearce et al. 2007, Rossi et al. 2005), or cobblestone assays (de Haan et al. 1997, de Haan and Van Zant 1999). The age-dependent expansion of HSC observed in the steady state was shown to be a transplantable, cell autonomous property of HSC ageing, so that even upon transplantation into young recipients, HSC from aged donors still had a greater propensity to self-renew and give rise to copies of themselves than HSC from young donors (Pearce et al. 2007, Rossi et al. 2005). A second surprising aspect of HSC biology is that these stem cells can be serially transplanted through successive generations of recipients as demonstrated in seminal experiments by Harrison and colleagues, who showed that HSC from C57Bl/6 mice could function for over eight years and thus far exceed the lifetime of the original donor (Harrison 1979). These experiments thereby established that replicative HSC exhaustion does not occur during periods of normal ageing in at least some strains of inbred mice.

It should be noted, however, that the case may be different in other experimental settings. For example, several forms of marrow toxins (chemotherapy, ionizing radiation) severely limit long-term HSC function and hasten stem cell exhaustion (Boccadoro et al. 2002, Gardner et al. 1997, Ito et al. 2004, Ito et al. 2006, Knudsen et al. 1999, Meng et al. 2003, Wang et al. 2006). Furthermore, as mentioned, certain strains of mice do not show expansion of the HSC compartment with ageing, and a correlation between stem cell expansion and longevity has been suggested (de Haan et al. 1997, Henckaerts et al. 2004). Thus, it is possible that genetically outbred mice in the wild would exhibit more pronounced stem cell attrition upon greater exposure to environmental stresses that are not normally encountered by laboratory-housed strains under experimental conditions optimized for longevity and easy breeding.

Additionally, even in strains where stem cell exhaustion does not occur with ageing, HSC still exhibit profound signs of intrinsic ageing. Indeed numerous studies have documented considerable impact on HSC function with advancing age, including alterations in their mobilization properties (Xing et al. 2006), diminished reconstitution and homing capability (Liang et al. 2005, Morrison et al. 1996, Rossi et al. 2005, Sudo et al. 2000, Xing et al. 2006), and a loss of lymphoid lineage potential in favor of myeloid lineage commitment with age (Kim et al. 2003, Liang et al. 2005, Rossi et al. 2005, Sudo et al. 2000). The molecular basis for this cell intrinsic stem cell ageing in HSC is not clearly established, but it is noteworthy that numerous genes involved in higher-order chromosome dynamics, chromatin remodeling, and epigenetic regulation of gene expression are differentially expressed during HSC ageing (Rossi et al. 2007, Rossi et al. 2005).

2.2 *Ageing in Other Tissue-Specific Stem Cells*

Analyses of stem cells in nonhematopoietic tissues is complicated by the fact that assays used to evaluate stem cell function, such as transplantation, can be technically difficult, if not impossible, to perform for certain types of stem cells. Moreover, while hematopoietic stem and progenitor cells can be enriched to near purity through a combination of cell surface phenotype and fluorescence-activated cell sorting, strategies for enriching other tissue-specific stem cells are only just beginning to be developed. With this said, numerous lines of evidence suggest that stem cells from a variety of tissues suffer functional decline with advancing age.

For example, a decline in the number of new neurons produced by neural stem cells (NSC) in the dentate gyrus with age has been well documented in rodents (Kuhn et al. 1996), while the numbers and proliferation of NSC in the subventricular zone diminishes (Maslov et al. 2004, Molofsky et al. 2006) along with their neurogenic potential, resulting in an impairment in fine olfactory discrimination with age (Enwere et al. 2004). Similar effects of ageing on pancreatic β-cell proliferation have been noted (Krishnamurthy et al. 2006, Sone and Kagawa 2005). In an even more blatant example of stem cell functional decline, hair greying in aged mice and humans has been linked to incomplete melanocyte stem cell maintenance (Lang et al. 2005, Nishimura et al. 2005). The lack of facile transplantation models hinders the determination of which cell or compartment is failing with ageing in these settings. For example, in the β-cell system, ageing could occur in the β-cells themselves (manifesting as decreased "replication"), in a putative pancreatic stem cell (manifesting as decreased "neogenesis"), or in another tissue that influences islet proliferation in a cell nonautonomous manner.

Indeed, non-cell-autonomous factors have been shown to impact stem cell ageing in other systems (see also Gutierrez and Ju, this volume). For example, Rando and colleagues have shown that the ageing of muscle satellite cells is influenced by cell extrinsic factors through the demonstration that satellite cells of aged mice that were surgically joined to young mice through heterochronic parabiosis to establish a conjoined circulatory system, were more capable of responding to injury than in the aged setting alone (Conboy et al. 2005). These experiments suggest that the aged milieu, hormonal or otherwise, significantly influences satellite cell ageing, although there may also be cell-autonomous components to satellite cell ageing. By the same token, while much of the functional decline that characterizes HSC ageing has been shown to be cell autonomous (Rossi et al., 2005), the influence of the stem cell niche and bone marrow microenvironment on HSC biology is undeniable, and thus it would be surprising if the aged microenvironment did not also influence HSC ageing to some degree. Recent studies in telomerase knockout mice provide further evidence that telomere dysfunction induces niche and perhaps other environmental defects, which limits function and engraftment of HSCs (Ju et al. in press; see also Gutierrez and Ju, this volume). The ability to conditionally target genes in a tissue-specific manner will facilitate the determination of which cells in which compartments are ageing intrinsically and which are ageing by "proxy."

3 Activation of $p16^{INK4a}$ Contributes to Stem Cell Ageing

The mechanisms whereby old self-renewing cells remember their prior proliferative history are beginning to be elucidated. In particular, a role for the senescence-promoting *Ink4a/Arf* locus and its downstream targets (Fig. 9.2) has been suggested. Senescence (see Allsopp, this volume; Passos, this volume; Chang, this volume; and Gutierrez and Ju, this volume) is a specialized form of terminal differentiation induced by a variety of stimuli, including alterations of telomere length and structure, some forms of DNA damage (for example, oxidative stress; see Passos, this volume), and activation of certain oncogenes (reviewed in (Campisi 2001, Wright and Shay 2002). Senescence requires activation of the retinoblastoma (Rb) and/or p53 protein and expression of their regulators such as $p16^{INK4a}$ and $p14^{ARF}$ (Alcorta et al. 1996, Kamijo et al. 1997, Sage et al. 2003, Stein et al. 1999) (Fig. 9.2). These proteins were first identified by their importance in tumor suppression, and the evidence suggesting senescence prevents cancer is unimpeachable (reviewed in (Campisi 2005, Kim and Sharpless 2006). Senescence is characterized by a few markers, such as senescence associated-β-galactosidase (SA-β-gal) (Dimri et al. 1995) and senescence-associated heterochromatic foci (SAHF) (Narita et al., 2003), although the in vivo study of this process has been slowed by the lack of reliable markers that work across tissues and species.

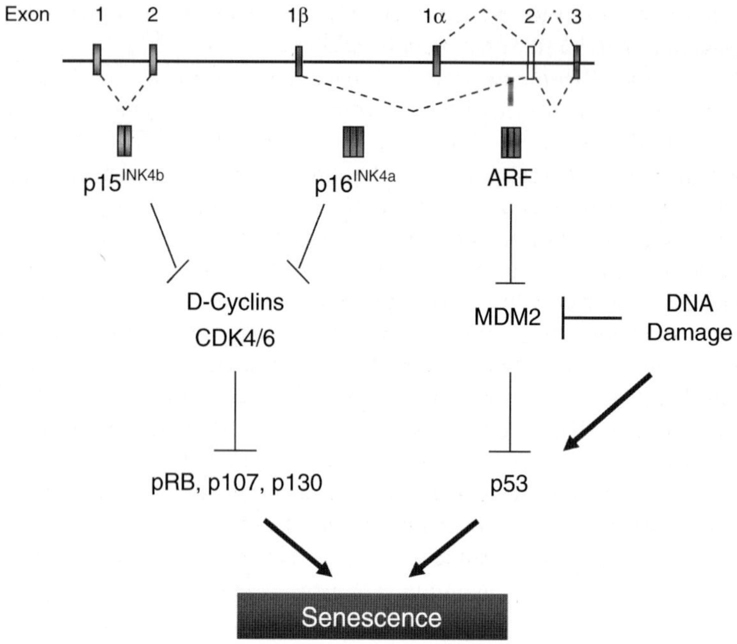

Fig. 9.2 The *INK4a/ARF/INK4b* locus and downstream targets. Three tumor suppressor genes are encoded in a common 40 kb stretch of human chromosome 9p21: two highly related CDK inhibitor proteins (p15^{INK4b} and p16^{INK4a}) and ARF, a regulator of p53 stability. Expression of p16^{INK4a} and p15^{INK4b} inhibits CDK4/6 activity, leading to hypophosphorylation of Rb-family proteins (Rb, p107, p130) and growth arrest. Expression of ARF inhibits the ubiquitin ligase activity of MDM2, thereby stabilizing p53. ARF and p16^{INK4a} share exons 2/3, but are encoded in alternate reading frames and therefore have no amino acid homology. DNA damage inhibits p53 degradation by MDM2 in an ARF-independent manner, and can promote senescence independent of *INK4a/ARF* activation. DNA damage, however, likely also activates p16^{INK4a}; although this relationship is complex (see section 5.2) (*See Color Plate*)

3.1 Senescence and p16^{INK4a} Expression with Ageing

The expression of several markers of senescence increases with ageing. Campisi and colleagues have shown an increase in SA-β-gal activity in skin with ageing in humans (Dimri et al. 1995). Similar increases in SA–β-gal have been seen in ageing rodent and human tissues, including heart, kidney, pancreatic islet, atherosclerotic plaques, and lung (Krishnamurthy et al. 2004, Matthews et al. 2006, Melk et al. 2003, Melk et al. 2004, Sone and Kagawa 2005, Tsuji et al. 2006, Urbanek et al. 2005). Likewise, Sherr and colleagues reported a marked increase in the expression of p16^{INK4a}, an effector of senescence, in many tissues with ageing (Zindy et al. 1997). Several groups have extended this finding in multiple mammalian species to a large number of ageing tissues in health and disease (Enomoto et al. 2006,

Gray-Schopfer et al. 2006, Herbig et al. 2006, Krishnamurthy et al. 2004, Melk et al. 2004, Menzel et al. 2006, Michaloglou et al. 2005, Nielsen et al. 1999, Ressler et al. 2006, Sasaki et al. 2006, Tsuji et al. 2006, Urbanek et al. 2005). Given the large change in $p16^{INK4a}$ expression with ageing (>10-fold in many tissues) and the relative ease of quantitating its expression, this finding has led to the proposal that $p16^{INK4a}$ expression could be used as a biomarker of physiological, as opposed to chronological, age (Krishnamurthy et al. 2004).

Although the expression of senescence markers clearly accompanies mammalian ageing, some evidence suggests that senescence mechanisms actually contribute to ageing. Caloric restriction (CR) potently retards ageing in many species, including mammals (see Dillin and Karlseder, this volume). Correspondingly, CR or other dietary changes retards or even abolishes the age-induced increase in the expression of senescence markers such as SA-β-gal and $p16^{INK4a}$ expression (Edwards et al. 2007, Krishnamurthy et al. 2004, Sone and Kagawa 2005). Provocatively, CR, like $p16^{INK4a}$ deficiency (Janzen et al. 2006), has been suggested to enhance HSC function with ageing (Chen et al. 2003). Therefore CR, the most potent means to delay ageing in mammals, is associated with a preservation of replicative stem cell function in at least some tissues, with a concomitant slowing in the rate of the age-induced increase in $p16^{INK4a}$ expression.

3.2 *Murine Studies of $p16^{INK4a}$ and Ageing*

Studies of $p16^{INK4a}$-deficient and overexpressing mice have suggested that expression of $p16^{INK4a}$ with ageing in these self-renewing compartments is not merely correlative, but actually plays a causal role in stem cell ageing. In particular, an effect on self-renewal was suggested in HSC, neural stem cells, and pancreatic β-cells (Janzen et al. 2006, Krishnamurthy et al. 2006, Molofsky et al. 2006, Stepanova and Sorrentino 2005) (e.g., Fig. 9.3). For example, a decline in islet proliferation and regenerative capacity with ageing has been well documented in rodent models, and a relative failure of β-cell mass appears to play a role in human type II diabetes (Butler et al. 2003, Yoon et al. 2003). More than 1% of β-cells in the islets of young mice are proliferating under steady-state conditions, but this frequency declines by more than 10-fold after a year of ageing (middle-aged for a mouse) (Fig. 9.3). β-cell proliferation is known to require cdk4 activity (Rane et al. 1999, Tsutsui et al. 1999), the biochemical target of $p16^{INK4}$, which significantly accumulates with ageing in the islet (Krishnamurthy et al. 2006, Nielsen et al. 1999). The finding that germ line $p16^{INK4a}$ deficiency largely abrogates the replicative failure of β-cells with ageing suggests that the age-induced expression of $p16^{INK4a}$ in part plays a causal role in this process (Krishnamurthy et al. 2006).

In all three cell types tested (islet, NSC, and HSC), $p16^{INK4a}$ deficiency partially abrogated an age-induced decline in proliferation and functional decline in each tissue (Janzen et al. 2006, Krishnamurthy et al. 2006, Molofsky et al. 2006). The effects of $p16^{INK4a}$ loss were consistent across these self-renewing tissues of vastly

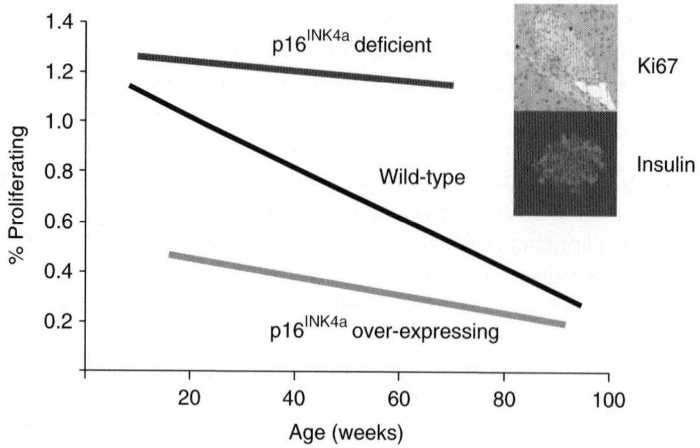

Fig. 9.3 β-cell proliferation with age. A decline in β-cell proliferation is modulated by the expression of the p16^{INK4a} tumor suppressor. Excess p16^{INK4a} expression correlates with reduced proliferation, while decreased p16^{INK4a} affords a resistance to β-cell ageing. The top panel in the inset shows an H&E image of an adult murine islet with immunohistochemical detection of Ki67, a marker of proliferation (brown), indicating proliferation in a rare subset of β-cells. The bottom panel shows an islet immunofluorescently labeled with insulin (red) and a nuclear stain (blue). Original magnification = 20X of both images (*See Color Plate*)

different biological properties–in true stem cells (HSC and NSC) as well as unipotent progenitors (pancreatic β-cells). Therefore, p16^{INK4a} appears capable of promoting ageing in disparate tissues that are developmentally distinct. In no organ examined in these studies (Janzen et al. 2006, Krishnamurthy et al. 2006, Molofsky et al. 2006), however, did p16^{INK4a} loss completely abrogate the effects of ageing, indicating p16^{INK4a}-independent ageing occurs in these compartments.

It is also worth noting, however, that similar evidence supports a role for p53, another potent effector of senescence in mammalian ageing. Although it has been difficult to show an increase in p53 activation with ageing, modest increases in p53 target transcripts such as p21^{CIP} have been reported in ageing tissues (Enomoto et al. 2006, Krishnamurthy et al. 2006, Matthews et al. 2006, Menzel et al. 2006, Tsuji et al. 2006). Moreover, genetically engineered animals possessing augmented p53 function also support a role for this archetypal tumor suppressor in stem cell ageing (Maier et al. 2004, Tyner et al. 2002). These results are informed by recent studies showing that p53 expression in HSC modulates their function with ageing, indicating that p53 activation per se compromises the HSC compartment (Dumble et al. 2007, TeKippe et al. 2003). Specifically, mice lacking p53 demonstrate enhanced HSC activity, and mice with increased p53 demonstrate decreased HSC activity. Therefore, p53 appears to exert age-promoting effects in concert with its well-established, beneficial anti-cancer activity.

It should be noted, however, that these data do not establish that p16^{INK4a} and p53 promote ageing through the induction of senescence. For example, the effect of p16^{INK4a} on HSC ageing may be unrelated to senescence or other effects on the cell cycle since purified HSC from aged mice have an equal capacity to enter cycle and give rise to progeny as HSC from young mice in rigorous clonal assays (Morrison et al. 1996, Sudo et al. 2000) (D.J.Rossi, personal communication). So although the induction of senescence in some tissues seems likely, it is also possible these proteins merely decrease the frequency of cell cycle entry in the absence of senescence or, in the case of p53, by inducing apoptosis rather than senescence. The distinction between these possible mechanisms is important for a few reasons, perhaps most importantly in that senescence is expected to be irreversible, whereas a decrease in cell cycle entry possibly could be reversed.

3.3 Human Genomic Studies of p16^{INK4a} and Age-Associated Phenotypes

With completion of the human haplotype map (HapMap), the discovery of genetic polymorphisms that correlate with aspects of human ageing has been greatly facilitated. Screens for loci that correlate with extraordinary longevity have been countenanced, but this approach suffers from differences between the control groups (young patients) and experimental groups (old patients). Recently, however, screens for genetic loci associated with two important age-associated phenotypes, frailty and Type II diabetes mellitus (T2DM), have suggested a likely role for p16^{INK4a} in these aspects of human ageing. In both studies, single nucleotide polymorphisms (SNPs) near the *INK4/ARF* locus were associated with the phenotype of interest. As there are no other plausible candidates in this genomic region, these results suggest that noncoding elements that influence expression of p16^{INK4a}, p15^{INK4b} and/or ARF might play a role in the onset of these human conditions.

Frailty scoring is a highly validated and reproducible instrument commonly used in clinical geriatrics as an indicator of age-related fitness in the elderly (Fried et al. 1998, Fried et al. 2001). Frail elderly are less able to care for themselves, and demonstrate a marked increase in mortality. Frailty is associated with expression of a few biomarkers of ageing (e.g., increased serum IL-6) as well as age-related diseases such as T2DM and heart disease. While one might expect that patients with subclinical neoplasms would be more likely to exhibit frailty, provocatively, frailty has recently been shown to inversely correlate with cancer incidence. Ershler and colleagues abstracted data from 48,000 Medicare beneficiaries and showed that patients with clinical features of frailty demonstrated a marked reduction in cancer incidence compared to nonfrail controls (W. Ershler et al., Personal Communication). The effect was the strongest in the oldest patients, with frail 85+ year olds enjoying a fivefold reduction in cancer incidence. Although this epidemiological association can be explained in several ways, it is tempting to speculate that patients with the strongest tumor suppressor mechanisms, such as senescence, experience less cancer but at the cost of accelerated ageing.

Against this backdrop, Melzer and colleagues performed a candidate screen for common SNPs associated with genes involved in the cell cycle and senescence

through a direct transcriptional effect of Ets transcription factors (Huot et al. 2002, Ohtani et al. 2001). A possible role for Polycomb Group complexes (PcG) with *Ink4a/Arf* regulation with ageing deserves particular consideration.

4.1 Polycomb Group Complexes

PcG complexes appear to repress transcription through covalent modifications of histones (reviewed in Pasini et al. 2004, Sparmann and van Lohuizen 2006). The relationship of PcG complexes to *INK4a/ARF/INK4b* expression has been particularly provocative, as Bmi1 is required for stem cell maintenance of HSC (Park et al. 2003) and NSC (Molofsky et al. 2005). The ability of these complexes to create heritable epigenetic marks might suggest a nongenetic mechanism whereby cells could carry the memory of prior ageing-related exposures. At least four PcG proteins (Bmi1, Cbx7, Cbx8, and Mel18) have been reported to repress transcripts encoded by the *Ink4a/Arf/Ink4b* locus ($p16^{Ink4a}$, $p15^{Ink4b}$, and *ARF*) (Dietrich et al. 2007, Gil et al. 2004, Itahana et al. 2003, Jacobs et al. 1999). Loss of *Bmi1* in mice is associated with failure to maintain diverse self-renewing stem cells (e.g., HSC and NSC), which can in large part be rescued by *Ink4a/Arf* deficiency (Bruggeman et al. 2005, Jacobs et al. 1999, Molofsky et al. 2005, Molofsky et al. 2003, Oguro et al. 2006, Park et al. 2003). In contrast, homeotic transformations of the axial skeleton noted in mice lacking *Bmi1* are not rescued by loss of *Ink4a/Arf* (Jacobs et al. 1999). These results prove that $p16^{INK4a}$ and Arf are two intrinsic mediators of the stem cell exhaustion phenotype in Bmi1-deficient mice, although there is experimental evidence for additional effectors (van Lohuizen, Keystone Stem cell meeting 2007). Intriguingly, enforced expression of another PcG protein, Ezh2, in HSC has been shown to greatly augment their performance in serial transplant (Kamminga et al. 2006), further suggesting PcG complexes are limiting during stem cell ageing, at least under the stressful conditions of serial transplant.

Most recently, employing chromatin immunoprecipitation, two groups have shown that Bmi1 and Ezh2 bind to the $p16^{INK4a}$ promoter in human and murine cells (Bracken et al. 2007, Kotake et al. 2007). Decreased occupancy of the $p16^{INK4a}$ promoter by PcG proteins coincided with increased $p16^{INK4a}$ expression during senescence (Bracken et al. 2007). Additionally, the Bmi1 interaction with the $p16^{INK4a}$ promoter required Rb-family proteins (Kotake et al. 2007), suggesting an explanation for the requirement for mitogenic stimuli in the induction of senescence. In both studies and in contrast to murine cells, in human cells Bmi1 only appears to regulate the expression of $p16^{INK4a}$, but not that of ARF or $p15^{INK4b}$. Therefore, perhaps this mouse and human difference with regard to PcG proteins explains the marked increase with ageing in *Arf* expression in mice but not humans. It remains to be seen if changes in $p16^{INK4a}$ promoter occupancy by PcG proteins occur with mammalian ageing.

4.2 *p16[INK4a] induction with Ageing: Telomere-Independent?*

A particularly important question relates to the relationship of telomeres and p16[INK4a]. As discussed, some data suggest that sufficient telomere attrition occurs in human stem cells during normal ageing to activate an intrinsic DNA damage response with attendant compromise of stem cell replicative function. Humans who harbor short telomeres because of congenital deficiencies of the telomerase complex develop an age-related failure of bone marrow function and other self-renewing compartments (Mitchell et al. 1999, Vulliamy et al. 2001, Yamaguchi et al. 2005) (see also Du et al., this volume). Moreover, telomere shortening has been shown to precede the development in overt cirrhosis in patients with chronic hepatitis of various etiologies (Kitada et al. 1995, Miura et al. 1997, Urabe et al. 1996, Wiemann et al. 2002). Lastly, some studies have demonstrated a relationship between telomere length in peripheral blood lymphocytes (PBL) and the onset of certain diseases associated with ageing. Such studies in non-neoplastic diseases have shown that PBL telomere lengths can provide predictive information on the risk of developing atherosclerosis (Obana et al. 2003, Samani et al. 2001) and overall mortality (Cawthon et al. 2003). Therefore, telomere dysfunction appears likely to occur with ageing in some human tissues.

Telomere dysfunction appears to activate the classic tumor suppressor mechanisms of senescence and apoptosis. Clearly, telomere dysfunction is sensed in a biochemically similar manner to a DSB, to potently induce p53. Telomere dysfunction in cultured human and mouse cells has been shown to induce p53-mediated senescence and other checkpoint responses such as apoptosis. Although more controversial, telomere dysfunction also appears to induce p16[INK4a] albeit with greatly delayed kinetics compared to p53 (Jacobs and de Lange 2004). For example, p16[INK4a] expression in human fibroblasts undergoing telomere shortening-induced senescence occurs 7 to 14 days later than p53 activation, presumably in response to the same DNA-damage-like stimulus. On the other hand, the placement of human epithelial cells and murine cells in culture has been shown to rapidly induce p16[INK4a] in the absence of telomere shortening (Foster et al. 1998, Huschtscha et al. 1998, Kiyono et al. 1998, Zindy et al. 1998). Likewise, because of differences in telomere biology between rodents and man (see Chang, this volume), telomere dysfunction is not likely to explain the massive increase in p16[INK4a] expression seen during rodent ageing. Therefore, p16[INK4a] appears to be induced as a result of telomere-independent and -dependent stimuli, although the relative contribution of each in unknown in human ageing.

5 Conclusions

An increasing body of evidence suggests that at least some aspects of mammalian ageing may be underwritten by a progressive stem cell functional decline that results from activation of the tumor suppressor machinery associated with senescence (Fig. 9.4).

In particular, expression of p16^{INK4a} as a result of stresses such as oncogene activation, telomere shortening, and DNA damage appears to limit the function of a few types of self-renewing cells in the adult mammal. Intriguingly, ageing or cancer appears to represent either the success or failure, respectively, of senescence. These observations predict a simple way of determining physiological age. As increased p16^{INK4a} expression, likely as a surrogate for senescence, is easily measured with ageing, these results suggest a means to determine human physiological, as opposed to chronological, age. Such a precise marker of molecular age should find clinical utility in that it permits the measure of the effects of stresses or would-be anti-ageing therapies on the rate of age-induced compromise of stem cell function. In addition, for elderly patients, an ever-increasing population, ageing markers may facilitate therapeutic decision making regarding surgery, chemotherapy, or other aggressive therapies that require self-renewal or regenerative capacity of patient tissues. Perhaps even more importantly, this enhanced understanding of the role of p16^{INK4a} and senescence in ageing may allow for the design of rational therapies designed to retard the process.

References

Ahmed S and Hodgkin J (2000) MRT-2 checkpoint protein is required for germ line immortality and telomere replication in C. elegans. Nature 403: 159–64

Akashi K, Traver D, Miyamoto T, Weissman IL (2000) A clonogenic common myeloid progenitor that gives rise to all myeloid lineages. Nature 404: 193–97

Alcorta DA, Xiong Y, Phelps D, Hannon G, Beach D, Barrett JC (1996) Involvement of the cyclin-dependent kinase inhibitor p16 (INK4a) in replicative senescence of normal human fibroblasts. Proc Natl Acad Sci USA 93: 13742–47

Appelbaum FR, Gundacker H, Head DR, Slovak ML, Willman CL, Godwin JE, Anderson JE, Petersdorf SH (2006) Age and acute myeloid leukemia. Blood 107″ 3481–85

Ash RC, Horowitz MM, Gale RP, van Bekkum DW, Casper JT, Gordon-Smith EC, Henslee PJ, Kolb HJ, Lowenberg B, Masaoka T, et al. (1991) Bone marrow transplantation from related donors other than HLA-identical siblings: effect of T cell depletion. Bone Marrow Tr 7: 443–52

Aslanian A, Iaquinta PJ, Verona R, Lees JA (2004) Repression of the Arf tumor suppressor by E2F3 is required for normal cell cycle kinetics. Genes Dev 18: 1413–22

Beghe C, Wilson A, Ershler WB (2004) Prevalence and outcomes of anemia in geriatrics: a systematic review of the literature. Am J Med 116 Suppl 7A: 3S–10S

Boccadoro M, Palumbo A, Bringhen S, Merletti F, Ciccone G, Richiardi L, Rus C, Bertola A, Giaccone L, Omede P, et al. (2002) Oral melphalan at diagnosis hampers adequate collection of peripheral blood progenitor cells in multiple myeloma. Haematologica 87, 846–50

Bracken AP, Kleine-Kohlbrecher D, Dietrich N, Pasini D, Gargiulo G, Beekman C, Theilgaard-Monch K, Minucci S, Porse BT, Marine JC, et al. (2007) The Polycomb group proteins bind throughout the INK4A-ARF locus and are disassociated in senescent cells. Genes Dev 21: 525–30

Bradford GB, Williams B, Rossi R, Bertoncello I (1997) Quiescence, cycling, and turnover in the primitive hematopoietic stem cell compartment. Exp Hematol 25: 445–53

Bruggeman SW, Valk-Lingbeek ME, van der Stoop PP, Jacobs JJ, Kieboom K, Tanger E, Hulsman D, Leung C, Arsenijevic Y, Marino S, et al. (2005) Ink4a and Arf differentially affect cell proliferation and neural stem cell self-renewal in Bmi1-deficient mice. Genes Dev 19: 1438–43

Brunello A, Basso U, Pogliani C, Jirillo A, Ghiotto C, Koussis H, Lumachi F, Iacobone M, Vamvakas L, Monfardini S (2005) Adjuvant chemotherapy for elderly patients (> or =70 years) with early high-risk breast cancer: a retrospective analysis of 260 patients. Ann Oncol 16: 1276–82

Bryder D, Rossi DJ, Weissman IL (2006) Hematopoietic stem cells: the paradigmatic tissue-specific stem cell. Am J Pathol 169, 338–46

Buckner CD, Clift RA, Sanders JE, Stewart P, Bensinger WI, Doney KC, Sullivan KM, Witherspoon RP, Deeg HJ, Appelbaum FR, et al. (1984) Marrow harvesting from normal donors. Blood 64: 630–04

Bulavin DV, Phillips C, Nannenga B, Timofeev O, Donehower LA, Anderson CW, Appella E, Fornace AJ Jr (2004) Inactivation of the Wip1 phosphatase inhibits mammary tumorigenesis through p38 MAPK-mediated activation of the p16(Ink4a)-p19(Arf) pathway. Nat Genet 36: 343–50

Butler AE, Janson J, Bonner-Weir S, Ritzel R, Rizza RA, Butler PC (2003) Beta-cell deficit and increased beta-cell apoptosis in humans with type 2 diabetes. Diabetes 52: 102–10

Campisi J (2001) Cellular senescence as a tumor-suppressor mechanism. Trends Cell Biol 11: S27–31

Campisi J (2003) Cellular senescence and apoptosis: how cellular responses might influence ageing phenotypes. Exp Gerontol 38: 5–11

Campisi J (2005) Suppressing cancer: the importance of being senescent. Science 309: 886–87

Castro-Malaspina H, Harris RE, Gajewski J, Ramsay N, Collins R, Dharan B, King R, Deeg HJ (2002) Unrelated donor marrow transplantation for myelodysplastic syndromes: outcome analysis in 510 transplants facilitated by the National Marrow Donor Program. Blood 99: 1943–51

Cawthon RM, Smith KR, O'Brien E, Sivatchenko A, Kerber RA (2003). Association between telomere length in blood and mortality in people aged 60 years or older. Lancet 361: 393–95

Chen J (2004) Senescence and functional failure in hematopoietic stem cells. Exp Hematol 32: 1025–32

Chen J, Astle CM, Harrison DE (2003) Hematopoietic senescence is postponed and hematopoietic stem cell function is enhanced by dietary restriction. Exp Hematol 31:1097–1103

Cheshier SH, Morrison SJ, Liao X, Weissman IL (1999) In vivo proliferation and cell cycle kinetics of long-term self-renewing hematopoietic stem cells. Proc Natl Acad Sci USA 96: 3120–25

Conboy IM, Conboy MJ, Wagers AJ, Girma ER, Weissman IL, Rando TA (2005) Rejuvenation of aged progenitor cells by exposure to a young systemic environment. Nature 433: 760–64

de Haan G, Nijhof W, Van Zant G (1997) Mouse strain-dependent changes in frequency and proliferation of hematopoietic stem cells during ageing: correlation between lifespan and cycling activity. Blood 89: 1543–50

de Haan G, Van Zant G (1999) Dynamic changes in mouse hematopoietic stem cell numbers during ageing. Blood 93: 3294–3301

DeGregori J, Leone G, Miron A, Jakoi L, Nevins JR (1997) Distinct roles for E2F proteins in cell growth control and apoptosis. Proc Natl Acad Sci USA 94: 7245–50

Dietrich N, Bracken AP, Trinh E, Schjerling CK, Koseki H, Rappsilber J, Helin K, Hansen KH (2007) Bypass of senescence by the polycomb group protein CBX8 through direct binding to the INK4A-ARF locus. EMBO J 26: 1637–48

Dimri GP, Lee X, Basile G, Acosta M, Scott G, Roskelley C, Medrano EE, Linskens M, Rubelj I, Pereira-Smith O, et al. (1995) A biomarker that identifies senescent human cells in culture and in ageing skin in vivo. Proc Natl Acad Sci USA 92: 9363–67

Dumble M, Moore L, Chambers SM, Geiger H, Van Zant G, Goodell MA, Donehower LA (2007) The impact of altered p53 dosage on hematopoietic stem cell dynamics during ageing. Blood. 109: 1736–42

Edwards MG, Anderson RM, Yuan M, Kendziorski CM, Weindruch R, Prolla TA (2007) Gene expression profiling of ageing reveals activation of a p53-mediated transcriptional program. BMC genomics 8: 80

Enomoto K, Mimura T, Harris DL, Joyce NC (2006) Age differences in cyclin-dependent kinase inhibitor expression and rb hyperphosphorylation in human corneal endothelial cells. Invest Ophthalmol Vis Sci 47: 4330–40

Enwere E, Shingo T, Gregg C, Fujikawa H, Ohta S, Weiss S (2004) Ageing results in reduced epidermal growth factor receptor signaling, diminished olfactory neurogenesis, and deficits in fine olfactory discrimination. J Neurosci 24: 8354–65

Foster SA, Wong DJ, Barrett MT, Galloway DA (1998) Inactivation of p16 in human mammary epithelial cells by CpG island methylation. Mol Cell Biol 18: 1793–1801

Fried LP, Kronmal RA, Newman AB, Bild DE, Mittelmark MB, Polak JF, Robbins JA, Gardin JM (1998) Risk factors for 5-year mortality in older adults: the Cardiovascular Health Study. JAMA 279: 585–92

Fried LP, Tangen CM, Walston J, Newman AB, Hirsch C, Gottdiener J, Seeman T, Tracy R, Kop WJ, Burke G, et al. (2001) Frailty in older adults: evidence for a phenotype. J Gerontol 56″ M146–56

Gardner RV, Astle CM, Harrison DE (1997) Hematopoietic precursor cell exhaustion is a cause of proliferative defect in primitive hematopoietic stem cells (PHSC) after chemotherapy. Exp Hematol 25: 495–501

Gil J, Bernard D, Martinez D, Beach D (2004) Polycomb CBX7 has a unifying role in cellular lifespan. Nat Cell Biol 6: 67–72

Gray-Schopfer VC, Cheong SC, Chong H, Chow J, Moss T, Abdel-Malek ZA, Marais R, Wynford-Thomas D, Bennett DC (2006) Cellular senescence in naevi and immortalisation in melanoma: a role for p16? Br J Cancer 95: 496–505

Guralnik JM, Eisenstaedt RS, Ferrucci L, Klein HG, Woodman RC (2004) Prevalence of anemia in persons 65 years and older in the United States: evidence for a high rate of unexplained anemia. Blood 104: 2263–68

Harrison DE (1979) Mouse erythropoietic stem cell lines function normally 100 months: loss related to number of transplantations. Mech Ageing Dev 9: 427–33

Henckaerts E, Langer JC, Snoeck HW (2004) Quantitative genetic variation in the hematopoietic stem cell and progenitor cell compartment and in lifespan are closely linked at multiple loci in BXD recombinant inbred mice. Blood 104: 374–79

Herbig U, Ferreira M, Condel L, Carey D, Sedivy JM (2006) Cellular senescence in ageing primates. Science 311: 1257

Hodgson GS, Bradley TR (1984) In vivo kinetic status of hematopoietic stem and progenitor cells as inferred from labeling with bromodeoxyuridine. Exp Hematol 12: 683–87

Huot TJ, Rowe J, Harland M, Drayton S, Brookes S, Gooptu C, Purkis P, Fried M, Bataille V, Hara E, et al. (2002) Biallelic mutations in p16(INK4a) confer resistance to Ras- and Ets-induced senescence in human diploid fibroblasts. Mol Cell Biol 22: 8135–43

Huschtscha LI, Noble JR, Neumann AA, Moy EL, Barry P, Melki JR, Clark SJ, Reddel RR (1998) Loss of p16INK4 expression by methylation is associated with lifespan extension of human mammary epithelial cells. Cancer Res 58: 3508–12

Itahana K, Zou Y, Itahana Y, Martinez JL, Beausejour C, Jacobs JJ, Van Lohuizen M, Band V, Campisi J, Dimri GP (2003) Control of the replicative life span of human fibroblasts by p16 and the polycomb protein Bmi-1. Mol Cell Biol 23: 389–401

Ito K, Hirao A, Arai F, Matsuoka S, Takubo K, Hamaguchi I, Nomiyama K, Hosokawa K, Sakurada K, Nakagata N, et al. (2004) Regulation of oxidative stress by ATM is required for self-renewal of haematopoietic stem cells. Nature 431: 997–1002

Ito K, Hirao A, Arai F, Takubo K, Matsuoka S, Miyamoto K, Ohmura M, Naka K, Hosokawa K, Ikeda Y, et al. (2006) Reactive oxygen species act through p38 MAPK to limit the lifespan of hematopoietic stem cells. Nat Med 12: 446–51

Jacobs JJ, de Lange T (2004) Significant role for p16INK4a in p53-independent telomere-directed senescence. Curr Biol 14: 2302–08

Jacobs JJ, Kieboom K, Marino S, DePinho RA, van Lohuizen M (1999) The oncogene and Polycomb-group gene bmi-1 regulates cell proliferation and senescence through the ink4a locus. Nature 397: 164–68

Janzen V, Forkert R, Fleming HE, Saito Y, Waring MT, Dombkowski DM, Cheng T, DePinho RA, Sharpless NE, Scadden DT (2006) Stem-cell ageing modified by the cyclin-dependent kinase inhibitor p16INK4a. Nature 443: 421–26

Kamijo T, Zindy F, Roussel MF, Quelle DE, Downing JR, Ashmun RA, Grosveld G, Sherr CJ (1997) Tumor suppression at the mouse INK4a locus mediated by the alternative reading frame product p19ARF. Cell 91: 649–59

Kamminga LM, Bystrykh LV, de Boer A, Houwer S, Douma J, Weersing E, Dontje B, de Haan G (2006) The Polycomb group gene Ezh2 prevents hematopoietic stem cell exhaustion. Blood 107: 2170–79

Kim M, Moon HB, Spangrude GJ (2003) Major age-related changes of mouse hematopoietic stem/progenitor cells. Ann N Y Acad Sci 996: 195–208

Kim WY, Sharpless NE (2006) The regulation of INK4/ARF in cancer and ageing. Cell 127: 265–75

Kitada T, Seki S, Kawakita N, Kuroki T, Monna T (1995) Telomere shortening in chronic liver diseases. Biochem Biophys Res Commun 211: 33–39

Kiyono T, Foster SA, Koop JI, McDougall JK, Galloway DA, Klingelhutz AJ (1998) Both Rb/p16INK4a inactivation and telomerase activity are required to immortalize human epithelial cells. Nature 396: 84–88

Knudsen LM, Rasmussen T, Jensen L, Johnsen HE (1999) Reduced bone marrow stem cell pool and progenitor mobilisation in multiple myeloma after melphalan treatment. Med Oncol 16: 245–54

Kollman C, Howe CW, Anasetti C, Antin JH, Davies SM, Filipovich AH, Hegland J, Kamani N, Kernan NA, King R, et al. (2001) Donor characteristics as risk factors in recipients after transplantation of bone marrow from unrelated donors: the effect of donor age. Blood 98: 2043–51

Kondo M, Weissman IL, Akashi K (1997) Identification of clonogenic common lymphoid progenitors in mouse bone marrow. Cell 91: 661–72

Kotake Y, Cao R, Viatour P, Sage J, Zhang Y, Xiong Y (2007) pRB family proteins are required for H3K27 trimethylation and Polycomb repression complexes binding to and silencing p16INK4alpha tumor suppressor gene. Genes & development 21: 49–54

Krishnamurthy J, Ramsey MR, Ligon KL, Torrice C, Koh A, Bonner-Weir S, Sharpless NE (2006). p16INK4a induces an age-dependent decline in islet regenerative potential. Nature 443: 453–57

Krishnamurthy J, Torrice C, Ramsey MR, Kovalev GI, Al-Regaiey K, Su L, Sharpless NE (2004) Ink4a/Arf expression is a biomarker of ageing. J Clin Invest 114: 1299–1307

Kuhn HG, Dickinson-Anson H, Gage FH (1996) Neurogenesis in the dentate gyrus of the adult rat: age-related decrease of neuronal progenitor proliferation. J Neurosci 16: 2027–33

Lang D, Lu MM, Huang L, Engleka KA, Zhang M, Chu EY, Lipner S, Skoultchi A, Millar SE, Epstein JA (2005) Pax3 functions at a nodal point in melanocyte stem cell differentiation. Nature 433: 884–87

Lenhoff S, Hjorth M, Westin J, Brinch L, Backstrom B, Carlson K, Christiansen I, Dahl IM, Gimsing P, Hammerstrom J, et al. (2006). Impact of age on survival after intensive therapy for multiple myeloma: a population-based study by the Nordic Myeloma Study Group. Br J Haematol 133: 389–96

Liang Y, Van Zant G, Szilvassy SJ (2005) Effects of ageing on the homing and engraftment of murine hematopoietic stem and progenitor cells. Blood 106: 1479–87

Lichtman MA, Rowe JM (2004) The relationship of patient age to the pathobiology of the clonal myeloid diseases. Semin Oncol 31: 185–97

Linton PJ, Dorshkind K (2004) Age-related changes in lymphocyte development and function. Nat Immunol 5: 133–39

Maier B, Gluba W, Bernier B, Turner T, Mohammad K, Guise T, Sutherland A, Thorner M, Scrable H (2004) Modulation of mammalian life span by the short isoform of p53. Genes Dev 18: 306–19

Maslov AY, Barone TA, Plunkett RJ, Pruitt SC (2004) Neural stem cell detection, characterization, and age-related changes in the subventricular zone of mice. J Neurosci 24: 1726–33

Matthews C, Gorenne I, Scott S, Figg N, Kirkpatrick P, Ritchie A, Goddard M, Bennett M (2006) Vascular smooth muscle cells undergo telomere-based senescence in human atherosclerosis: effects of telomerase and oxidative stress. Circ Res 99: 156–64

Meier B, Clejan I, Liu Y, Lowden M, Gartner A, Hodgkin J, Ahmed S (2006) trt-1 is the Caenorhabditis elegans catalytic subunit of telomerase. PLOS Genetics 2: e18
Melk A, Kittikowit W, Sandhu I, Halloran KM, Grimm P, Schmidt BM, Halloran PF (2003) Cell senescence in rat kidneys in vivo increases with growth and age despite lack of telomere shortening. Kidney Int 63: 2134–43
Melk A, Schmidt BM, Takeuchi O, Sawitzki B, Rayner DC, Halloran PF (2004) Expression of p16INK4a and other cell cycle regulator and senescence associated genes in ageing human kidney. Kidney Int 65: 510–20
Melzer D, Frayling TM, Murray A, Hurst AJ, Harries LW, Song H, Khaw K, Luben R, Surtees PG, Bandinelli SS (2007) A common variant of the p16(INK4a) genetic region is associated with physical function in older people. Mech Ageing Dev 128: 370–77
Meng A, Wang Y, Van Zant G, Zhou D (2003) Ionizing radiation and busulfan induce premature senescence in murine bone marrow hematopoietic cells. Cancer Res 63: 5414–19
Menzel O, Migliaccio M, Goldstein DR, Dahoun S, Delorenzi M, Rufer N (2006) Mechanisms regulating the proliferative potential of human CD8+ T lymphocytes overexpressing telomerase. J Immunol 177: 3657–68
Michaloglou C, Vredeveld LC, Soengas MS, Denoyelle C, Kuilman T, van der Horst CM, Majoor DM, Shay JW, Mooi WJ, Peeper DS (2005) BRAFE600-associated senescence-like cell cycle arrest of human naevi. Nature 436: 720–24
Mitchell JR, Wood E, Collins K (1999) A telomerase component is defective in the human disease dyskeratosis congenita. Nature 402: 551–55
Miura N, Horikawa I, Nishimoto A, Ohmura H, Ito H, Hirohashi S, Shay JW, Oshimura M (1997) Progressive telomere shortening and telomerase reactivation during hepatocellular carcinogenesis. Cancer Genet Cytogenet 93: 56–62
Molofsky AV, He S, Bydon M, Morrison SJ, Pardal R (2005) Bmi-1 promotes neural stem cell self-renewal and neural development but not mouse growth and survival by repressing the p16Ink4a and p19Arf senescence pathways. Genes Dev 19: 1432–37
Molofsky AV, Pardal R, Iwashita T, Park IK, Clarke MF, Morrison SJ (2003) Bmi-1 dependence distinguishes neural stem cell self-renewal from progenitor proliferation. Nature 425: 962–67
Molofsky AV, Slutsky SG, Joseph NM, He S, Pardal R, Krishnamurthy J, Sharpless NE, Morrison SJ (2006) Increasing p16INK4a expression decreases forebrain progenitors and neurogenesis during ageing. Nature 443: 448–52
Moore KA, Lemischka IR (2006) Stem cells and their niches. Science 311: 1880–85
Morrison SJ, Wandycz AM, Akashi K, Globerson A, Weissman IL (1996) The ageing of hematopoietic stem cells. Nat Med 2: 1011–16
Narita M, Nunez S, Heard E, Lin AW, Hearn SA, Spector DL, Hannon GJ, Lowe SW (2003) Rb-mediated heterochromatin formation and silencing of E2F target genes during cellular senescence. Cell 113: 703–16
Nielsen GP, Stemmer-Rachamimov AO, Shaw J, Roy JE, Koh J, Louis DN (1999) Immunohistochemical survey of p16INK4A expression in normal human adult and infant tissues. Lab Invest 79: 1137–43
Nishimura EK, Granter SR, Fisher DE (2005) Mechanisms of hair graying: incomplete melanocyte stem cell maintenance in the niche. Science 307: 720–24
Obana N, Takagi S, Kinouchi Y, Tokita Y, Sekikawa A, Takahashi S, Hiwatashi N, Oikawa S, Shimosegawa T (2003) Telomere shortening of peripheral blood mononuclear cells in coronary disease patients with metabolic disorders. Intern Med 42: 150–53
Oguro H, Iwama A, Morita Y, Kamijo T, van Lohuizen M, Nakauchi H (2006) Differential impact of Ink4a and Arf on hematopoietic stem cells and their bone marrow microenvironment in Bmi1-deficient mice. J Exp Med 203: 2247–53
Ohtani N, Zebedee Z, Huot TJ, Stinson JA, Sugimoto M, Ohashi Y, Sharrocks AD, Peters G, Hara E (2001) Opposing effects of Ets and Id proteins on p16INK4a expression during cellular senescence. Nature 409: 1067–70
Osawa M, Hanada K, Hamada H, Nakauchi H (1996) Long-term lymphohematopoietic reconstitution by a single CD34-low/negative hematopoietic stem cell. Science 273: 242–45
Park IK, Qian D, Kiel M, Becker MW, Pihalja M, Weissman IL, Morrison SJ, Clarke MF (2003) Bmi-1 is required for maintenance of adult self-renewing haematopoietic stem cells. Nature 423: 302–05

Pasini D, Bracken AP, Helin K (2004) Polycomb group proteins in cell cycle progression and cancer. Cell Cycle 3: 396–400

Passegue E, Wagers AJ, Giuriato S, Anderson WC, Weissman IL (2005) Global analysis of proliferation and cell cycle gene expression in the regulation of hematopoietic stem and progenitor cell fates. J Exp Med 202: 1599–1611

Pearce DJ, Anjos-Afonso F, Ridler CM, Eddaoudi A, Bonnet D (2007) Age dependent increase in SP distribution within hematopoiesis: implications for our understanding of the mechanism of ageing. Stem Cells 25: 828–35

Rane SG, Dubus P, Mettus RV, Galbreath EJ, Boden G, Reddy EP, Barbacid M (1999) Loss of Cdk4 expression causes insulin-deficient diabetes and Cdk4 activation results in beta-islet cell hyperplasia. Nat Genet 22: 44–52

Ressler S, Bartkova J, Niederegger H, Bartek J, Scharffetter-Kochanek K, Jansen-Durr P, Wlaschek M (2006) p16INK4A is a robust in vivo biomarker of cellular ageing in human skin. Ageing Cell 5: 379–89

Reya T, Morrison SJ, Clarke MF, Weissman IL (2001) Stem cells, cancer, and cancer stem cells. Nature 414: 105–11

Rossi DJ, Bryder D, Weissman IL (2007) Hematopoietic stem cell ageing: mechanism and consequence. Exp Gerontol 42: 385–90

Rossi DJ, Bryder D, Zahn JM, Ahlenius H, Sonu R, Wagers AJ, Weissman IL (2005) Cell intrinsic alterations underlie hematopoietic stem cell ageing. Proc Natl Acad Sci USA 102: 9194–99

Sage J, Miller AL, Perez-Mancera PA, Wysocki JM, Jacks T (2003) Acute mutation of retinoblastoma gene function is sufficient for cell cycle re-entry. Nature 424: 223–28

Samani NJ, Boultby R, Butler R, Thompson JR, Goodall AH (2001) Telomere shortening in atherosclerosis. Lancet 358: 472–73

Sasaki M, Ikeda H, Sato Y, Nakanuma Y (2006) Decreased expression of Bmi1 is closely associated with cellular senescence in small bile ducts in primary biliary cirrhosis. Am J Pathol 169: 831–45

Satyanarayana A, Greenberg RA, Schaetzlein S, Buer J, Masutomi K, Hahn WC, Zimmermann, S, Martens U, Manns MP, Rudolph KL (2004) Mitogen stimulation cooperates with telomere shortening to activate DNA damage responses and senescence signaling. Mol Cell Biol 24: 5459–74

Saxena R, Voight BF, Lyssenko V, Burtt NP, de Bakker PI, Chen H, Roix JJ, Kathiresan S, Hirschhorn JN, Daly MJ (2007) Genome-wide association analysis identifies loci for type 2 diabetes and triglyceride levels. Science 316: 1331–36.

Schofield R (1978) The relationship between the spleen colony-forming cell and the haemopoietic stem cell. Blood Cells 4″ 7–25

Scott LJ, Mohlke KL, Bonnycastle LL, Willer CJ, Li Y, Duren WL, Erdos MR, Stringham HM, Chines PS, Jackson AU (2007) A genome-wide association study of type 2 diabetes in Finns detects multiple susceptibility variants. Science 316: 1341–45

Sharpless NE (2005) INK4a/ARF: a multifunctional tumor suppressor locus. Mutat Res 576: 22–38

Sone H, Kagawa Y (2005) Pancreatic beta cell senescence contributes to the pathogenesis of type 2 diabetes in high-fat diet-induced diabetic mice. Diabetologia 48: 58–67

Sparmann A, van Lohuizen M (2006. Polycomb silencers control cell fate, development and cancer. Nat Rev Cancer 6: 846–56

Stein GH, Drullinger LF, Soulard A, Dulic V (1999) Differential roles for cyclin-dependent kinase inhibitors p21 and p16 in the mechanisms of senescence and differentiation in human fibroblasts. Mol Cell Biol 19: 2109–17

Stepanova L, Sorrentino BP (2005) A limited role for p16Ink4a and p19Arf in the loss of hematopoietic stem cells during proliferative stress. Blood 106: 827–32

Sudo K, Ema H, Morita Y, Nakauchi H (2000) Age-associated characteristics of murine hematopoietic stem cells. J Exp Med 192: 1273–80

TeKippe M, Harrison DE, Chen J (2003) Expansion of hematopoietic stem cell phenotype and activity in Trp53-null mice. Exp Hematol 31″ 521–27

Tsuji T, Aoshiba K, Nagai A (2006) Alveolar cell senescence in patients with pulmonary emphysema. Am J Respir Crit Care Med 174: 886–93

Tsutsui T, Hesabi B, Moons DS, Pandolfi PP, Hansel KS, Koff A, Kiyokawa H (1999) Targeted disruption of CDK4 delays cell cycle entry with enhanced p27(Kip1) activity. Mol Cell Biol 19: 7011–19

Tyner SD, Venkatachalam S, Choi J, Jones S, Ghebranious N, Igelmann H, Lu X, Soron G, Cooper B, Brayton C, et al. (2002). p53 mutant mice that display early ageing-associated phenotypes. Nature 415: 45–53

Urabe Y, Nouso K, Higashi T, Nakatsukasa H, Hino N, Ashida K, Kinugasa N, Yoshida K, Uematsu S, Tsuji T (1996) Telomere length in human liver diseases. Liver 16: 293–97

Urbanek K, Torella D, Sheikh F, De Angelis A, Nurzynska D, Silvestri F, Beltrami CA, Bussani R, Beltrami AP, Quaini F, et al. (2005) Myocardial regeneration by activation of multipotent cardiac stem cells in ischemic heart failure. Proc Natl Acad Sci USA 102: 8692–97

von Zglinicki T, Saretzki G, Ladhoff J, d'Adda di Fagagna F, Jackson SP (2005) Human cell senescence as a DNA damage response. Mech Ageing Dev 126: 111–17

Vulliamy T, Marrone A, Goldman F, Dearlove A, Bessler M, Mason PJ, Dokal I (2001) The RNA component of telomerase is mutated in autosomal dominant dyskeratosis congenita. Nature 413: 432–35

Wang W, Chen JX, Liao R, Deng Q, Zhou JJ, Huang S, Sun P (2002) Sequential activation of the MEK-extracellular signal-regulated kinase and MKK3/6–p38 mitogen-activated protein kinase pathways mediates oncogenic ras-induced premature senescence. Mol Cell Biol 22: 3389–3403

Wang Y, Schulte BA, Larue AC, Ogawa M, Zhou D (2006) Total body irradiation selectively induces murine hematopoietic stem cell senescence. Blood 107: 358–66

Weissman IL (2000) Stem cells: units of development, units of regeneration, and units in evolution. Cell 100: 157–68

Wiemann SU, Satyanarayana A, Tsahuridu M, Tillmann HL, Zender L, Klempnauer J, Flemming P, Franco S, Blasco MA, Manns MP, et al. (2002) Hepatocyte telomere shortening and senescence are general markers of human liver cirrhosis. Faseb J 16: 935–42

Wright WE, Shay JW (2002) Historical claims and current interpretations of replicative ageing. Nat Biotechnol 20: 682–88

Xing Z, Ryan MA, Daria D, Nattamai KJ, Van Zant G, Wang L, Zheng Y, Geiger H (2006) Increased hematopoietic stem cell mobilization in aged mice. Blood 108: 2190–97

Yakoub-Agha I, Mesnil F, Kuentz M, Boiron JM, Ifrah N, Milpied N, Chehata S, Esperou H, Vernant JP, Michallet M, et al. (2006) Allogeneic marrow stem-cell transplantation from human leukocyte antigen-identical siblings versus human leukocyte antigen-allelic-matched unrelated donors (10/10) in patients with standard-risk hematologic malignancy: a prospective study from the French Society of Bone Marrow Transplantation and Cell Therapy. J Clin Oncol 24: 5695–5702

Yamaguchi H, Calado RT, Ly H, Kajigaya S, Baerlocher GM, Chanock SJ, Lansdorp PM, Young NS (2005) Mutations in TERT, the gene for telomerase reverse transcriptase, in aplastic anemia. N Engl J Med 352: 1413–24

Yamazaki S, Iwama A, Takayanagi S, Morita Y, Eto K, Ema H, Nakauchi H (2006) Cytokine signals modulated via lipid rafts mimic niche signals and induce hibernation in hematopoietic stem cells. Embo J 25: 3515–23

Yoon KH, Ko SH, Cho JH, Lee JM, Ahn YB, Song KH, Yoo SJ, Kang MI, Cha BY, Lee KW, et al. (2003) Selective beta-cell loss and alpha-cell expansion in patients with type 2 diabetes mellitus in Korea. J Clin Endocrinol Metab 88: 2300–08

Zeggini E, Weedon MN, Lindgren CM, Frayling TM, Elliott KS, Lango H, Timpson NJ, Perry JR, Rayner NW, Freathy RM (2007) Replication of genome-wide association signals in UK samples reveals risk loci for type 2 diabetes. Science 316: 1336–41

Zhou S, Schuetz JD, Bunting KD, Colapietro AM, Sampath J, Morris JJ, Lagutina I, Grosveld GC, Osawa M, Nakauchi H, et al. (2001) The ABC transporter Bcrp1/ABCG2 is expressed in a wide variety of stem cells and is a molecular determinant of the side-population phenotype. Nat Med 7: 1028–34

Zhu J, Woods D, McMahon M, Bishop JM (1998) Senescence of human fibroblasts induced by oncogenic Raf. Genes Dev 12: 2997–3007

Zindy F, Eischen CM, Randle DH, Kamijo T, Cleveland JL, Sherr CJ, Roussel MF (1998) Myc signaling via the ARF tumor suppressor regulates p53-dependent apoptosis and immortalization. Genes Dev 12: 2424–33

Zindy F, Quelle DE, Roussel MF, Sherr CJ (1997) Expression of the p16INK4a tumor suppressor versus other INK4 family members during mouse development and ageing. Oncogene 15: 203–11

Chapter 10
Telomerase as a Potential Regulator of Tissue Progenitor Cells

Steven E. Artandi

Abstract Telomerase, the reverse transcriptase that synthesizes telomeres, comprises two minimal subunits, TERT, the protein subunit, and TERC, the RNA subunit that provides the template for telomere addition (Blackburn 2001, Smogorzewska and de Lange 2004). Telomerase is intimately associated with progenitor cell compartments in vivo and with cancers. This association between telomerase and actively proliferating normal and malignant cells is likely due in part to the need for telomere synthesis to offset the end replication problem. However, recent data implicate telomerase as contributing in a more direct fashion to the proliferation of tissue progenitor cells, and this action of telomerase in facilitating progenitor cell proliferation is independent of its function in adding telomere repeats.

1 Telomerase Is Expressed in Tissue Progenitor Cells in Mouse and Human

One means of gaining insight into telomerase function is to assess how telomerase is expressed in mammalian tissues. This seemingly straightforward question has been complicated by the low abundance of the telomerase enzyme in cells and tissues. TERT protein and TERT mRNA are therefore difficult to detect, which explains why most studies have relied on the PCR-based TRAP assay for telomerase activity. Using the TRAP assay, telomerase is detected in many, and perhaps all, self-renewing tissues, but its expression may be restricted to certain cell compartments within tissues, particularly those enriched in progenitor cells (Forsyth et al. 2002). For example, in human hair follicles, telomerase is expressed most highly in the bulb region, the portion of the follicle that harbors actively dividing progenitor cells during anagen, the active growth phase of the hair follicle cycle (Ramirez et al. 1997). Telomerase was expressed at much lower levels in the bulge region, which contains quiescent multipotent stem cells. In human epidermis, telomerase is expressed

Steven E. Artandi
Department of Medicine, Cancer Biology Program, Stanford, California, United States.
e-mail: sartandi@stanford.edu

K.L. Rudolph (ed.), *Telomeres and Telomerase in Ageing, Disease, and Cancer.*

in the basal layer, which contains dividing progenitor cells that differentiate to regenerate the skin. More differentiated layers of skin did not express telomerase (Harle-Bachor and Boukamp 1996). Similarly, dissection of gastrointestinal tissue showed that telomerase activity was highest in the crypts, which contain the rapidly proliferating progenitor cells that replenish the population of epithelial cells that migrate along the crypt-villus axis and are continuously shed into the lumen of the gut (Bachor et al. 1999, Tahara et al. 1999). Therefore, telomerase is expressed in normal self-renewing human epithelial tissues, particularly in compartments that contain proliferating progenitor cells.

In the hematopoietic system, telomerase expression patterns have been investigated extensively (Lansdorp 2005). In mouse, telomerase is expressed in purified hematopoietic stem cells (HSCs)(Morrison et al. 1996). Similarly, in humans, telomerase is detected in primitive HSCs, although these levels were found to be low when the stem cells were quiescent. Stimulation of HSCs in culture led to cell cycle entry and to a significant increase in the levels of telomerase activity (Chiu et al. 1996, Yui et al. 1998). These results could mean that in vivo telomerase is switched on as HSCs become activated to self-renew or to differentiate. One confounding set of observations that makes a simple model of telomerase regulation in HSC unlikely relates to the fact that telomeres shorten in HSC and in differentiated hematopoietic cells with advancing age (Vaziri et al. 1994, Rufer et al. 1999). Therefore, although telomerase may be activated as HSCs enter cycle, either the levels or the action of telomerase is insufficient to prevent the subsequent telomere shortening that accompanies cell division. A better understanding of telomere dynamics in hematopoiesis will require detailed analyses of telomerase regulation in more committed hematopoietic progenitors as well as in HSCs.

Telomerase activity is detected in mouse embryonic stem (ES) cells (Niida et al. 1998) and human ES cells (Thomson et al. 1998). Telomerase activity markedly diminishes with differentiation of mouse ES cells, consistent with the general paradigm that telomerase is shut off in differentiated cells. Through the use of a TERT promoter fragment driving green fluorescent protein (GFP) in mouse ES cells, silencing of the TERT promoter was shown to be the primary variable underlying the loss of telomerase activity with ES cell differentation (Armstrong et al. 2000). This phenomenon of TERT transcriptional silencing with terminal differentiation has been seen in many cell culture experiments, including cancer cells that retain the ability to differentiate under appropriate conditions (Greenberg et al. 1998). Under some circumstances, telomerase activity is reduced with cell cycle exit, for example, with serum starvation of NIH3T3 cells (Holt et al. 1996). This paradigm of telomerase silencing is recapitulated during embryogenesis in some tissues. For example, in brain, TERT mRNA and telomerase are well expressed at embryonic day 13 in mouse, but diminish during embryogenesis and the first few days of postnatal life. Similar patterns of telomerase expression were seen within the brain, in cerebral cortex, hippocampus, and brain stem (Klapper et al. 2001).

There are many commonalities between mouse and human in terms of patterns of telomerase expression. In both species, telomerase is not expressed in tissues comprised largely of terminally differentiated tissues, such as brain, muscle, and heart (Chadeneau et al. 1995, Prowse and Greider 1995, Artandi et al. 2002). As in human, telomerase is detected in gastrointestinal epithelium, anagen hair follicles (Sarin et al.

2005), and hematopoietic progenitor cells (Morrison et al. 1996) from mouse. Telomerase is detected in lymphocytes from both species and increases with cytokine stimulation of lymphocytes (Weng et al. 1996). There is some evidence that the increase in telomerase activity occurs at a posttranslational level. TERT protein was found to be cytoplasmic in resting human T cells, but with cytokine stimulation TERT translocated to the nucleus (Liu et al. 1999). Some notable differences between mouse and human include fibroblasts and liver. Mouse embryonic fibroblasts are telomerase positive, whereas human fibroblast lines typically derived from embryonic tissues or from foreskin are telomerase negative, at least after adaptation to culture. BJ fibroblasts were found to express low levels of telomerase transiently during S-phase, perhaps reflecting the fact that some human fibroblasts can express telomerase (Masutomi et al. 2003). Mouse liver is telomerase positive, whereas human liver is telomerase negative (Kim et al. 1994, Prowse and Greider 1995).

In many cases, it is unclear which cell populations express telomerase in a complex tissue comprising multiple different cell types. This question of which cells express telomerase would be greatly aided by in situ techniques that allow detection of telomerase components. These approaches have proven challenging because TERT mRNA and TERT protein are nonabundant. One study analyzed TERT mRNA by RNA in situ hybridization in human tissues and found TERT mRNA in the basal layer of the epidermis and the gastrointestinal crypts (Kolquist et al. 1998). Additional investigation of telomerase expression patterns by RNA in situ hybridization in human and mouse is clearly warranted. Detection of TERT protein has been similarly challenging. Our understanding of TERT regulation and the role that TERT plays in tissue homeostasis would be significantly advanced if it could reliably detected by immunohistochemistry or Western blot. In addition to the limitations associated with the low abundance of TERT, the specificity of certain anti-TERT antibodies has been problematic (Wu et al. 2006). Taken together, telomerase appears restricted in its expression pattern in mammalian tissues and is enriched in cycling tissue progenitor cells.

2 TERT Overexpression Causes Progenitor Cells to Proliferate

One role that telomerase serves in these cycling progenitor cells is to synthesize telomere repeats, offsetting the end-replication problem and helping to keep telomeres sufficiently long and stable (see Gutierrez and Ju, this volume). However, TERT overexpression experiments show that TERT can enhance progenitor cell proliferation, raising the possibility that TERT contributes to the proliferative state of progenitor cells. A clear connection between TERT and progenitor cell proliferation was revealed through analysis of the effect of TERT overexpression in skin and hair follicle. Conditional expression of TERT protein in adult mice was achieved using a tetracycline-regulated bi-transgenic approach (Gossen and Bujard 1992, Furth et al. 1994). In this system, the mouse TERT cDNA is placed under control of a tetracycline-inducible promoter comprising a TATA binding site and a heptamerized tet-repressor binding element. The second transgene used a β-actin promoter, CMV enhancer element to direct expression of the chimeric transcription factor rtTA (reverse tetracycline transactivator), which binds to the tet-repressor elements only in the presence of tetraycline or its analog doxycycline. This strategy avoids potential effects of TERT overexpression during development and enables the expression of TERT in adult tissues with temporal control.

Hair follicles cycle between two states, anagen, the active phase, and telogen, the resting phase (Blanpain and Fuchs 2006). The initiation of a new anagen cycle requires the activation of a small number of multipotent stem cells that reside in a niche termed the bulge region (Cotsarelis et al. 1990, Braun et al. 2003, Blanpain et al. 2004, Tumbar et al. 2004). These activated stem cells give rise to the newly generated portion of the anagen follicle, including cycling progenitor cells (matrix cells) that produce the cells of the growing hair shaft (Fig. 10.1). Hair follicle cycles are synchronized in mouse for the first 60–70 days of life, which makes it possible to switch on TERT at specific points in the hair follicle cycle (Muller-Rover et al. 2001). Induction of TERT expression in inducible TERT (i-TERT) mice during telogen triggered the initiation of a new anagen cycle, a developmental phenotype that had not been previously linked to telomerase (Sarin et al. 2005). Using classic label-retaining cell (LRC) analysis, TERT was shown to cause otherwise quiescent bulge stem cells to enter cell cycle. LRC analysis requires injecting mice with the thymidine analog BrdU to serially label proliferating stem cells during early life,

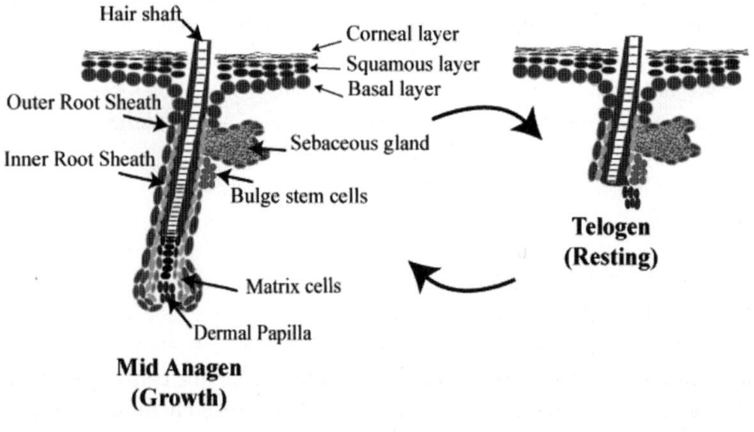

Fig. 10.1 Multipotent epidermal stem cells in the hair follicle bulge region. Hair follicles cycle between two states: telogen, the resting phase, and anagen, the active phase. Quiescent stem cells reside in the bulge region just below the sebaceous gland. A small number of these stem cells become activated, enter cell cycle, and give rise to the lower portion of the new anagen follicle. In anagen, proliferating progenitor cells (matrix cells) surround the dermal papilla and differentiate to produce the growing hair, which is pushed upward through the root sheaths. With injury to the skin, bulge stem cells can also repair interfollicular skin, although the epidermis of the skin is likely maintained by progenitors in the basal layer. Conditional expression of TERT in skin is sufficient to initiate a new anagen cycle and causes quiescent bulge stem cells to proliferate (*See Color Plate*)

followed by a chase period of several weeks, after which time BrdU label is restricted to the stem cells in the bulge region. Switching on TERT in i-TERT mice caused a marked reduction in the number of stem cells in the bulge region that retained detectable BrdU, indicating that TERT caused these cells to divide, thus diluting the label (Sarin et al. 2005). Similarly, expressing TERT constitutively in skin in K5-TERT transgenic mice enhanced the proliferative effect of phorbol esters on quiescent stem cells. After Brdu injection and a chase period, K5-TERT mice and controls were treated with phorbol esters, which causes keratinocyte proliferation. BrdU label was significantly more depleted from the bulge stem cell population in K5-TERT mice compared to controls (Flores et al., 2005). Thus, expression of TERT causes quiescent epidermal stem cells to proliferate and induction of a new anagen hair follicle cycle.

This connection between TERT and progenitor cells may underlie the pro-oncogenic effect of TERT. Overexpression of TERT in mouse tissues has shown a modest but clearly enhanced sensitivity to cancer development. Expressing TERT under control of a keratin-5 promoter, which is active in the basal layer if the skin and the outer root sheath of the hair follicle, led to an approximately twofold increase in the number of papillomas formed when these mice were treated with a classic DMBA, phorbol ester chemical carcinogen protocol (Gonzalez-Suarez et al. 2001). A TERT transgenic mouse expressing TERT from a ubiquitous promoter resulted in spontaneous invasive breast cancers at long latency and to increased numbers of premalignant breast lesions (Artandi et al. 2002). Transgenic expression of TERT in T lymphocytes resulted in lymphomas that were more widely disseminated than controls (Canela et al. 2004). In human cells, overexpression of TERT was required, together with activated H-Ras, to transform human ALT fibroblasts, which maintain their telomeres through a recombination-based mechanism (Stewart et al. 2002). In these experiments, TERT was thought to provide a second activity that supported tumorigenesis because telomeres are not typically limiting in laboratory mice and in ALT fibroblasts. To dissect this further, an epitope-tagged TERT protein that cannot elongate telomeres in human cells substituted effectively for wild-type TERT in the human ALT fibroblasts transformation experiment, supporting the idea that TERT may enhance tumor formation through a telomere length-independent mechanism (Stewart et al. 2002).

TERT's only well-understood molecular function is as a reverse transcriptase adding telomere repeats to chromosome ends. Although critical telomere shortening impairs cell proliferation, the converse is not true, making it unlikely that TERT's ability to enhance proliferation of progenitor cells depends on telomere elongation. Nonetheless, it is crucial to rigorously determine if TERT's telomere lengthening activity is required for its effect on cell proliferation. In i-TERT mice, TERT's ability to activate the hair follicle cycle and induce anagen was completely independent of the telomerase component TERC. Induction of TERT in a TERC-/- background caused initiation of anagen, showing that the anagen-promoting effect of TERT is genetically separable from its role in adding telomere repeats (Sarin et al. 2005). However, in constitutive K5-TERT transgenic mice, the effects of TERT on keratinocytes were found to require TERC. The increased number of papillomas in K5-TERT mice was lost in a TERC-/- background, and the increased size of keratinocyte colonies in culture was also found to depend on TERC (Cayuela et al. 2005, Flores et al. 2005).

The dependence on TERC in the setting of constitutive TERT transgenic expression may implicate TERT's role in telomere elongation. However, no change in overall telomere length was detected in keratinocytes with constitutive TERT overexpression, making this explanation less likely (Flores et al. 2005). Alternatively, the demonstrated need for TERC in some contexts may reflect the fact that TERT and TERC physically interact, with TERC stabilizing TERT or altering its conformation. In the conditional TERT experiments, TERC is clearly dispensable, evidence that any potential TERT-stabilizing effect is not required for TERT's effects on progenitor cells in this conditional context (Sarin et al. 2005). Therefore, TERT is sufficient to cause quiescent epidermal stem cells to enter cycle, initiating the anagen cycle through a mechanism that does not require TERC.

The connection between TERT and progenitor cell proliferation is an exciting new area of telomerase biology, but important questions remain to be addressed. Although TERT clearly has a potent effect on skin progenitors, it is not known if this activity extends to other tissue compartments. While TERT initiates a regenerative program in hair follicle, the consequence of TERT overexpression on other tissues remains to be determined. Furthermore, to what extent does this new TERT activity contribute to cancer in mice or humans? TERT can clearly facilitate progenitor cell proliferation in a gain-of-function context, but it will be important to determine if such an activity can be demonstrated and studied in loss-of-function experiments. Finally, what is the molecular mechanism underlying this function of TERT? Despite intense interest, TERT protein has been largely refractory to biochemical analysis at the endogenous level, in part due to its low expression level and restricted compartments of expression. Further investigations of TERT protein will be critical for revealing what role TERT plays in normal progenitor cells and in cancer.

References

Armstrong L, Lako M, Lincoln J, Cairns PM, Hole N (2000) mTert expression correlates with telomerase activity during the differentiation of murine embryonic stem cells. Mech Dev 97: 109–16

Artandi SE, Alson S, Tietze MK, Sharpless N, Ye S, Greenberg R, Castrillion D, Horner J, Weiler S, Carrasco R, DePinho RA (2002) Constitutive telomerase expression promotes mammary carcinomas in aging mice. Proc Natl Acad Sci USA 99: 8191–96

Bachor C, Bachor OA, Boukamp P (1999) Telomerase is active in normal gastrointestinal mucosa and not up-regulated in precancerous lesions. J Cancer Res Clin Oncol 125: 453–60

Blackburn EH (2001) Switching and signaling at the telomere. Cell 106: 661–73

Blanpain C, Fuchs E (2006) Epidermal stem cells of the skin. Annu Rev Cell Dev Biol 22: 339–73

Blanpain C, Lowry WE, Geoghegan A, Polak L, Fuchs E (2004) Self-renewal, multipotency, and the existence of two cell populations within an epithelial stem cell niche. Cell 118: 635–48

Braun KM, Niemann C, Jensen UB, Sundberg JP, Silva-Vargas V, Watt FM (2003) Manipulation of stem cell proliferation and lineage commitment: visualisation of label-retaining cells in wholemounts of mouse epidermis. Development 130: 5241–55

Canela A, Martin-Caballero J, Flores JM, Blasco MA (2004) Constitutive expression of tert in thymocytes leads to increased incidence and dissemination of T-cell lymphoma in Lck-Tert mice. Mol Cell Biol 24: 4275–93

Cayuela ML, Flores JM, Blasco MA (2005) The telomerase RNA component Terc is required for the tumour-promoting effects of Tert overexpression. EMBO Rep 6: 268–74

Chadeneau C, Siegel P, Harley CB, Muller WJ, Bacchetti S (1995) Telomerase activity in normal and malignant murine tissues. Oncogene 11: 893–98

Chiu CP, Dragowska W, Kim NW, Vaziri H, Yui J, Thomas TE, Harley CB, Lansdorp PM (1996) Differential expression of telomerase activity in hematopoietic progenitors from adult human bone marrow. Stem Cells 14: 239–48

Cotsarelis G, Sun TT, Lavker RM (1990) Label-retaining cells reside in the bulge area of pilosebaceous unit: implications for follicular stem cells, hair cycle, and skin carcinogenesis. Cell 61: 1329–37

Flores I, Cayuela ML, Blasco MA (2005) Effects of telomerase and telomere length on epidermal stem cell behavior. Science 309: 1253–56

Forsyth NR, Wright WE, Shay JW (2002) Telomerase and differentiation in multicellular organisms: turn it off, turn it on, and turn it off again. Differentiation 69: 188–97

Furth PA, St Onge L, Boger H, Gruss P, Gossen M, Kistner A, Bujard H, Hennighausen L (1994) Temporal control of gene expression in transgenic mice by a tetracycline-responsive promoter. Proceedings of the National Academy of Sciences, USA 91: 9302–06

Gonzalez-Suarez E, Samper E, Ramirez A, Flores JM, Martin-Caballero J, Jorcano JL, Blasco MA (2001) Increased epidermal tumors and increased skin wound healing in transgenic mice overexpressing the catalytic subunit of telomerase, mTERT, in basal keratinocytes. Embo J 20: 2619–30

Gossen M, Bujard H (1992) Tight control of gene expression in mammalian cells by tetracycline-responsive promoters. Proc Natl Acad Sci USA 89: 5547–51

Greenberg RA, Allsopp RC, Chin L, Morin GB, DePinho RA (1998) Expression of mouse telomerase reverse transcriptase during development, differentiation and proliferation. Oncogene 16: 1723–30

Harle-Bachor C, Boukamp P (1996) Telomerase activity in the regenerative basal layer of the epidermis inhuman skin and in immortal and carcinoma-derived skin keratinocytes. Proc Natl Acad Sci USA 93: 6476–81

Holt SE, Wright WE, Shay JW (1996) Regulation of telomerase activity in immortal cell lines. Mol Cell Biol 16: 2932–39

Kim NW, Piatyszek MA, Prowse KR, Harley CB, West MD, Ho PL, Coviello GM, Wright WE, Weinrich SL, Shay JW (1994) Specific association of human telomerase activity with immortal cells and cancer. Science 266: 2011–15

Klapper W, Shin T, Mattson MP (2001) Differential regulation of telomerase activity and TERT expression during brain development in mice. J Neurosci Res 64: 252–60

Kolquist KA, Ellisen LW, Counter CM, Meyerson M, Tan LK., Weinberg RA., Haber D A, Gerald WL (1998). Expression of TERT in early premalignant lesions and a subset of cells in normal tissues. Nat Genet 19: 182–86

Lansdorp PM (2005) Role of telomerase in hematopoietic stem cells. Ann NY Acad Sci 1044: 220–27

Liu K, Schoonmaker MM, Levine BL, June CH, Hodes RJ, Weng NP (1999) Constitutive and regulated expression of telomerase reverse transcriptase (hTERT) in human lymphocytes. Proc Natl Acad Sci USA 96: 5147–52

Masutomi K, Yu EY, Khurts S, Ben-Porath I, Currier JL, Metz GB, Brooks MW, Kaneko S, Murakami S, DeCaprio JA, et al. (2003) Telomerase maintains telomere structure in normal human cells. Cell 114, 241–53

Morrison SJ, Prowse KR, Ho P, Weissman IL (1996) Telomerase activity in hematopoietic cells is associated with self-renewal potential. Immunity 5: 207–16

Muller-Rover S, Handjiski B, van der Veen C, Eichmuller S, Foitzik K, McKay IA, Stenn KS, Paus R (2001) A comprehensive guide for the accurate classification of murine hair follicles in distinct hair cycle stages. J Invest Dermatol 117: 3–15

Niida H, Matsumoto T, Satoh H, Shiwa M, Tokutake Y, Furuichi Y, Shinkai Y (1998) Severe growth defect in mouse cells lacking the telomerase RNA component. Nature Genetics 19: 203–06

Prowse KR, Greider CW (1995) Developmental and tissue-specific regulation of mouse telomerase and telomere length. Proc Natl Acad Sci USA 92: 4818–22

Ramirez RD, Wright WE, Shay JW, Taylor RS (1997) Telomerase activity concentrates in the mitotically active segments of human hair follicles. J Invest Dermatol 108: 113–17

Rufer N, Brummendorf TH, Kolvraa S, Bischoff C, Christensen K, Wadsworth L, Schulzer M, Lansdorp PM (1999) Telomere fluorescence measurements in granulocytes and T lymphocyte subsets point to a high turnover of hematopoietic stem cells and memory T cells in early childhood. J Exp Med 190: 157–67

Sarin KY, Cheung P, Gilison D, Lee E, Tennen RI, Wang E, Artandi MK, Oro AE, Artandi SE (2005) Conditional telomerase induction causes proliferation of hair follicle stem cells. Nature 436: 1048–52

Smogorzewska A, de Lange T (2004) Regulation of telomerase by telomeric proteins. Annu Rev Biochem 73: 177–208

Stewart SA, Hahn WC, O'Connor BF, Banner EN, Lundberg AS, Modha P, Mizuno H, Brooks MW, Fleming M, Zimonjic DB, et al. (2002) Telomerase contributes to tumorigenesis by a telomere length-independent mechanism. Proc Natl Acad Sci USA 99: 12606–11

Tahara H, Yasui W, Tahara E, Fujimoto J, Ito K, Tamai K, Nakayama J, Ishikawa F, Ide T (1999) Immuno-histochemical detection of human telomerase catalytic component, hTERT, in human colorectal tumor and non-tumor tissue sections. Oncogene 18: 1561–67

Thomson JA, Itskovitz-Eldor J, Shapiro SS, Waknitz MA, Swiergiel JJ, Marshall VS, Jones, J. M. (1998). Embryonic stem cell lines derived from human blastocysts. Science 282: 1145–47

Tumbar T, Guasch G, Greco V, Blanpain C, Lowry WE, Rendl M, Fuchs E (2004) Defining the epithelial stem cell niche in skin. Science 303: 359–63

Vaziri H, Dragowska W, Allsopp RC, Thomas TE, Harley CB, Lansdorp PM (1994) Evidence for a mitotic clock in human hematopoietic stem cells: loss of telomeric DNA with age. Proc Natl Acad Sci USA 91: 9857–60

Weng NP, Levine BL, June CH, Hodes RJ (1996) Regulated expression of telomerase activity in human T lymphocyte development and activation. J Exp Med 183: 2471–79

Wu YL, Dudognon C, Nguyen E, Hillion J, Pendino F, Tarkanyi I, Aradi J, Lanotte M, Tong JH, Chen GQ, Segal-Bendirdjian E (2006) Immunodetection of human telomerase reverse-transcriptase (hTERT) re-appraised: nucleolin and telomerase cross paths. J Cell Sci 119: 2797–2806

Yui J, Chiu CP, Lansdorp PM (1998) Telomerase activity in candidate stem cells from fetal liver and adult bone marrow. Blood 91: 3255–62

Part III
Telomeres, DNA Damage and Cancer

Chapter 11
Telomere Shortening and Telomerase Activation during Cancer Formation

K. Lenhard Rudolph

Abstract Telomere shortening limits the proliferative capacity of human cells by inducing cell cycle arrest/senescence or apoptosis. Telomere shortening and the activation of these checkpoints represent tumor suppressor mechanisms in humans. In contrast, telomere shortening can induce chromosomal instability during ageing and chronic disease. This correlates with an increased cancer risk. This chapter summarizes experimental data from mouse models as well as data from primary human cancer demonstrating that telomere shortening and telomerase activation have ambiguous functions in cancer formation.

1 Telomere Shortening and Telomerase Activation Characterize Human Tumorigenesis

Most somatic tissues in adult humans lack telomerase activity. The absence of telomerase expression and the end replication problem of DNA-polymerase lead to a continuous shortening of telomeres in replicating cells (see Allsopp, this volume). In addition, the accumulation of reactive oxygen species (ROS) can accelerate telomere shortening (see Passos et al., this volume). During ageing significant telomere shortening occurs in most human tissues and telomere shortening is accelerated in chronic diseases that increase the rate of cell turnover (for review see Djojosubroto et al. 2003).

The telomere hypothesis of cellular ageing indicates that telomere shortening represents a tumor suppressor mechanism. In response to critical shortening, telomeres lose capping function and dysfunctional telomeres induce cell cycle arrest, senescence, or apoptosis (Wright and Shay 1992, Chin et al. 1999, d'Adda di Fagagna et al. 2003, Herbig et al. 2004, Choudhury et al. 2007; see also Gutierrez and Ju, this volume). Telomere shortening and the activation of these checkpoints constrain cell proliferation and prevent immortal proliferation of cells (Wright and Shay 1992). Telomere shortening and the activation of these checkpoints. This mechanism

K. Lanhard Rudolph
Institute of Moleculare Medicine and Max-Planck-Research-Group on Stem Cell Aging, University of Ulm, Albert-Einstein-Allee 11, 89081 Ulm, Germany
e-mail: Lenhard.Rudolph@uni-ulm.de

K.L. Rudolph (ed.), *Telomeres and Telomerase in Ageing, Disease, and Cancer.*

protects organisms from cancer, since the proliferative capacity of transformed cancer cells that arise during a lifetime is limited by telomere shortening not allowing tumor mass growth or metastasis. In agreement with this hypothesis, over 80% of human cancers show an activation of telomerase (for review see Satyanarayana et al. 2004, Ju and Rudolph 2006) – the enzyme, which stabilizes telomeres thus facilitating immortal proliferation of cells (see Artandi, this volume). The activation of telomerase is a general phenomenon in different types of human cancer, thus making it a promising target for cancer therapy (Shay and Wright 2002, see also Keith and Bilsland, this volume). In human cancer the activation of telomerase often correlates with tumor progression. Premalignant tumors show no or very little telomerase activity, but an increase in activity is seen in malignant tumors and metastasis (for review see Satyanarayana et al. 2004). A subset of human cancers is telomerase negative. It has been postulated that these tumors exhibit alternative mechanisms of telomere lengthening (ALT). ALT has been demonstrated in experimental cell culture systems and appears to involve recombination of telomeric DNA (Muntoni and Reddel 2005). Together, we can conclude that an activation of telomere-maintenance mechanisms is a prerequisite for human cancer formation. This assumption was proven by cell culture experiments showing that the transformation of human cells into tumor-forming cancer cells requires the activation of telomerase (Hahn et al. 1999).

The above data indicate that suppression of telomerase in most somatic human tissues represents a tumor suppressor mechanism. According to this hypothesis, telomere shortening and loss of regenerative reserve during ageing and in response to chronic disease may represent a "side effect" of this tumor suppressor mechanism. This model is complicated by the fact that certain stem cell compartments in humans express telomerase activity such as the hematopoietic stem cells (HSC) as well as intestinal stem cells in the basal crypts (Yui et al. 1998, Tahara et al. 1999). An emerging hypothesis in molecular oncology indicates that stem cells represent the cell of origin of different types of cancers (Clarke et al. 2006). One could speculate that telomerase suppression is not a rate-limiting step in tumor formation since it is already expressed in stem cells (Armanios and Greider 2005). However, there is evidence that the level of telomerase expression in stem cells is not sufficient to maintain telomeres stable. It has been shown that $CD34^+$ hematopoietic cells show significant telomere shortening during human ageing (Vaziri et al. 1994). Therefore, it seems likely that telomere shortening limits the proliferative capacity of human stem cells during ageing. In addition, these data suggest that transformed stem cells need to upregulate telomerase to achieve immortality. In agreement with this assumption, an upregulation of telomerase occurs in most malignancies of the hematopoietic system, which originate from telomerase-positive hematopoietic stem and progenitor cells (for review see Ju and Rudolph, 2006).

Another finding complicates the hypothesis that telomere shortening simply acts as a tumor suppressor mechanism: telomere shortening correlates with increased cancer risk. As summarized above most human tissues show significant telomere shortening during ageing. Telomere shortening during ageing is not associated with improved tumor suppression, but the cancer risk sharply increases during ageing (Hayat et al. 2007). Similarly, telomere shortening is associated with chronic diseases, which accelerate the rate of cell turnover, e.g., colitis ulcerosa, chronic hepatitis. In these diseases, critical telomere shortening correlates with the

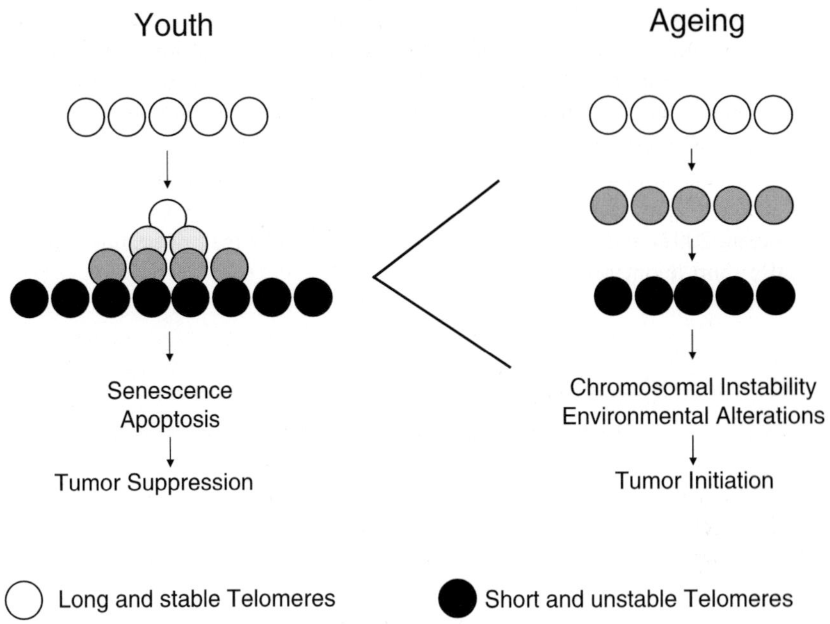

Fig. 11.1 Telomere shortening has a dual role in cancer formation. At young age telomere shortening acts as a tumor suppressor: Telomeres are long in most organ cells. If at this stage a single cell transforms into a cancer cell, clonal expansion of the transformed cell leads to telomere shortening and checkpoint activation (senescence and apoptosis) inhibiting the progression of the tumor. In contrast, at old age and at the end stage of chronic disease (e.g. liver cirrhosis), a growing number of organ cells show critically short telomeres. This increases the tumor risk by inducing chromosomal instability and environmental alteration (cytokine secretion). It is possible that during ageing and chronic disease, these mechanisms of tumor initiation become dominant leading to a net increase in overall tumor risk.

progression of disease and an increasing cancer risk (O'Sullivan et al. 2002, Wiemann et al. 2002). Moreover, mutations in dyskerin or telomerase induce telomere shortening, and this is again associated with an increased cancer risk in patients with the disease (see Du et al., this volume). Together, these data indicate that telomere shortening may in fact increase the cancer risk during ageing and at the end stage of chronic diseases, whereas telomere shortening can also limit the transformation of human cells and cancer growth (see above). A possible explanation for this apparent paradox is that telomere shortening protects from cancer in young organisms when the average telomere length of organ cells is long and telomeres in most cells are stable. In this situation an emerging tumor cell clone will experience telomere shortening as a consequence of cell division and critical telomere shortening will eventually stop expansion of this clone by induction of senescence or apoptosis. In contrast, when telomeres become critically short on whole organ level (during ageing or disease) the organ loses regenerative capacity and telomeres become instable in an increasing number of organ cell. This instability then increases the rate of cancer initiation, possibly outweighing the tumor suppressor effects of telomere shortening on emerging tumor clones (Fig. 11.1).

In agreement with the assumption that telomere shortening can induce cancer initiation, it has been observed that most human cancers exhibit critically short telomeres, much shorter than the surrounding nontransformed tissues (for review see Satyanarayana et al. 2004). This has been observed for many kinds of cancers, including hematopoietic malignancies, colorectal cancer, liver cancer, breast cancer, pancreatic cancer, and others. In addition, telomere shortening is seen already at early tumor stages (van Heek et al. 2002, Meeker et al. 2004, Chin et al. 2004, Plentz et al. 2007), indicating that tumors arise from cells (cell of origin) that have critically short telomeres.

2 Telomere Shortening Increases Tumor Initiation but Suppresses Tumor Progression in Telomerase-Deficient Mice

Similar to the results in human cancer, studies in telomerase knockout mice have revealed a dual role of telomere shortening in cancer formation.

Telomerase knockout mice lack the RNA component of telomerase (mTERC$^{-/-}$, Blasco et al. 1997; see also Blasco, this volume). First-generation (G1) mTERC$^{-/-}$ mice do not show an obvious organ phenotype due to long telomere reserves in laboratory mouse strains (Prowse and Greider 1995). When G1 mTERC$^{-/-}$ mice are crossed with each other, significant telomere shortening is seen in the second generation of knockout mice (G2). Using this mating scheme, successive telomere shortening occurs from one generation to the next, resulting in critically short telomeres and telomere dysfunction in late-generation animals (G3-G6, Blasco et al. 1997). The onset of telomere dysfunction depends on the initial telomere length in individual mouse strains. Mouse strains with relatively short telomeres reach a critically short length in earlier generations compared to mouse strains with longer telomeres (Herrera et al. 1999). Late generation mTERC$^{-/-}$ mice show a variety of premature ageing phenotypes predominantly affecting stem cell compartments and organ systems with high rates of cell turnover (Rudolph et al. 1999, Herrera et al. 1999, Choudhury et al. 2007; see also Chang, this volume; Gutierrez and Ju, this volume; Artandi, this volume; Blasco, this volume).

In line with the model that telomere shortening represents a tumor suppressor mechanism, late-generation mTERC$^{-/-}$ mice show an impaired development of macroscopic cancers in response to carcinogen treatment (Gonzalez-Suarez et al. 2000, Farazi et al. 2003) or when crossed with tumor-prone, genetic mouse models of cancer (Greenberg at al. 1999a, Rudolph et al. 2001). In contrast to its role in tumor suppression, telomere shortening in mTERC$^{-/-}$ mice is also associated with increased rates of cancer initiation. mTERC$^{-/-}$ mice on a mixed genetic background showed increased rates of spontaneous cancer formation during ageing (Rudolph et al. 1999). Moreover, APC-mutant mice exhibited an increased rate of intestinal tumor initiation (microadenoma and dysplastic crypts) when crossed with mTERC$^{-/-}$ mice (Rudolph et al. 2001). Similarly, the rate of liver tumor initiation in response to carcinogen-treatment was increased in late-generation $mTERC^{-/-}$ mice compared to $mTERC^{+/+}$ mice (Farazi et al. 2003). Moreover, mTERC deficiency accelerated cancer formation in response to destabilization of telomeres induced by TRF2 overexpression (Blanco et al. 2007; see also Blasco, this volume). Together, these data indicate that telomere dysfunction suppresses tumor progression in tumor-prone mice but, in contrast, can increase tumor initiation during ageing and in response to genetic lesions or carcinogen exposure.

There is evidence that loss of checkpoint function can accelerate tumor formation in response to telomere dysfunction, specifically the loss of p53 checkpoint function (Chin et al. 1999). Moreover, heterozygous deletion of p53 cooperated with telomere dysfunction and increased the incidence of epithelial cancers in ageing $mTERC^{-/-}$, $p53^{+/-}$ mice (Artandi et al. 2000). This finding is interesting since an increase in epithelial cancer is a hallmark of human ageing, which is characterized by telomere shortening, and often the tumors show defects in p53 checkpoint function (see above and Satyanarayana et al. 2004, Lowe et al. 2004).

3 Molecular Mechanisms of Tumor Initiation in Response to Telomere Shortening

The main function of telomeres is to cap chromosomal ends and to stabilize chromosomes (McEachern et al. 2000). Specifically, telomere capping is necessary to distinguish the chromosome ends from DNA breaks within the genome. While DNA breaks induce checkpoint responses and DNA repair of broken ends, the natural ends of the chromosome should not provoke such responses. Thus, telomere capping prevents the inappropriate activation of DNA repair pathways and the induction of chromosomal fusion. For capping function, telomeres need to have a minimum length and, in addition, need to form three-dimensional structures (e.g., telomere loops, g-quadruplexes) that are stabilized by specific telomere-binding proteins (de Lange 2004; see also Blasco, this volume; Dillin and Karlseder, this volume; Chang, this volume).

Most human cancers that evolve during ageing and at the end stage of chronic diseases are associated with high rates of chromosomal instability (CIN). A current concept in molecular oncology indicates that CIN induces genetic lesions that promote the step-wise progression of altered cells to transform into cancer cells (Michor et al. 2005). Since telomere shortening is associated with an increased cancer risk during ageing and chronic disease, it is conceivable that loss of telomere capping function contributes to the induction of chromosomal instability and cancer initiation (Fig. 11.2). In agreement with this assumption, telomere shortening in humans correlates with the evolution of CIN in chronic diseases associated with an increased cancer risk, e.g., ulcerative colitis and liver cirrhosis (O'Sullivan et al. 2002), as well as with the evolution of aneuploidy on the single-cell level in human hepatocellular carcinoma (Plentz et al. 2005).

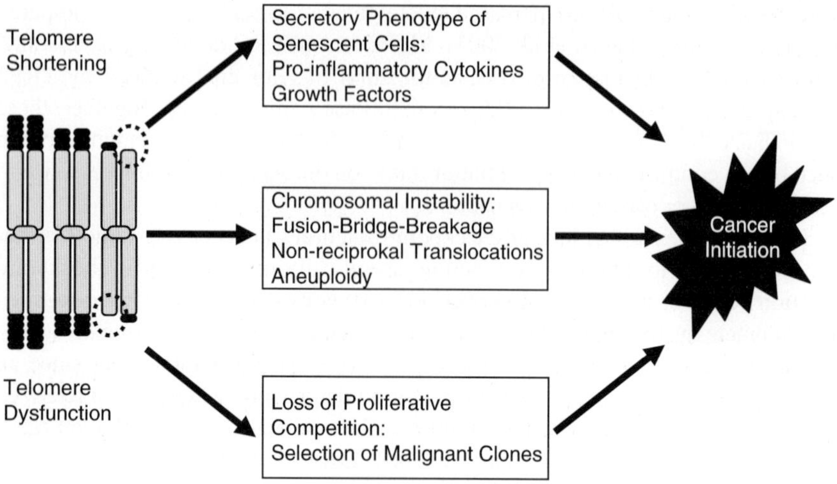

Fig. 11.2 Telomere shortening increases tumor initiation. Different mechanisms can increase the tumor risk in response to telomere shortening and telomere dysfunction: (i) Telomere dysfunction induces chromosomal instability by inducing fusion-bridge-breakage cycles resulting in aneuploidy and non-reciprocal tranlocations; (ii) cell with dysfunctional telomeres show an activation of stress signalling pathways (p38) leading to a an increased secretion of cytokines and growth factors from senescent cells; (iii) telomere shortening limits the proliferation of stem and progenitor cells in organs. This loss of proliferative competition leads to a selection of malignant stem and progenitor cells that have an increased or abnormal proliferation capacity

Experimental support for the telomere hypothesis of cancer initiation has come from studies in mTERC$^{-/-}$ mice. Increased tumor initiation in mTERC$^{-/-}$ mice was associated with increased rates of chromosomal instability in tumor cells (Artandi et al. 2000). These studies have also shown that heterozygous deletion of p53 can increase CIN and tumor initiation specifically in epithelial compartments (Artandi et al. 2000). The molecular basis of increased cancer formation in response to telomere dysfunction and checkpoint abrogation remains to be defined. Virtually all tumors in mTERC$^{-/-}$, p53$^{+/-}$ mice showed a loss of heterozygosity of the remaining p53 allele, indicating that telomere dysfunction can increase LOH, which then promotes cancer formation. In addition, the tumors showed high rates of CIN, including nonreciprocal translocations – a hallmark of human cancer during ageing. It is an interesting area of investigation to map these chromosomal lesions aiming to identify recurrent lesions that are functionally involved in carcinogenesis (O'Hagan et al. 2002).

In addition to its role in initiation of CIN, telomere dysfunction could promote tumor formation by inducing environmental alterations. Primary human fibroblasts develop a secretory phenotype in response to telomere dysfunction and senescence (Parrinello et al. 2005). Senescent fibroblasts secrete elevated amounts of proinflammatory cytokines and matrix-modulating proteins. Co-culture experiments indicate that these secreted factors alter proliferation and differentiation of epithelial cells (Parrinello et al. 2005). Studies in mTERC$^{-/-}$ mice have provided evidence that the secretory phenotype occurs in response to telomere dysfunction in vivo. Telomere dysfunction in late-generation mTERC$^{-/-}$ mice induces environmental alterations that impair the function and engraftment of hematopoietic stem cells (Ju et al. 2007; see also Gutierrez and Ju, this volume). It is conceivable that such environmental alterations would influence the selective outgrowth of (pre-) malignant cell clones in ageing organs and tissues, thus promoting cancer formation (Fig. 11.2). Moreover it is possible that loss of proliferative competition itself promotes cancer formation in response to telomere shortening. It has been shown that the loss of replicative competition of hematopoietic stem and progenitor cells promotes selective outgrowth of malignant hematopoietic clones and leukemogenesis (Bilousova et al. 2005). It is conceivable that telomere shortening would induce the same process, since it impairs the proliferative capacity of stem and progenitor cell compartments (Choudhury et al. 2007).

3 Molecular Mechanisms of Tumor Suppression in Response to Telomere Shortening

A current concept indicates that DNA damage checkpoints impair cell proliferation in response to telomere dysfunction thereby suppressing tumor formation. In agreement with this model, inactivation of the p53 and/or Rb pathways allows human fibroblast to bypass the M1-senescent checkpoint (Wright and Shay 1992, Gire and Wynford-Thomas 1998; see also Allsopp, this volume; Gutierrez and Ju, this volume). In addition, there is a second mortality stage (M2) that limits survival of cells that bypassed senescence. M2 is characterized by very short telomeres, high levels of CIN, and massive cell death. This stage has therefore been called crisis (Wright and Shay 1992). The crisis checkpoint is much less defined than the senescence checkpoint. Both crisis and senescence represent potent barriers protecting the organisms against cancer. Vice versa, defects in these checkpoints may allow proliferation of cells with critically short telomeres and could cooperate with telomere dysfunction during initiation of CIN and cancer.

In agreement with this concept, p53 deletion cooperates with telomere dysfunction to induce chromosomal instability and initiation of epithelial cancer in ageing telomerase knockout mice (Artandi et al. 2000; see above). However, the cooperation between loss of DNA damage checkpoints and telomere dysfunction appears to be more complex and needs to be further elucidated. In contrast to p53 deletion, the deletion of ATM does not lead to an increase in cancer formation in mTERC$^{-/-}$ mice but accelerated ageing in the double mutant mice (Wong et al. 2003). These data indicate that ATM-independent mechanisms can activate checkpoints in response to telomere dysfunction possibly involving ATR, which has been implicated in replicative senescence of human fibroblasts (d'Adda di Fagagna et al. 2003). In addition, the data suggest that loss of ATM may increases DNA damage in telomere dysfunctional mice, thus accelerating ageing phenotypes.

It is possible that the role of ATM in telomere capping (Verdun and Karlseder 2006) or detoxification of reactive oxygen species (Barlow et al. 1999) is responsible for increased DNA/telomere damage and accelerated ageing of mTERC$^{-/-}$, ATM$^{-/-}$ double mutant mice.

In contrast to ATM deletion, the knockout of p21 rescues premature ageing and life span of mTERC$^{-/-}$ mice (Choudhury et al. 2007). p21 is one of the primary targets of p53-inducing senescence of human cells in response to telomere dysfunction (Brown et al. 1997). Notably, p21 deletion did not increase the rate of cancer formation in ageing mTERC$^{-/-}$ mice. A possible explanation is that p21 deletion rescues cell cycle arrest and stem cell depletion in mTERC$^{-/-}$ mice but leaves apoptosis checkpoints intact (Choudhury et al. 2007). Apoptosis checkpoints could remove genetically instable cells with high rates of CIN, thus preventing the accumulation of CIN and cancer initiation in p21$^{-/-}$, TERC$^{-/-}$ mice. In agreement with this hypothesis, the inhibition of p53-dependent apoptosis in p53 mutant mice leads to an increase of cancer (Cosme-Blanco et al. 2007). However, when apoptosis-deficient, p53 mutant mice were crossed with mTERC$^{-/-}$ mice tumor induction was impaired, correlating with a strong activation of p21 and senescence associated β -galactosidase in mouse tissues (Cosme-Blanco et al. 2007; see also Chang, this volume). Together, these data suggest that both pathways, p53-dependent apoptosis and p53-dependent cell cycle arrest, can compensate each other to suppress tumor formation in response to telomere dysfunction (Fig. 11.3).

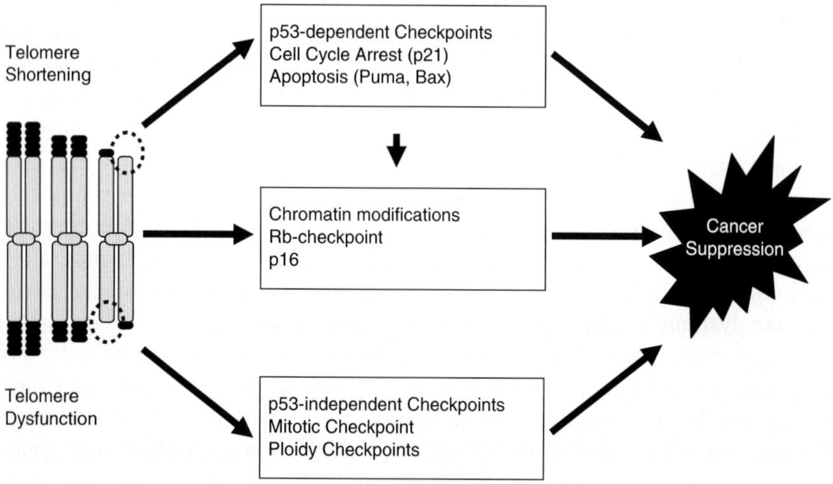

Fig. 11.3 Telomere shortening suppresses tumor progression: (i) Telomere dysfunction induces p53-dependent checkpoints limiting cell proliferation (p21) and this checkpoint could also involve p53 target genes including apoptosis (PUMA, Bax); (ii) In addition, telomere dysfunction can limit cell survival by inducing p53-independent checkpoints. The nature of these checkpoints remains to be analyzed but could involve mitotic checkpoints or ploidy checkpoints. (iii) Telomere dysfunction and the activation of checkpoint responses induce chromatin modification and the activation of the Rb-checkpoint leading to cell cycle arrest

In agreement with the pivotal role of p53 in human carcinogenesis, tumor suppression in telomere-dysfunctional mice is associated with an activation of p53 in tumor cells as well as with increased rates of tumor cell apoptosis and impaired tumor cell proliferation (Rudolph et al. 2001). In addition, there is emerging evidence that p53-independent checkpoints can impair cancer formation in response to telomere dysfunction and chromosomal instability in vivo. Acute induction of telomere dysfunction by inhibition of the telomere-binding protein TRF2 leads to p53-independent apoptosis in liver cells of $p53^{-/-}$ mice (Lechel et al. 2005). In addition, p53-independent checkpoints impair proliferation and induce apoptosis in $p53^{-/-}$ tumors with dysfunctional telomeres (Lechel et al. 2007). These findings are in line with the observation that two checkpoints (M1 and M2) can limit proliferation of primary human cells in response to telomere shortening. In human fibroblasts the senescence checkpoint is p53-dependent, but the crisis checkpoint is p53-independent (Wright and Shay 1992; see also Allsopp, this volume). While the above studies have provided experimental evidence that p53-independent checkpoints can impair tumor progression in mouse models, the molecular components of this checkpoint and its relevance to human carcinogenesis need to be explored in greater detail. Studies in mice and human cells have shown that the activation of p53-independent checkpoints in response to telomere dysfunction correlates with the accumulation of DNA damage and the evolution of massive chromosomal instability (Wright and Shay 1992, Lechel et al. 2007, Fig. 11.3), suggesting that these checkpoints involve DNA damage signaling and ploidy checkpoints.

5 Molecular Mechanisms Promoting Progression of Telomere-Dysfunctional Tumors

As outlined above, most human tumors exhibit critically short telomeres, suggesting that tumors arise from genetically instable cells with short, dysfunctional telomeres. However, telomere dysfunction induces p53-dependent and p53-indepenedent checkpoints that limit tumor progression. These circumstances indicate that tumors need to activate mechanisms that limit telomere dysfunction, thus allowing tumor progression. Telomerase activation represents the most common pathway for stabilizing telomeres in human tumors (see Artandi, this volume; Keith and Bilsland, this volume, Fig. 11.4).

Telomerase consists of two essential components: (1) telomerase reverse transcriptase (TERT), the catalytic active component of the enzyme (Lingner et al. 1997), and (2) telomerase RNA component (TERC), a functional RNA serving as a template for telomere synthesis (Greider and Blackburn 1989; see also Dillin and Karlseder, this volume; Blasco, this volume). Most human cells and organs lack telomerase activity due the suppression of TERT expression (Meyerson et al. 1997). TERT is expressed in stem and progenitor cell compartments of some human tissues and organs and correlates with detectable levels of telomerase activity

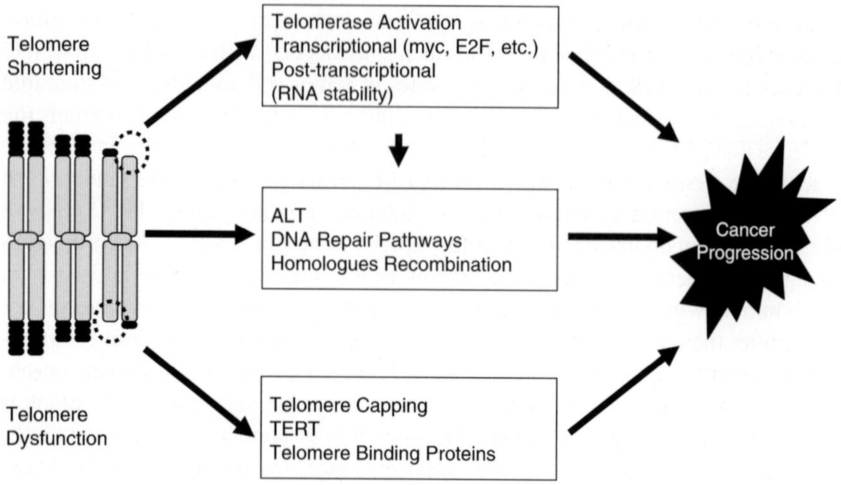

Fig. 11.4 Telomere stabilization induces tumor progression. (i) Telomerase activations is the most common pathway of telomere stabilization in human cancers an >80% of human cancers are telomerase positive; (ii) 10–20% of human cancers are telomerase negative and stabilize telomeres by activation of telomerase-independent telomere lengthening mechanism (ALT) involving DNA homologous recombination, (iii) It is possible that an altered expression of telomere binding proteins and TERT contributes to telomere stabilization in tumor cells independent of telomerase activity by improving telomere capping function

in these compartments (see Artandi, this volume). There is experimental evidence that the expression of TERT itself is required for normal function of stem and progenitor cells (see Artandi, this volume). In addition, telomerase expression could be necessary to ensure that stem and progenitor cells have a higher replicative potential compared to somatic cells (Blasco, this volume; Gutierrez and Ju, this volume; Zimmermann and Martens, this volume). However, as outlined above, the expression level of TERT and the level of telomerase activity in stem and progenitor cells is not sufficient to maintain a stable telomere length in this compartment during ageing. It can be assumed that cancer stem cells also experience telomere shortening and need to activate TERT expression levels to allow immortal proliferation and tumor progression. The same could apply for TERC expression. Although TERC is more ubiquitously expressed than TERT, its expression level increases in most human cancers compared to nontransformed tissue (see Keith and Bilsland, this volume). There is experimental evidence that overexpression of TERT and TERC can cooperate to increase telomere length in human cells (Cristofari and Lingner 2006). In addition, TERT expression can stabilize telomere function without increasing the length of the telomeres, suggesting that TERT itself can improve telomere capping (Zhu et al. 1999, Fig. 11.4).

Studies on telomerase-deficient mice and studies on inhibition of telomerase in human cancer cells have proven the functional relevance of telomerase for tumor progression. Impaired tumor progression in response to telomerase deletion or in response to telomerase inhibition is associated with increased telomere dysfunction, accumulation of DNA damage, and activation of p53-dependent and p53-independent checkpoint responses (see above). In addition, TERT expression was necessary for the experimental transformation of primary human cells (Hahn et al. 1999). Together, these data underscore that telomerase represents a valuable target for anticancer therapies (see Keith and Bilsland, this volume). An active area of cancer research is to analyze molecular pathways that reactivate telomerase in cancer cells. Overexpression of *myc* represents the first oncogenic pathway, which was shown to activate telomerase through binding to the TERT promoter (Greenberg et al. 1999b, Wang et al. 1998). Experimental data from mouse models have shown that activation of telomerase represents a functional relevant mechanism by which *myc* promotes tumorigenesis (Flores et al. 2006). A variety of additional transcription factors have been identified that regulate TERT expression (see, for example, Alonso et al. 2006). In addition, there is experimental evidence for posttranscriptional regulation of TERT-expression in cancer cells (Emerald et al. 2007). However, the functional relevance of these regulatory networks for cancer progression needs to be evaluated. In contrast to the growing understanding on the regulation of TERT, the regulatory units controlling TERC expression and its upregulation in tumor cells are much less defined. A growing understanding of the molecular pathways controlling telomerase reactivation in cancer cells will likely lead to the development of new therapies aiming to impair telomerase activity in cancer cells, thereby blocking tumor progression. Of note, first clinical trials to explore the use of telomerase inhibitors or TERT vaccines for anticancer therapy have been initiated (see Keith and Bilsland, this volume).

In addition to an activation of telomerase, an improvement in telomere capping function could allow proliferation and survival of tumor cells with critically short telomeres. It has been proposed that overexpression of TERT improves telomere capping independent of telomere elongation (Zhu et al. 1999). It is known that telomere-binding proteins control the formation and stabilization of telomere loops that are necessary for telomere capping function. In principle, it is possible that alterations in telomere binding proteins could optimize capping function of critically short telomeres. Vice versa, altered expression of telomere-binding proteins could also destabilize capping function of telomeres and could increase chromosomal instability and tumor initiation. In agreement with this assumption, overexpression of the telomere-binding protein TRF2 increases telomere dysfunction and tumor initiation in mTERC$^{-/-}$ mice (Blanco et al. 2007; see also Blasco, this volume). A variety of studies have demonstrated an altered expression of telomere-binding proteins in human cancer compared to nontransformed cells and tissues (see for example Bellon et al. 2006). A detailed understanding of the equilibrium of telomere-binding proteins and its function in controlling telomere capping could point to therapeutic targets of cancer prevention and treatment.

Ten percent to 10-20% of human malignancies do not express telomerase but stabilize telomeres by an alternative mechanism of telomere elongation, which has been named ALT (Muntoni and Reddel 2005, Fig. 11.4). Compared to telomerase, the molecular basis of ALT is less well defined, but experimental evidence suggests that ALT involves DNA repair pathways (Jiang et al. 2007). It is also unknown why certain tumors choose ALT instead of telomerase reactivation for telomere maintenance. Overexpression of TERT in ALT cancer cell lines is sufficient to reactivate telomerase (Greenberg and Rudolph, unpublished observation). This observation suggests that ALT cancers can in principle reactivate telomerase but fail to upregulate TERT expression during transformation. Human sarcomas represent tumors that are often telomerase-negative and show an activation of ALT in 20%–60% of the cases (Johnson et al. 2005). Sarcomas can derive from mesenchymal stem cells (MSCs), which in humans show little or no telomerase expression (Zimmermann et al. 2003, Parsch et al. 2004). In this regard MSCs are different from other stem cell compartments in humans that express telomerase (see above). The suppression of telomerase in MSCs could provide a plausible explanation why sarcomas derived from MSCS more often fail to upregulate telomerase compared to other cancer types derived from telomerase-positive stem cell compartments. One could speculate that ALT might be a more frequent mechanism of immortalization in tumors deriving from telomerase-negative cells compared to tumors deriving from telomerase-positive cells. In addition, ALT could represent an important escape mechanism of telomerase-positive tumors treated with telomerase inhibitors (see Keith and Bilsland, this volume). Together, a further analysis of the molecular mechanisms of ALT is clearly warranted to further strengthen therapeutic approaches aiming to target telomeres for anticancer treatment.

6 Conclusions

Telomeres and telomerase have a dual role in cancer initiation and progression. Telomere shortening can induce chromosomal instability, environmental alterations, and cancer initiation. In contrast, telomere shortening induces checkpoints that limit tumor progression, and the stabilization of telomeres represents an essential step during tumorigenesis. The disclosure of molecular mechanisms downstream of telomere dysfunction, as well as of mechanisms that can stabilize telomere function, will likely be of clinical value to improve cancer treatment, prevention, and screening.

References

Alonso MM, Fueyo J, Yung WK, Gomez-Manzano C (2006) E2F1 and telomerase: alliance in the dark side. Cell Cycle 5: 930–35

Armanios M, Greider CW (2005) Telomerase and cancer stem cells. Cold Spring Harb Symp Quant Biol. 70: 205–08

Artandi SE, Chang S, Lee SL, Alson S, Gottlieb GJ, Chin L, DePinho RA (2000) Telomere dysfunction promotes non-reciprocal translocations and epithelial cancers in mice. Nature 406: 641–45

Barlow C, Dennery PA, Shigenaga MK, Smith MA, Morrow JD, Roberts LJ 2nd, Wynshaw-Boris A, Levine RL (1999) Loss of the ataxia-telangiectasia gene product causes oxidative damage in target organs. Proc Natl Acad Sci USA 96(17): 9915–19

Bellon M, Datta A, Brown M, Pouliquen JF, Couppie P, Kazanji M, Nicot C (2006) Increased expression of telomere length regulating factors TRF1, TRF2 and TIN2 in patients with adult T-cell leukemia. Int J Cancer 119: 2090–97

Bilousova G, Marusyk A, Porter CC, Cardiff RD, DeGregori J (2005) Impaired DNA replication within progenitor cell pools promotes leukemogenesis. PLOS Biol 3: e401

Blanco R, Munoz P, Flores JM, Klatt P, Blasco MA (2007) Telomerase abrogation dramatically accelerates TRF2-induced epithelial carcinogenesis. Genes Dev 21: 206–20

Blasco MA, Lee HW, Hande MP, Samper E, Lansdorp PM, DePinho RA, Greider CW (1997) Telomere shortening and tumor formation by mouse cells lacking telomerase RNA. Cell 91: 25–34

Brown JP, Wei W, Sedivy JM (1997) Bypass of senescence after disruption of p21CIP1/WAF1 gene in normal diploid human fibroblasts. Science 277: 831–34

Chin L, Artandi SE, Shen Q, Tam A, Lee SL, Gottlieb GJ, Greider CW, DePinho RA (1999) p53 deficiency rescues the adverse effects of telomere loss and cooperates with telomere dysfunction to accelerate carcinogenesis. Cell 97: 527–38

Chin K, de Solorzano CO, Knowles D, Jones A, Chou W, Rodriguez EG, Kuo WL, Ljung BM, Chew K, Myambo K, Miranda M, Krig S, Garbe J, Stampfer M, Yaswen P, Gray JW, Lockett SJ (2004) In situ analyses of genome instability in breast cancer. Nat Genet. 36: 984–88

Choudhury AR, Ju Z, Djojosubroto MW, Schienke A, Lechel A, Schaetzlein S, Jiang H, Stepczynska A, Wang C, Buer J, Lee HW, von Zglinicki T, Ganser A, Schirmacher P, Nakauchi H, Rudolph KL (2007) Cdkn1a deletion improves stem cell function and lifespan of mice with dysfunctional telomeres without accelerating cancer formation. Nat Genet 39: 99–105

Clarke MF, Dick JE, Dirks PB, Eaves CJ, Jamieson CH, Jones DL, Visvader J, Weissman IL, Wahl GM (2006) Cancer stem cells–perspectives on current status and future directions: AACR workshop on cancer stem cells. Cancer Res 66: 9339–44

Cosme-Blanco W, Shen MF, Lazar AJ, Pathak S, Lozano G, Multani AS, Chang S (2007) Telomere dysfunction suppresses spontaneous tumorigenesis in vivo by initiating p53-dependent cellular senescence. EMBO Rep 8(5): 497–503

Cristofari G, Lingner J (2006) Telomere length homeostasis requires that telomerase levels are limiting. EMBO J 25: 565–74

d'Adda di Fagagna F, Reaper PM, Clay-Farrace L, Fiegler H, Carr P, Von Zglinicki T, Saretzki G, Carter NP, Jackson SP (2003) A DNA damage checkpoint response in telomere-initiated senescence. Nature 426: 194–98

de Lange T (2004) T-loops and the origin of telomeres. Nat Rev Mol Cell Biol 5: 323–29

Djojosubroto MW, Choi YS, Lee HW, Rudolph KL (2003) Telomeres and telomerase in ageing, regeneration and cancer. Mol Cells 15(2): 164–75

Emerald BS, Chen Y, Zhu T, Zhu Z, Lee KO, Gluckman PD, Lobie PE (2007) AlphaCP1 mediates stabilization of hTERT mRNA by autocrine human growth hormone. J Biol Chem 282: 680–90

Farazi PA, Glickman J, Jiang S, Yu A, Rudolph KL, DePinho RA (2003) Differential impact of telomere dysfunction on initiation and progression of hepatocellular carcinoma. Cancer Res 63(16): 5021–27

Flores I, Evan G, Blasco MA (2006) Genetic analysis of myc and telomerase interactions in vivo. Mol Cell Biol 26: 6130–38

Gire V, Wynford-Thomas D (1998) Reinitiation of DNA synthesis and cell division in senescent human fibroblasts by microinjection of anti-p53 antibodies. Mol Cell Biol 18(3): 1611–21

Gonzalez-Suarez E, Samper E, Flores JM, Blasco MA (2000) Telomerase-deficient mice with short telomeres are resistant to skin tumorigenesis. Nat Genet 26: 114–17

Greenberg RA, Chin L, Femino A, Lee KH, Gottlieb GJ, Singer RH, Greider CW, DePinho RA (1999a) Short dysfunctional telomeres impair tumorigenesis in the INK4a(delta2/3) cancer-prone mouse. Cell 97: 515–25

Greenberg RA, O'Hagan RC, Deng H, Xiao Q, Hann SR, Adams RR, Lichtsteiner S, Chin L, Morin GB, DePinho RA (1999b) Telomerase reverse transcriptase gene is a direct target of c-Myc but is not functionally equivalent in cellular transformation. Oncogene 18: 1219–26

Greider CW, Blackburn EH (1989) A telomeric sequence in the RNA of Tetrahymena telomerase required for telomere repeat synthesis. Nature 337: 331–37

Hahn WC, Counter CM, Lundberg AS, Beijersbergen RL, Brooks MW, Weinberg RA (1999) Creation of human tumour cells with defined genetic elements. Nature 400: 464–68

Hayat MJ, Howlader N, Reichman ME, Edwards BK (2007) Cancer statistics, trends, and multiple primary cancer analyses from the Surveillance, Epidemiology, and End Results (SEER) Program. Oncologist 12: 20–37

Herbig U, Jobling WA, Chen BP, Chen DJ, Sedivy JM (2004) Telomere shortening triggers senescence of human cells through a pathway involving ATM, p53, and p21(CIP1), but not p16(INK4a). Mol Cell 14: 501–13

Herrera E, Samper E, Martin-Caballero J, Flores JM, Lee HW, Blasco MA (1999) Disease states associated with telomerase deficiency appear earlier in mice with short telomeres. EMBO J 18(11): 2950–60

Jiang WQ, Zhong ZH, Henson JD, Reddel RR (2007) Identification of candidate alternative lengthening of telomeres genes by methionine restriction and RNA interference. Oncogene 26(32):4635–47.

Johnson JE, Varkonyi RJ, Schwalm J, Cragle R, Klein-Szanto A, Patchefsky A, Cukierman E, von Mehren M, Broccoli D (2005) Multiple mechanisms of telomere maintenance exist in liposarcomas. Clin Cancer Res 11: 5347–55

Ju Z, Rudolph KL (2006) Telomeres and telomerase in cancer stem cells. Eur J Cancer 42(9): 1197–203

Ju Z, Jiang H, Jaworski M, Gompf A, Rathinam C, Klein C, Trumpp A, Rudolph KL (2007) Telomere dysfunction induces environmental defects limiting hematopoietic stem/progenitor cell function and engraftment. Nature Med 13(6):742–7.

Lechel A, Satyanarayana A, Ju Z, Plentz RR, Schaetzlein S, Rudolph C, Wilkens L, Wiemann SU, Saretzki G, Malek NP, Manns MP, Buer J, Rudolph KL (2005) The cellular level of telomere dysfunction determines induction of senescence or apoptosis in vivo. EMBO Rep 6: 275–81

Lechel A, Holstege H, Begus Y, Schienke A, Kamino K, Lehmann U, Kubicka S, Schirmacher P, Jonkers J, Rudolph KL (2007) Telomerase deletion limits progression of p53-mutant hepatocellular carcinoma with short telomeres in chronic liver disease. Gastroenterology, 132(4):1465–75.

Lingner J, Hughes TR, Shevchenko A, Mann M, Lundblad V, Cech TR (1997) Reverse transcriptase motifs in the catalytic subunit of telomerase. Science 276: 561–67

Lowe SW, Cepero E, Evan G (2004) Intrinsic tumour suppression. Nature 432: 307–15

McEachern MJ, Krauskopf A, Blackburn EH (2000) Telomeres and their control. Annu Rev Genet 34: 331–58

Meeker AK, Hicks JL, Iacobuzio-Donahue CA, Montgomery EA, Westra WH, Chan TY, Ronnett BM, De Marzo AM (2004) Telomere length abnormalities occur early in the initiation of epithelial carcinogenesis. Clin Cancer Res 10: 3317–26

Meyerson M, Counter CM, Eaton EN, Ellisen LW, Steiner P, Caddle SD, Ziaugra L, Beijersbergen RL, Davidoff MJ, Liu Q, Bacchetti S, Haber DA, Weinberg RA (1997) hEST2, the putative human telomerase catalytic subunit gene, is up-regulated in tumor cells and during immortalization. Cell 90: 785–95

Michor F, Iwasa Y, Vogelstein B, Lengauer C, Nowak MA (2005) Can chromosomal instability initiate tumorigenesis? Semin Cancer Biol 15: 43–49

Muntoni A, Reddel RR (2005) The first molecular details of ALT in human tumor cells. Hum Mol Genet 14: R191–96

O'Hagan RC, Chang S, Maser RS, Mohan R, Artandi SE, Chin L, DePinho RA (2002) Telomere dysfunction provokes regional amplification and deletion in cancer genomes. Cancer Cell 2: 149–55

O'Sullivan JN, Bronner MP, Brentnall TA, Finley JC, Shen WT, Emerson S, Emond MJ, Gollahon KA, Moskovitz AH, Crispin DA, Potter JD, Rabinovitch PS (2002) Chromosomal instability in ulcerative colitis is related to telomere shortening. Nat Genet 32: 280–84

Parrinello S, Coppe JP, Krtolica A, Campisi J (2005) Stromal-epithelial interactions in ageing and cancer: senescent fibroblasts alter epithelial cell differentiation. J Cell Sci 118(Pt 3): 485–96

Parsch D, Fellenberg J, Brummendorf TH, Eschlbeck AM, Richter W (2004) Telomere length and telomerase activity during expansion and differentiation of human mesenchymal stem cells and chondrocytes. J Mol Med 82: 49–55

Plentz RR, Schlegelberger B, Flemming P, Gebel M, Kreipe H, Manns MP, Rudolph KL, Wilkens L (2005) Telomere shortening correlates with increasing aneuploidy of chromosome 8 in human hepatocellular carcinoma. Hepatology 42(3): 522–26

Plentz RR, Park YN, Lechel A, Kim H, Nellessen F, Langkopf BH, Wilkens L, Destro A, Fiamengo B, Manns MP, Roncalli M, Rudolph KL (2007) Telomere shortening and inactivation of cell cycle checkpoints characterize human hepatocarcinogenesis. Hepatology 45: 968–76

Prowse KR, Greider CW (1995) Developmental and tissue-specific regulation of mouse telomerase and telomere length. Proc Natl Acad Sci USA 92: 4818–22

Rudolph KL, Millard M, Bosenberg MW, DePinho RA (2001) Telomere dysfunction and evolution of intestinal carcinoma in mice and humans. Nat Genet 28: 155–59

Rudolph KL, Chang S, Lee HW, Blasco M, Gottlieb GJ, Greider C, DePinho RA (1999) Longevity, stress response, and cancer in ageing telomerase-deficient mice. Cell 96(5): 701–12

Satyanarayana A, Manns MP, Rudolph KL (2004) Telomeres and telomerase: a dual role in hepatocarcinogenesis. Hepatology 40: 276–83

Shay JW, Wright WE (2002) Telomerase: a target for cancer therapeutics. Cancer Cell 2(4): 257–65

Tahara H, Yasui W, Tahara E, Fujimoto J, Ito K, Tamai K, Nakayama J, Ishikawa F, Tahara E, Ide T (1999) Immuno-histochemical detection of human telomerase catalytic component, hTERT, in human colorectal tumor and non-tumor tissue sections. Oncogene 18: 1561–67

van Heek NT, Meeker AK, Kern SE, Yeo CJ, Lillemoe KD, Cameron JL, Offerhaus GJ, Hicks JL, Wilentz RE, Goggins MG, De Marzo AM, Hruban RH, Maitra A (2002) Telomere shortening is nearly universal in pancreatic intraepithelial neoplasia. Am J Pathol 161: 1541–47

Vaziri H, Dragowska W, Allsopp RC, Thomas TE, Harley CB, Lansdorp PM (1994) Evidence for a mitotic clock in human hematopoietic stem cells: loss of telomeric DNA with age. Proc Natl Acad Sci USA 91: 9857–60

Verdun RE, Karlseder J (2006) The DNA damage machinery and homologous recombination pathway act consecutively to protect human telomeres. Cell 127: 709–20

Wang J, Xie LY, Allan S, Beach D, Hannon GJ (1998) Myc activates telomerase. Genes Dev 12: 1769–74

Wiemann SU, Satyanarayana A, Tsahuridu M, Tillmann HL, Zender L, Klempnauer J, Flemming P, Franco S, Blasco MA, Manns MP, Rudolph KL (2002) Hepatocyte telomere shortening and senescence are general markers of human liver cirrhosis. FASEB J 16: 935–42

Wong KK, Maser RS, Bachoo RM, Menon J, Carrasco DR, Gu Y, Alt FW, DePinho RA (2003) Telomere dysfunction and Atm deficiency compromises organ homeostasis and accelerates ageing. Nature 421: 643–48

Wright WE, Shay JW (1992) The two-stage mechanism controlling cellular senescence and immortalization. Exp Gerontol 27(4): 383–89

Yui J, Chiu CP, Lansdorp PM (1998) Telomerase activity in candidate stem cells from fetal liver and adult bone marrow. Blood 91: 3255–62

Zhu J, Wang H, Bishop JM, Blackburn EH (1999) Telomerase extends the lifespan of virus-transformed human cells without net telomere lengthening. Proc Natl Acad Sci USA 96: 3723–28

Zimmermann S, Voss M, Kaiser S, Kapp U, Waller CF, Martens UM (2003) Lack of telomerase activity in human mesenchymal stem cells. Leukemia 17: 1146–49

Chapter 12
Telomere Binding Proteins and Disease

Maria A. Blasco

Abstract An increasingly growing number of human diseases including cancer and age-related diseases have been associated to telomeric dysfunction. The progressive telemore loss that occurs with aging in humans is proposed to contribute to the pathobiology of age-related diseases. Evidence for this comes from the study of mouse models that reproduce the corresponding full-blown human pathology only when in the setting of critically short telemores. Moreover, telomerase itself, as well as several telomere-binding proteins, has been found altered in human pathologies. Here, we will review recent findings suggesting that the length and function of telomeres are a biological determinant in the development of certain human diseases associated with aging, particularly from the perspective of mouse models for telomere dysfunction.

1 Chromatin Structure of Mammalian Telomeres

1.1 Telomeric DNA

Vertebrate telomeres end in a single-stranded 3′overhang of the G-rich DNA strand (de Lange et al. 2002; see Dillin and Karlseder, this volume), which is thought to be generated by mechanisms that involve the postreplicative processing of the leading-strand telomere (Bailey et al. 2001) and provides the substrate for telomerase (Blackburn 2001, Chan and Blackburn 2002). The overhang is bound by a single-strand telomere binding protein known as protection of telomeres 1 (Pot1) (Fig. 12.1), which has been proposed to regulate the access of telomerase to the telomere (Baumann and Cech 2001). Pot1 interacts with TPP1 and the two proteins are thought to form a heterodimer that is involved in telomere length regulation (Ye et al. 2004, Liu et al. 2004). In addition, the G-strand overhang can fold back and invade the double-stranded region of the telomere, generating a looped structure known as the telomere loop or t-loop (Griffith

Maria A. Blasco
Telomeres and Telomerase Group, Molecular Oncology Program, Spanish National Cancer Centre (CNIO), 3 Melchor Fernández Almagro, Madrid E-28029, Spain.
e-mail: mblasco@cnio.es

K.L. Rudolph (ed.), *Telomeres and Telomerase in Ageing, Disease, and Cancer*.

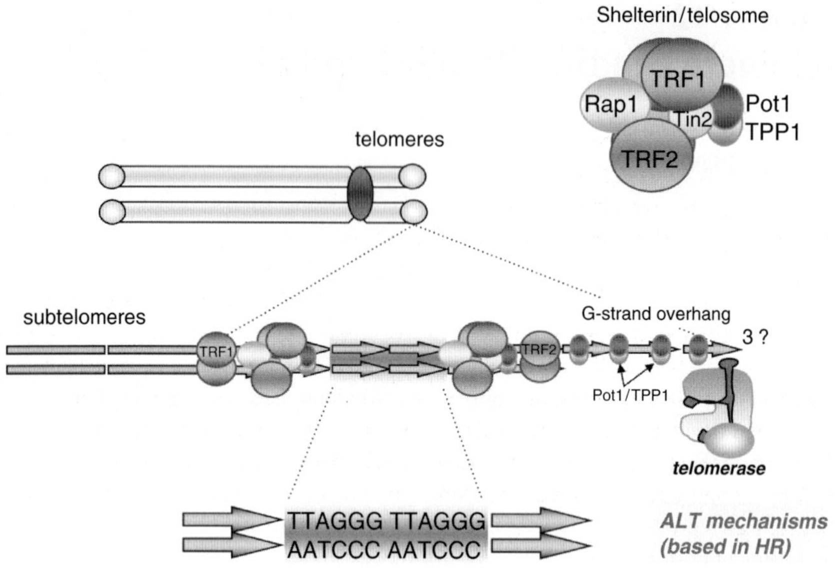

Fig. 12.1 Structure of mammalian telomeres. Double-stranded telomeric repeats are bound by a multiprotein complex known as *shelterin* or the *telosome*, which comprises TRF1, TRF2, Tin2, Rap1, Pot1, and TPP1. The G-strand overhang is also bound by the Pot1/TPP1 heterodimer. Telomerase is able to recognize the 3′end of the G-strand overhang and to elongate telomeres. In telomerase-negative cells, telomeres can be maintained by mechanisms involving homologous recombination between telomeric repeats, the so-called alternative lengthening of telomeres or ALT (*See Color Plate*)

et al. 1999), which provides telomere protection by sequestering the 3′ end of telomeres from repair and degradation activities. t-loops resemble intermediates of homologous recombination and may be processed by activities involved in this DNA repair pathway, such as the XRCC3 protein (de Lange 2004, Wang et al. 2004).

1.2 Telomere Repeat Binding Factors

Telomere repeats are bound by the so-called shelterin multiprotein complex (de Lange et al. 2005; see also Chang, this volume). Shelterin contains factors that bind directly to both single-strand repeats, such as the Pot1/PTOP heterodimer, or to double-strand telomeric repeats such as the telomere binding factors TRF1 and TRF2 and their interacting proteins Rap1 (**r**epressor/**a**ctivator **p**rotein 1) and Tin2 (**T**RF1-**I**nteracting **N**uclear Factor 2) (Fig. 12.1) (de Lange et al. 2005). TRF1 also interacts with the TANK1 and TANK2 poly(ADP)-ribosylases (also known as tankyrases) (Smith et al. 1998). TRF1 and the TRF1-interacting proteins have been proposed to regulate telomere length, probably by controlling the access of telomerase to the telomere (Smogorzewska et al. 2000).

TRF2 and Pot1 have additional roles in telomere protection, as indicated by the fact that they prevent end-to-end chromosome fusions (van Steensel et al. 1998, Celli et al. 2005, Hockemeyer et al. 2006). The mechanisms underlying the roles of TRF2 and Pot1 in telomere protection may be related to their interplay with DNA-damage signaling and repair factors (reviewed in Blasco 2005). In particular, TRF2 can interact with components of the Mre11 complex (**m**eiotic **r**ecombination 11) (Zhu et al. 2000), which is central to both nonhomologous end joining (NHEJ) and homologous recombination (HR). TRF2 has been proposed to bind to ATM and to cancel the ATM-triggered damage response, suggesting that TRF2 has a role in preventing a DNA damage response at telomeres (Karlseder et al. 2004, Bradshaw et al. 2005). Pot 1 has recently been proposed to provide a similar function, as indicated by the fact that Pot1-deficient mice show increased DNA damage signaling at telomeres (Hockemeyer et al. 2006; see also Chang, this volume).

In addition to the shelterin proteins, components of the NHEJ (Ku and DNA-PKcs) and HR (Rad51D, Rad54) pathways have been also shown to have roles in telomere length regulation and telomere capping (Bailey et al. 2001, Samper et al. 2000, Tarsounas et al. 2004, Jaco et al. 2003).

1.3 Histone Modifications at Mammalian Telomeres

Telomeres and subtelomeres are bound by nucleosomes (de Lange et al. 1990, Makarov et al. 1993) that are enriched in histone marks characteristic of constitutive heterochromatin domains (Garcia-Cao et al. 2004, Gonzalo et al. 2005, Gonzalo et al. 2006). Two important histone marks at telomeric and subtelomeric domains are trimethylation of H3K9 and H4K20 (Garcia-Cao et al. 2004, Gonzalo et al. 2005, Gonzalo et al. 2006) (Fig. 12.2), which are carried out by the histone methyltransferases suppressor of variegation 3-9 homologues (Suv39h1 and h2) and suppressor of variegation 4-20 homologues (Suv4-20h1 and h2), respectively (Peters et al. 2001, Schotta et al. 2004, Kourmouli et al. 2004). In addition, proteins of the retinoblastoma family of tumor suppressors (Rb, p107, and p130) interact with the Suv420h1 and h2 HMTases to maintain trimethylation of H4K20, both at telomeric and pericentric chromatin (Gonzalo et al. 2005) (Fig. 12.2). Both telomeres and subtelomeres are also enriched in CBX1, CBX3, and CBX5 (the homologues of the *Drosophila* heterochromatin proteins HP1β, HP1γ and HP1α, respectively) (Fig. 12.2). These proteins are recruited to chromatin through their affinity for trimethylated H3K9 residues and are important for chromatin compaction at heterochromatic domains (Garcia-Cao et al. 2004, Gonzalo et al. 2005, Gonzalo et al. 2006, Lachner et al. 2001). In addition, telomeric and subtelomeric repeats are characterized by low levels of acetylated H3 (AcH3) and H4 (AcH4) (Benetti et al. 2007), in agreement with increased silencing in these regions.

These epigenetic marks are characteristic of compacted heterochromatin domains, such as those found at pericentric heterochromatin, suggesting that telomeric and pericentric chromatin domains share a similar chromatin architecture (Peters et al. 2001, Schotta et al. 2004, Kourmouli et al. 2004, Lachner et al. 2001).

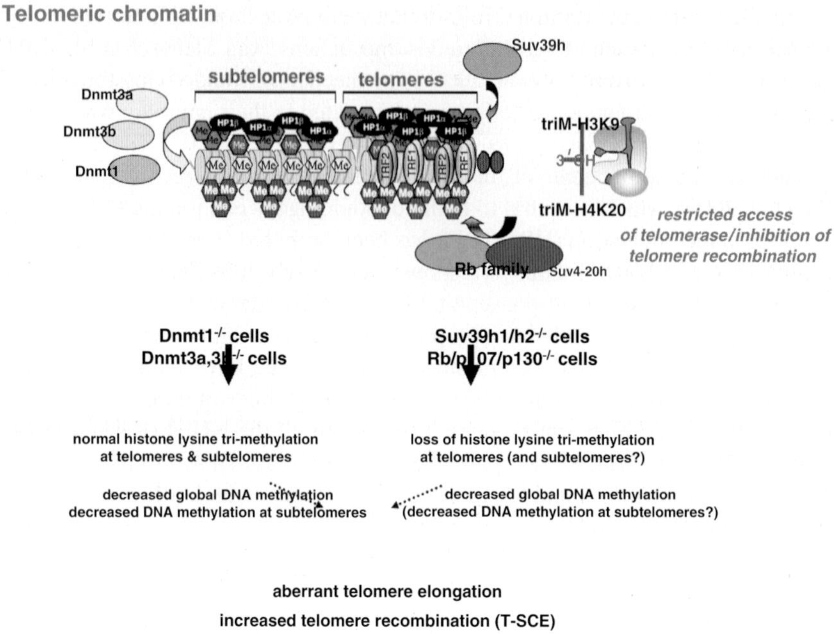

Fig. 12.2 Epigenetic modifications at mammalian pericentromeric and telomeric chromatin domains. Both telomeric and subtelomeric chromatin regions are enriched in trimethylated H3K9 and H4K20 and HP1 isoforms. In addition, subtelomeric DNA is heavily methylated by the DNMT1 and DNMT3a,3b enzymes. Both histone trimethylation and DNA methylation have been shown to independently act as negative regulators of telomere length, concomitant with their ability to inhibit telomere recombination (*See Color Plate*)

1.4 DNA Methylation at Mammalian Subtelomeres

DNA methylation is an abundant chromatin modification, with key roles in transcriptional regulation and in the definition of chromatin domains (Jones and Baylin 2002). Highly repetitive regions such as pericentric chromatin are heavily methylated, and this methylation has been proposed to be important to prevent the high level of homologous recombination that would otherwise be expected to take place at these domains (Dominguez-Bendala and McWhir 2004, Maloisel and Rossignol 1998, Bender 1998).

Telomere repeats cannot be methylated since they lack CpG sequences, which are the substrates for DNA methyltranfesases (DNMTases). However, mouse and human subtelomeric sequences have recently been shown to be methylated (Gonzalo et al. 2006, Steiner et al. 2004) (Fig. 12.2). There are three major DNMTs in mammalian cells: DNMT1, which functions as a maintenance DNMT that copies parental strand DNA methylation onto the daughter strand after DNA replication, and DNMT3a and DNMT3b, which are thought to function as de novo DNMTs (Okano et al. 1999, Okano et al. 1998). Cells that lack either DNMT1 or DNMT3a and DNMT3b isoforms show a marked reduction in DNA methylation at subtelomeric domains (Gonzalo et al. 2006), similar to that previously described for pericentric heterochromatin (Okano et al. 1999, Okano et al. 1998, Chen et al. 2004). Interestingly, this decreased DNA methylation does not affect the abundance of heterochromatic histone methylation marks (trimethylated H3K9 and H4K20) and of HP1 at subtelomeric chromatin (Gonzalo et al. 2006).

2 Telomere Maintenance Mechanisms in Mammals

2.1 *Telomerase*

The main mechanism for telomere elongation is the enzyme telomerase, which consists of a reverse transcriptase subunit (TERT) that is able to synthesize telomeric repeats de novo and add them to chromosome ends using an associated RNA molecule (TERC) as a template (Chan and Blackburn 2002; see also Dillin and Karlseder, this volume). The attrition of telomeric DNA that takes place during ageing is likely to result from limiting amounts of telomerase activity in the adult organism, which cannot compensate for the progressive telomere shortening that occurs as cells divide during tissue regeneration (Collins and Mitchell 2002, Blasco et al. 2005; see also Allsopp, this volume). In this regard, telomerase has been found to be crucial for stem cell function and proliferative potential (See Du et al., this volume, and Zimmerman and Martens, this volume), which anticipates its known role in cancer and ageing (See Rudolph, this volume). On one hand, short telomeres due to telomerase deficiency result in impaired stem cell functionality, defective tissue regeneration, and decreased tumorigenesis (Flores et al. 2005), while telomerase overexpression has opposite effects (Flores et al. 2005, Sarin et al. 2005). These findings suggest that telomere length and telomerase activity are important determinants of stem cell behavior and that it is in this context that they may influence cancer and ageing (see Gutierrez and Ju, this volume).

2.2 *Alternative Lengthening of Telomeres (ALT)*

Some immortal human and mouse cell lines that lack telomerase activity are still able to maintain or elongate their telomeres through a mechanism known as alternative lengthening of telomeres (ALT) (Muntoni and Reddel 2005, Dunham et al. 2000; see also Rudolph, this volume). In yeast and mammals, ALT has been shown to involve homologous recombination events between telomeric sequences (Lundblad 2002, Muntoni and Reddel 2005, Dunham et al. 2000). ALT-positive cells are characterized by heterogeneous telomere lengths, by the presence of both very short and very long telomeres, and by the co-localization of telomeres with a specific type of PML body, so-called ALT-associated PML bodies (APBs) (Muntoni and Reddel 2005, Dunham et al. 2000).

The telomerase-deficient mouse model has demonstrated telomere maintenance in the absence of telomerase both in cultured mouse embryonic fibroblasts (MEFs) (Hande et al. 1999, Chang et al. 2003) and ES cells (Niida et al. 2000), as well as in vivo in the context of B cell development, indicating that ALT takes place in situations other than immortalized cell lines (Herrera et al. 2000). However, while ALT can rescue the viability of telomerase-deficient yeast strains (Lundblad 2002), ALT mechanisms cannot rescue the viability of telomerase-deficient mice, suggesting that ALT mechanisms do not operate to rescue survival of multicellular organisms.

Of notice, components of shelterin complex such as Pot1 and TRF2, or TRF2-interacting proteins such as WRN, have been proposed as potential regulators of ALT, as they can influence telomere recombination (Blanco et al. 2007, Wu et al. 2006, Laud et al. 2005).

2.3 Epigenetic Regulation of Telomere Length and Telomere Recombination

Disruptions of histone and DNA modifications at telomeres result in loss of telomere length control (Garcia-Cao et al. 2004, Gonzalo et al. 2005, Garcia-Cao et al. 2002) (Fig. 12.2). In particular, cells lacking the Suv39h1 and Suv39h2 HMTases show decreased trimethylated H3K9 at telomeres, which is concomitant with aberrant telomere elongation (Garcia-Cao et al. 2004). A similar deregulation of telomere length is seen cells that lack all three members of the retinoblastoma family and show decreased levels of H4K20 trimethylation at telomeres (Gonzalo et al. 2005, Garcia-Cao et al. 2002). All together, these findings suggest an important role of histone modifications in mammalian telomere length control (Fig. 12.2).

Independent of histone methylation, decreased DNA methylation at subtelomeric regions is also accompanied by dramatically elongated telomeres in the absence of loss in heterochromatic histone methylation marks (Gonzalo et al. 2006) (Fig. 12.2). Indeed, aberrantly elongated telomeres in DNMT-deficient cells show a further increase in H3K9me3 and H4K20me3 density, as well as in HP1 binding, suggesting that the length of telomere repeats controls the formation of heterochromatin domains at telomeres when heterochromatinizing activities (Suv39h and Suv420 HMTases) are present.

The demethylation of subtelomeric regions in the absence of DNMTs is also concomitant with increased homologous recombination between telomeric sequences, which is corrected by reintroduction of DNMT3a,3b into cells that lack these enzymes results (Gonzalo et al. 2006) (Fig.12.2). These results suggest that DNA methylation at subtelomeric repeats may regulate ALT, which relies on homologous recombination.

All together, these results suggest that DNA and histone methylation at telomeric domains are an important mechanism for telomere length control. In turn, it has recently been shown that progressively shorter telomeres can also lead to epigenetic changes both in telomeric and subtelomeric chromatin (Benetti et al. 2007). In particular, short telomeres have a decreased density of heterochromatic histone marks such as H3K9me3 and H4K20me3 at telomeric repeats, as well as decreased binding of HP1. They are also characterized by an increased density of histone marks that are characteristic of more "open" or active chromatin domains, such as histone H3 and H4 acetylation (Benetti et al. 2007). In addition to these epigenetic changes at the TTAGGG repeats, subtelomeric regions show decreased histone methylation, increased histone acetylation, and a decrease in subtelomeric DNA methylation, suggesting that the length of the distal TTAGGG repeats influences the epigenetic status of subtelomeric chromatin (Benetti et al. 2007).

Of notice, similar to the findings described above for DNMT-deficient cells, the decreased DNA methylation seen at subtelomeric regions as a consequence of telomere shortening was also accompanied by increased telomere recombination (Benetti et al. 2007). This provides further support for a role of DNA methylation as a negative regulator of recombination at these regions, and might also provide an explanation for the fact that ALT is activated in the context of short telomeres in telomere-deficient cells and mice (Hande et al. 1999, Chang et al. 2003, Niida et al. 2000, Herrera et al. 2000).

The fact that short telomeres lead to a change in the epigenetic status of telomeric and subtelomeric chromatin may also explain the preferential elongation of short telomeres by telomerase. In particular, telomerase has been shown to specifically act on the shortest telomeres, both in budding yeast (Teixeira et al. 2004) and mammals (Samper et al. 2001, Hemann et al. 2001), raising the possibility that short telomeres have specific chromatin marks which are recognized by the telomerase complex. The precise nature of these marks is still unknown but may include increased H3 and H4 actelylation, as this mark is increased associated to telomere loss (Benetti et al. 2007).

3 Role of Mammalian Telomeres in Disease

3.1 The Telomerase-Deficient Mouse Model

The telomerase-deficient mouse model has been instrumental in demonstrating the impact of short telomeres in the context of the whole organism. Telomerase-deficient mice were first generated by elimination of the gene encoding the murine *Terc* gene (Blasco et al. 1995, Blasco et al. 1997). The long-term viability of the *Terc*$^{-/-}$ mouse strain is severely compromised, and only a limited number of generations can be derived, due to infertility and the progressive anticipation of pathologies associated with loss of telomeric repeats (Blasco et al. 1997, Lee et al. 1998, Blasco 2005, Herrera et al. 1999, Herrera et al. 2000; Franco et al. 2002; Ferron et al. 2004; Leri et al. 2003; Samper et al. 2002). These pathologies include loss of fertility, heart failure, immunosenescence, various tissue atrophies, and decreased tissue regeneration (of the digestive system, skin, hematopoietic system), among others (See

2000, Rudolph et al. 2001, Greenberg et al. 1999, Blanco et al. 2007). Therefore, DC is a human premature ageing syndrome that closely, but not completely, recapitulates the phenotype of the telomerase-deficient mouse model. A reason for this difference may be that DC patients still retain some telomerase activity, while the telomerase knockout mouse model lacks telomerase activity.

In addition to DC and aplastic anemia patients, who have defective telomerase activity and short telomeres, a number of additional human premature ageing syndromes are also characterized by an accelerated rate of telomere loss and chromosomal instability. Interestingly, these diseases are produced by mutations in DNA repair proteins such as Nbs1 (Nijmegen breakage syndrome), Mre11 (ataxia telangiectasia-like disorder), WRN (Werner syndrome; see Davis and Kipling, this volume), BLM (Bloom syndrome), ATM (ataxia telangiectasia) (ATM), and FANC genes (Fanconi anemia) (reviewed in Blasco 2005), many of which interact with the TRF2 telomere-binding protein (reviewed in Blasco 2005). Strikingly, when mice deficient for these proteins have been generated they do not faithfully reproduce premature ageing pathologies. In fact, the ageing pathologies associated with the Werner, Bloom, and ATM syndromes have been modeled in mice only when in combination with telomerase deficiency and short telomeres in the context of the telomerase-deficient mouse model (Chang et al. 2004, Wong et al. 2003, Du et al. 2004, Mochizuki et al. 2004), suggesting that short telomeres contribute to the pathobiology of these premature-ageing diseases.

3.3 Role of Telomere-Binding Proteins in Disease

A putative role for mammalian shelterin components in disease is suggested by the fact that TRF1, TRF2, and Tin2 are upregulated in some human carcinomas (Oh et al. 2005, Matsutani et al. 2001, Muñoz et al. 2005, Nakanishi et al. 2003, Bellon et al. 2006). TRF2 is one of the best-known components of the shelterin complex (de Lange 2005). In particular, TRF2 has been involved in telomere-length regulation, as well as in preventing end-to-end chromosome fusions (Smogorzewska et al. 2000, van Steensel et al. 1998, Celli and de Lange 2005). As mentioned above, TRF2 interacts with a number of factors involved in DNA repair, which also have been shown to have a role at telomeres (Blasco 2005 Fig. 12.3). The fact that many of the TRF2-interacting proteins are responsible for human chromosomal instability syndromes, together with the fact that TRF2 itself is upregulated in cancer, predicts that TRF2 itself may also have an important role in human disease.

In this regard, mice with increased TRF2 expression in stratified epithelia show increased UV-induced skin carcinogenesis, as well as a premature deterioration of the skin, consisting in hair loss, skin dryness, and skin hyperpigmentation (Muñoz et al. 2005). These skin phenotypes are reminiscent of the human xeroderma pigmentosum syndrome (de Boer et al. 2000). The increased epithelial carcinogenesis in TRF2 overexpressing mice is coincidental with increased chromosomal instability and telomere recombination (Muñoz et al. 2005, Blanco et al. 2007),

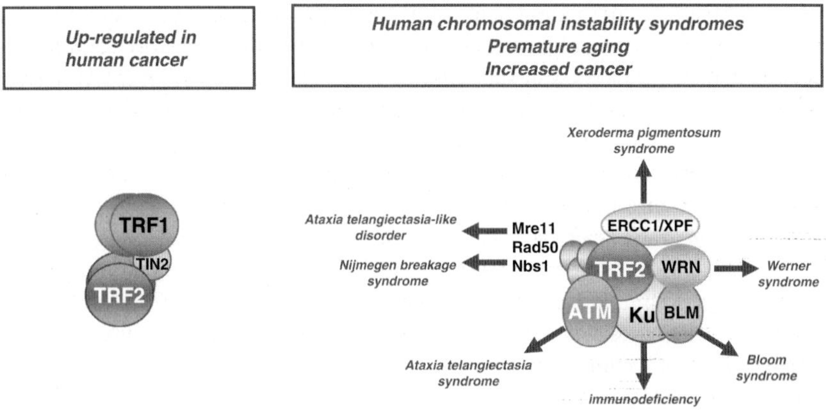

Fig. 12.3 Role of telomeric proteins in human disease. Components of the shelterin complex that are upregulated in human cancer are shown in the left panel. DNA repair and DNA damage signaling proteins that have been shown to interact with TRF2 are indicated in the right panel, together with the human chromosome instability syndromes where these proteins have been found to be mutated (*See Color Plate*)

suggesting that increased TRF2 expression is actively contributing to tumor development. It still remains to be tested, however, whether alteration in TRF2 and/or other telomere-binding protein impacts on cancer formation in human disease or ageing. In addition, TRF2 overexpressing mice show a rapid and dramatic loss of telomere sequences in skin areas exposed to light (Muñoz et al. 2005). Overexpression of telomerase cannot prevent telomere shortening in these mice, suggesting that short telomeres generated by TRF2 overexpression are not susceptible to rescue by telomerase (Muñoz et al. 2005). Short telomeres in TRF2 overexpressing mice were rescued, however, in the absence of the XPF nuclease, indicating that this enzyme is responsible for rapidly degrading telomeres in the presence of high TRF2 levels (Muñoz et al. 2005). The XPF/ERCC1 complex is involved in the repair of UV-induced DNA lesions through the nucleotide excision repair pathway (NER), as well as in DNA cross-link repair (de Laat et al. 1999). Interestingly, an interaction between TRF2 and XPF/ERCC1 and the localization of this complex to telomeres has been previously demonstrated (Zhu et al. 2003). Furthermore, XPF/ERCC1 has been shown to be able to degrade the single-stranded G-strand overhang at telomeres in cells with mutant TRF2, indicating that TRF2 can regulate the activity of the ERCC1/XPF complex at mammalian telomeres (Zhu et al. 2003). All together, these observations suggest that telomere shortening in TRF2 mice may be the result of an excess of XPF/ERCC1 nuclease activity at telomeres. In particular, increased TRF2 may sequester most XPF/ERCC1 at telomeres resulting in abnormal telomere degradation at the same time that results in defective NER at DNA lesions elsewhere in the genome. In support of this hypothesis, TRF2 mice not only show critically short telomeres and premature skin deterioration (hair loss, hyperpigmentation, etc.) but also show hypersensitivity to UV-induced DNA damage similarly to patients with xeroderma pigmentosum syndrome (Muñoz et al. 2005). Of note, these phenotypes are not observed in mice deficient for different NER components, suggesting they are likely to result from a combination of the NER defect together with the presence of critically short telomeres in K5-TRF2 mice. It is relevant to note that increased cancer susceptibility of K5-TRF2 mice with short telomeres is in marked contrast to telomerase-deficient mice also with short telomeres, which are resistant to skin carcinogenic treatments. Indeed, telomerase deficiency in the context of TRF2 overexpression dramatically accelerated epithelial carcinogenesis, further supporting the theory that short telomeres produced by TRF2 overexpression lead to increased tumorigenesis even in the absence of telomerase (Blanco et al. 2007). Finally, TRF2 conditional deletion in adult hepatocytes has no effect on their viability (Lazzerini Denchi et al. 2006), and further studies in additional cell types are required to understand the impact of TRF2 deficiency in organismal and cellular viability (Celli and de Lange 2005).

A role for the TRF1 shelterin components in human disease is less clear, other than the fact that this protein is also upregulated in some human tumors (Oh et al. 2005, Matsutani et al. 2001, Muñoz et al. 2005, Nakanishi et al. 2003, Bellon et al. 2006). Mice deficient for TRF1 show embryonic lethality in the absence of defective telomere length or loss of telomere protection (Karlseder et al. 2003). Interestingly, the TRF1-regulatory protein TANK1 is essential for the separation of sister chromatid telomeres during mitosis (Dynek et al. 2004), suggesting a role for this protein in cell division. Similarly, mice deficient for TIN2 show embryonic lethality independently of telomere length maintenance and telomerase activity (Chiang et al. 2004). Interestingly, TIN2 co-localizes to nontelomeric heterochromatin domains through its interaction with HP1γ (Kaminker et al. 2005), suggesting a putative role for this protein in heterochromatin assembly.

Finally, a role for the Pot1 shelterin component in human disease is suggested by recent findings showing that deletion of the Pot1 genes in mice result in chromosomal instability and increased telomere recombination, which in turn may impact on carcinogenesis (Wu et al. 2006, Hockmeyer et al. 2006). However, specific involvement of Pot1 in human disease remains to be demonstrated.

Acknowledgements M. A. Blasco's laboratory is funded by the MCyT (SAF2005-00277, GEN2001-4856-C13-08), by the Regional Government of Madrid (GR/SAL/0597/2004), European Union (TELOSENS FIGH-CT-2002-00217, INTACT LSHC-CT-2003-506803, ZINCAGE FOOD-CT-2003-506850, RISC-RAD FI6R-CT-2003-508842) and the Josef Steiner Award 2003.

References

Bailey SM et al. (2001) Strand-specific postreplicative processing of mammalian telomeres. Science 293: 2462–65

Baumann P, Cech TR (2001) Pot1, the putative telomere end-binding protein in fission yeast and humans. Science 292: 1171–75

Bellon et al. (2006) Increased expression of telomere length regulating factors TRF1, TRF2 and TIN2 in patients with adult T-cell leukemia. Int J Cancer 119: 2090–7

Bender J (1998) Cytosine methylation of repeated sequences in eukaryotes: the role of DNA pairing. Trends Biochem Sci 23: 252–56

Benetti R, Garcia-Cao M, Blasco MA (2007) Telomere length regulates the epigenetic status of mammalian telomeres and subtelomeres. Nature Genetics 39: 243–350

Blackburn EH (2001) Switching and signaling at the telomere. Cell 106: 661–73

Blanco R, Muñoz P, Klatt P, Flores JM, Blasco MA (2007) Telomerase abrogation dramatically accelerates TRF2-induced epithelial carcinogenesis. Genes Dev 21: 206–20

Blasco MA, Funk WD, Villeponteau B, Greider CW (1995) Functional characterization and developmental regulation of mouse telomerase RNA. Science 269: 1267–70

Blasco MA et al. (1997) Telomere shortening and tumor formation by mouse cells lacking telomerase RNA. Cell 91: 25–34

Blasco MA, Hahn WC (2003) Evolving views of telomerase and cancer. Trends Cell Biol 13: 289–94

Blasco MA (2005) Telomeres and human disease: ageing, cancer and beyond. Nat Rev Genet 6: 611–22

Bradshaw PS, Stavropoulos DJ, Meyn MS (2005) Human telomeric protein TRF2 associates with genomic double-strand breaks as an early response to DNA damage. Nat Genet 37: 193–97

Cawthon et al. (2003) Association between telomere length in blood and mortality in people aged 60 years or older. Lancet 361: 393–95

Celli GB, de Lange T (2005) DNA processing is not required for ATM-mediated telomere damage response after TRF2 deletion. Nat Cell Biol 7: 712

Chan SW, Blackburn EH (2002) New ways not to make ends meet: telomerase, DNA damage proteins and heterochromatin. Oncogene 21: 553–63

Chang et al. (2004) Essential role of limiting telomeres in the pathogenesis of Werner syndrome. Nat Genet 36: 877–82

Chang S, Khoo CM, Naylor ML, Maser RS, DePinho RA (2003) Telomere-based crisis: functional differences between telomerase activation and ALT in tumor progression. Genes Dev 17: 88–100

Chen T, Tsujimoto N, Li E (2004) The PWWP domain of Dnmt3a and Dnmt3b is required for directing DNA methylation to the major satellite repeats at pericentric heterochromatin. Moll Cell Biol 24: 9048–58

Chin et al. (1999) p53 deficiency rescues the adverse effects of telomere loss and cooperates with telomere dysfunction to accelerate carcinogenesis. Cell 97: 527–38

Chiang YJ, Kim SH, Tessarollo L, Campisi J, Hodes RJ (2004) Telomere-associated protein TIN2 is essential for early embryonic development through a telomerase-independent pathway.Mol Cell Biol 15: 6631–34

Collins K, Mitchell JR (2002) Telomerase in the human organism. Oncogene 21: 564–79

d'Adda di Fagagna et al. (2003) A DNA damage checkpoint response in telomere-initiated senescence. Nature 426: 194–98

de Laat WL, Jaspers NG, Hoeijmakers JH (1999) Molecular mechanism of nucleotide excision repair. Genes Dev 13: 768–85

de Lange T (2002) Protection of mammalian telomeres. Oncogene 21: 532–40

de Lange T (2004) T-loops and the origin of telomeres. Nat Rev Mol Cell Biol 5: 323–29

de Lange T (2005) Shelterin: the protein complex that shapes and safeguards human telomeres. Genes Dev 19: 2100–10

de Lange T, Shiue L, Myers RM, Cox DR, Naylor SL, Killery AM, Varmus HE (1990) Structure and variability of human chromosome ends. Mol Cell Biol 10: 518–27

Dominguez-Bendala J, McWhir J (2004) Enhanced gene targeting frequency in ES cells with low genomic methylation levels. Transgenic Res 13: 69–74

Du et al. (2004) Telomere shortening exposes functions for the mouse Werner and Bloom syndrome genes. Mol Cell Biol 24: 8437–46

Dunham MA, Neumann AA, Fasching CL, Reddel RR (2000) Telomere maintenance by recombination in human cells. Nat Genet 26: 447–50

Dynek JN, Smith S (2004) Resolution of sister telomere association is required for progression through mitosis. Science 304: 97–100

Epel et al. (2004) Accelerated telomere shortening in response to life stress. Proc Natl Acad Sci USA 101: 17312–15

Ferron et al. (2004) Telomere shortening and chromosomal instability abrogates proliferation of adult but not embryonic neural stem cells. Development 131: 4059–70

Flores I, Cayuela ML, Blasco MA (2005) Effects of telomerase and telomere length on epidermal stem cell behavior. Science 309: 1253–56

Franco S, Segura I, Riese H, Blasco, MA (2002) Decreased B16F10 melanoma growth and impaired vascularization in telomerase-deficient mice with critically short telomeres. Cancer Res 62: 552–59

Garcia-Cao M, Gonzalo S, Dean D, Blasco MA (2002) A role for the Rb family of proteins in controlling telomere length. Nat Genet 32: 415–19

Garcia-Cao M, O'Sullivan R, Peters AH, Jenuwein T, Blasco MA (2004) Epigenetic regulation of telomere length in mammalian cells by the Suv39h1 and Suv39h2 histone methyltransferases. Nat Genet 36: 94–99

Gonzalez-Suarez E, Samper E, Flores JM, Blasco MA (2000) Telomerase-deficient mice with short telomeres are resistant to skin tumorigenesis. Nat Genet 26: 114–17

Gonzalo et al. (2006) DNA methyltransferases control telomere length and telomere recombination in mammalian cells. Nature Cell Biol 8: 416

Gonzalo S et al. (2005) Role of the RB1 family in stabilizing histone methylation at constitutive heterochromatin. Nat Cell Biol 7: 420–28

Greenberg et al. (1999) Short dysfunctional telomeres impair tumorigenesis in the INK4a(delta2/3) cancer-prone mouse. Cell 97: 515–25

Griffith JD, Comeau L, Rosenfield S, Stansel RM, Bianchi A, Moss H, de Lange T (1999) Mammalian telomeres end in a large duplex loop. Cell 97: 503–14

Hande MP, Samper E, Lansdorp P, Blasco MA (1999) Telomere length dynamics and chromosomal instability in cells derived from telomerase null mice. J Cell Biol 44: 589–601

Hemann MT, Strong MA, Hao LY, Greider CW (2001) The shortest telomere, not average telomere length, is critical for cell viability and chromosome stability. Cell 107: 67–77

Herrera E, Martinez AC, Blasco MA (2000) Impaired germinal center reaction in mice with short telomeres. EMBO J 19: 472–81

Herrera E, Samper E, Martin-Caballero J, Flores JM, Lee HW, Blasco MA (1999) Disease states associated to telomerase deficiency appear earlier in mice with short telomeres. EMBO J 18: 2950–60

Hockemeyer D, Daniela JP, Takai H, de Lange T (2006) Recent expansion of the telomeric complex in rodents: Two distinct POT1 proteins protect mouse telomeres. Cell 126: 63–77

Jaco et al. (2003) Role of mammalian Rad54 in telomere length maintenance. Mol Cell Biol 23: 5572–80

Jones PA, Baylin SB (2002). The fundamental role of epigenetic events in cancer. Nat Rev Genet 3: 415–28

Kaminker et al. (2005) Higher-order nuclear organization in growth arrest of human mammary epithelial cells: a novel role for telomere-associated protein TIN2. J Cell Sci 183: 1321–30

Karlseder et al. (2003) Targeted deletion reveals an essential function for the telomere length regulator Trf1. Mol Cell Biol 23: 6533–41

Karlseder et al. (2004) The telomeric protein TRF2 binds the ATM kinase and can inhibit the ATM-dependent DNA damage response. PLOS Biol 2: E240

Kourmouli N et al. (2004) Heterochromatin and tri-methylated lysine 20 of histone 4 in mammals. J Cell Sci 117: 2491–2501

Lachner M, O'Carroll D, Rea S, Mechtler K, Jenuwein T (2001) Methylation of histone H3 lysine 9 creates a binding site for HP1 proteins. Nature 410: 116–20

Laud PR et al. (2005) Elevated telomere-telomere recombination in WRN-deficient, telomere dysfunctional cells promotes escape from senescence and engagement of the ALT pathway. Genes Dev 19: 2560–70

Lazzerini Denchi E, Celli G, de Lange T (2006) Hepatocytes with extensive telomere deprotection and fusion remain viable and regenerate liver mass through endoreduplication. Genes Dev 20: 2648–53

Lee H-W, Blasco MA, Gottlieb GJ, Greider CW, RA. DePinho. (1998) Essential role of mouse telomerase in highly proliferative organs. Nature 392: 569–74

Leri et al. (2003) Ablation of telomerase and telomere loss leads to cardiac dilatation and heart failure associated with p53 upregulation. EMBO J 22: 131–39

Liu D, Safari A, O'Connor MS, Chan DW, Laegeler A, Qin J, Songyang Z (2004) PTOP interacts with POT1 and regulates its localization to telomeres. Nat Cell Biol 6: 673–80

Lundblad V. (2002) Telomere maintenance without telomerase. Oncogene 21: 522–31

Maloisel L, Rossignol JL (1998) Suppression of crossing-over by DNA methylation in *Ascobolus*. Genes Dev 12: 1381–89

Makarov VL, Lejnine S, Bedoyan J, Langmore JP (1993) Nucleosomal organization of telomere-specific chromatin in rat. Cell 73: 775–87

Marrone et al. (2004) Heterozygous telomerase RNA mutations found in dyskeratosis congenita and aplastic anemia reduce telomerase activity via haploinsufficiency. Blood 104: 3936–42

Mason PJ, Wilson DB, Bessler M (2005) Dyskeratosis congenita – a disease of dysfunctional telomere maintenance. Curr Mol Med 5: 159–70

Matsutani et al. (2001) Expression of telomeric repeat binding factor 1 and 2 and TRF1-interacting nuclear protein 2 in human gastric carcinomas. Int J Oncol 19: 507–12

Mitchell JR, Wood E, Collins K (1999) A telomerase component is defective in the human disease dyskeratosis congenita. Nature 402: 551–55

Mochizuki et al. (2004) Mouse dyskerin mutations affect accumulation of telomerase RNA and small nucleolar RNA, telomerase activity, and ribosomal RNA processing. Proc Natl Acad Sci USA 101: 10756–761

Muñoz et al. (2005) XPF nuclease-dependent telomere loss and increased DNA damage in mice overexpressing TRF2 result in premature ageing and cancer. Nature Genetics 10: 1063

Muntoni A, Reddel RR (2005) The first molecular details of ALT in human tumor cells. Hum Mol Genet 14: 191–96

Nakanishi et al. (2003) Expression of mRNAs for telomeric repeat binding factor (TRF)-1 and TRF2 in atypical adenomatous hyperplasia and adenocarcinoma of the lung. Clin Cancer Res 9: 1105–11

Niida et al. (2000) Telomere maintenance in telomerase-deficient mouse embryonic stem cells: characterization of an amplified telomeric DNA. Mol Cell Biol 20: 4115–2417

Oh et al. (2003) Telomere attrition and Chk2 activation in human heart failure. Proc Natl Acad Sci USA 100: 5378–83

Oh B-K, Kim Y-J, Park C, Park YN (2005) Up-regulation fo telomere-binding proteins, TRF1, TRF2, and TIN2 is related to telomere shortening during human multistep hepatocarcinogenesis. Am J Pathol 166: 73–80

Okano M, Bell DW, Haber DA, Li E (1999) DNA methyltransferases Dnmt3a and Dnmt3b are essential for de novo methylation and mammalian development. Cell 99: 247–57

Okano M, Xie S, Li E (1998) Cloning and characterization of a family of novel mammalian DNA (cytosine-5) methyltransferases. Nat Genet 19: 219–20

O'Sullivan et al. (2002) Chromosomal instability in ulcerative colitis is related to telomere shortening. Nat Genet 32, 280–84

Peters AH et al. (2001) Loss of the Suv39h histone methyltransferases impairs mammalian heterochromatin and genome stability. Cell 107: 323–37

Poch et al. (2004) Short telomeres protect from diet-induced atherosclerosis in apolipoprotein E-null mice. FASEB J 18: 418–20

Prime G, Markie D (2005) The telomere repeat binding protein Trf1 interacts with the spindle checkpoint protein Mad1 and Nek2 mitotic kinase. Cell Cycle 4: 121–24

Rudolph KL, Millard M, Bosenberg MW, DePinho RA (2001) Telomere dysfunction and evolution of intestinal carcinoma in mice and humans. Nat Genet 28: 155–59

Samani et al. (2001) Telomere shortening in atherosclerosis. Lancet 358: 472–73

Samper E, Flores JM, Blasco MA (2001) Restoration of telomerase activity rescues chromosomal instability and premature ageing in Terc-/- mice with short telomeres. EMBO Rep 2: 800–07

Samper et al. (2002) Long-term repopulating ability of telomerase-deficient murine hematopoietic stem cells. Blood 99: 2767–75

Samper E, Goytisolo FA, Slijepcevic P, van Buul PP, Blasco MA (2000) Mammalian Ku86 protein prevents telomeric fusions independently of the length of TTAGGG repeats and the G-strand overhang. EMBO Rep 1L 244–52

Sarin KY et al. (2005) Conditional telomerase induction causes proliferation of hair follicle stem cells. Nature 436: 1048–52

Schotta G et al. (2004) A silencing pathway to induce H3-K9 and H4-K20 trimethylation at constitutive heterochromatin. Genes Dev 18: 1251–62

Shay JW, Wright WE (2006) Telomerase therapeutics for cancer: challenges and new directions. Nat Rev Drug Discov 5: 577–84

Smith S, Giriat I, Schmitt A, de Lange T (1998) Tankyrase, a poly(ADP-ribose) polymerase at human telomeres. Science 282: 1484–87

Smogorzewska A, van Steensel B, Bianchi A, Oelmann S, Schaefer MR, Schnapp G, de Lange T (2000) Control of human telomere length by TRF1 and TRF2. Mol Cell Biol 20: 1659–68

Steinert S, Shay JW, Wright WE (2004) Modification of subtelomeric DNA. Mol Cell Biol 24: 4571

Takai H, Smogorzewska A, de Lange T (2003) DNA damage foci at dysfunctional telomeres. Curr Biol 13: 1549–56

Tarsounas et al. (2004) Telomere maintenance requires the RAD51D recombination/repair protein. Cell 117: 337–47

Teixeira MT, Arneric M, Sperisen P, Lingner J (2004) Telomere length homeostasis is achieved via a switch between telomerase- extendible and -nonextendible states. Cell 117: 323–35

Valdes et al. (2005) Obesity, cigarette smoking, and telomere length in women.Lancet 366: 662–64

van Steensel B, Smogorzewska A, de Lange T (1998) TRF2 protects human telomeres from end-to-end fusions. Cell 92: 401–13

Vulliamy et al. (2001) The RNA component of telomerase is mutated in autosomal dominant dyskeratosis congenita. Nature 413: 432–35

Vulliamy et al. (2004) Disease anticipation is associated with progressive telomere shortening in families with dyskeratosis congenita due to mutations in TERC. Nat Genet 36: 447–49

Wang RC, Smogorzewska A, de Lange T (2004) Homologous recombination generates T-loop-sized deletions at human telomeres. Cell 119: 355–68

Wiemann et al. (2002) Hepatocyte telomere shortening and senescence are general markers of human liver cirrhosis. FASEB J 16: 935–42

Wong et al. (2003) Telomere dysfunction and Atm deficiency compromises organ homeostasis and accelerates ageing. Nature 421: 643–48

Wu et al. (2006) Pot1 deficiency initiates DNA damage checkpoint activation and aberrant homologous recombination at telomeres. Cell 126: 49–62

Yamaguchi et al. (2005) Mutations in TERT, the gene for telomerase reverse transcriptase, in aplastic anemia. N Engl J Med 352: 1413–24

Ye JZ, Hockemeyer D, Krutchinsky AN, Loayza D, Hooper SM, Chait BT, de Lange T (2004) POT1-interacting protein PIP1: a telomere length regulator that recruits POT1 to the TIN2/TRF1 complex. Genes Dev 18: 1649–54

Zhu XD, Kuster B, Mann M, Petrini JH, Lange T (2000) Cell-cycle-regulated association of RAD50/MRE11/NBS1 with TRF2 and human telomeres. Nat Genet 25: 347–52

Zhu et al. (2003) ERCC1/XPF removes the 3′ overhang from uncapped telomeres and represses formation of telomeric DNA-containing double minute chromosomes. Mol Cell 12: 1489–98

Part IV
Therapeutic Targets

Chapter 13
Targeting Telomerase: Therapeutic Options for Cancer Treatment

W. Nicol Keith(✉) and Alan E. Bilsland

Abstract Telomerase is a suitable target for drug development because of its high prevalence, selective expression, and mechanistic importance in many cancer types. Telomerase targeting strategies can be defined as immunotherapy, gene therapy, and inhibition either by oligonucleotides or small molecules. However, few published approaches have entered clinical testing. There are advantages and limitations to each approach, and their different mechanisms of action must be taken into account in order to design appropriate preclinical models and identify appropriate endpoints for clinical trials. In this chapter we review the development pipelines for each class of telomerase-directed therapeutics, summarizing future challenges, opportunities, and development gaps.

1 Target-Based Drug Discovery and Telomerase Validation

Despite significant advances in recent years in our understanding of the cellular and molecular mechanisms underpinning cancer progression, mortality rates remain unacceptably high for most common malignancies, and effective treatment options are often unavailable. Therefore, drug development remains a major goal of cancer research. Drug discovery begins with exploratory research to identify potential molecular targets, followed by preclinical target validation, assay development, compound screening, hit identification and confirmation, lead identification and optimization, and in vitro and in vivo efficacy and toxicity studies (Sams-Dodd 2005) (Fig. 13.1). Therefore, preclinical development alone requires significant financial, intellectual, and manpower investment. Assuming that a new lead meets acceptance criteria in all assays, and subject to relevant regulatory and ethical approval, clinical testing may begin. At this point in development, both costs and risk increase substantially as a candidate progresses through clinical trials (DiMasi and Grabowski 2007, DiMasi et al. 2003).

W. Nicol Keith
Centre for Oncology and Applied Pharmacology, University of Glasgow, Cancer Research UK Beatson Laboratories, Garscube Estate, Switchback Road, Bearsden, Glasgow, United Kingdom
email: n.keith@beatson.gla.ac.uk

2 Telomerase-Specific Adaptive Immunotherapy

2.1 Rationale

Telomerase-specific adaptive immunotherapy aims to stimulate a CD8+ cytotoxic T lymphocyte (CTL) response against hTERT overexpression in cancer cells. Clinical approaches to induce CTL responses against tumor "neoantigens" generally make use of the professional antigen-presenting function of dendritic cells (DC) pulsed with neoantigen mRNA or, in the case of peptide vaccination approaches, with synthetic peptides capable of binding MHC class I molecules (HLA-A, HLA-B, or HLA-C). Pulsed DC prime CTL to lyse target cells displaying endogenously synthesized antigenic peptides in the context of the class I presentation pathway which is present in all nucleated cells (Rock and Goldberg 1999). Although antigen-specific CTL responses can be raised by these methods, clinical efficacy is limited, and the importance of both inhibiting CTL suppressive signals and inducing antigen-specific CD4+ T-helper cell support is now widely recognized (Emens 2006). T helper cells secrete cytokines that both enhance CTL activity and induce humoral and innate immunity. However, CD4+ responses are mainly raised through the endocytic class II presentation pathway, and endogenous proteins are not normally displayed in this context unless routed to the lysosomal compartment (Chamuleau et al. 2006, Humphreys et al. 2004). Immunotherapy directed against cancer-associated neoantigens is a promising approach under intense investigation (Berzofsky et al. 2004). Unfortunately, most tumor-associated antigens are represented only in small subsets of tumors and are not broad-spectrum markers. In contrast, the overexpression of telomerase in most tumor cells provides a rationale for its investigation as a potential antigen. Telomerase-specific immunotherapy represents a direct cell-killing approach not subject to the phenotypic lag.

2.2 Preclinical Studies

Epitope deduction identified hTERT peptide I540 (^{540}ILAKFLHWL^{548}) as a potential high-affinity binding epitope for the MHC class I allele HLA-A2.1 (Vonderheide et al. 1999). CD8+ cells sorted from I540 pulsed peripheral blood mononuclear cells (PBMC) of both healthy donors and cancer patients lysed hTERT+, HLA-A2.1+ cancer cell lines and primary leukemia cells in the presence of I540, but not hTERT- or HLA-A2.1- targets or autologous CD34+ cells, which have low telomerase activity (Minev et al. 2000, Vonderheide et al. 1999). There is some controversy regarding the processing of endogenous I540, since these studies did not sort CTL on the basis of their antigen specificity. When enriched polyclonal lines and clones of I540-specific CTL were generated from melanoma patients, target cells expressing endogenous hTERT were lysed only when pulsed with I540 peptide. Further, I540 was not generated by preoteosomes from longer peptide fragments, suggesting that processing from endogenous hTERT may be inefficient (Ayyoub et al. 2001). However, two groups have reported direct elution of I540 from endogenous HLA-A2 complexes on cancer cells, and recombinant antibodies with specificity against purified I540/HLA-A2 complexes appear to recognize the endogenous epitope on cancer cells (Gannage et al. 2005, Hirano et al. 2006, Lev et al. 2002).

Fragments of hTERT that bind with high affinity to the MHC class I alleles HLA-A24, HLA-A3, HLA-A1, and HLA-B7 and stimulate appropriately restricted CTL responses against hTERT+ cells have also been described (Adotevi et al. 2006, Arai et al. 2001, Schreurs et al. 2005, Vonderheide et al. 2001). Additionally, since CTL specific for immunodominant epitopes may be subject to tolerance mechanisms, the MHC class I binding affinity of several low-affinity hTERT peptides was enhanced by mutation, resulting in appropriately restricted CTL responses against hTERT+ target cells expressing the wild-type epitope (Hernandez et al. 2002, Scardino et al. 2002). This approach may circumvent tolerance to peptide vaccination. Epitopes that bind to multiple MHC class II alleles, thereby stimulating CD4+ T helper cell proliferation, have also been identified (Schroers et al. 2002, Schroers et al. 2003). In an alternative approach, hTERT-specific CTL were generated using hTERT mRNA transduced DC (Nair et al. 2000). This strategy is nonbiased with respect to MHC allele restriction and has the advantage of presentation of multiple epitopes. Subsequently, the same group reported that transduction of DC with a chimeric hTERT/ lysosome-associated membrane protein 1 (LAMP-1) mRNA fusion activates both CD4+ and CD8+ responses, presumably by directing the product into the lysosomal compartment (Su et al. 2002).

In vivo models of hTERT immunotherapy using humanised HLA transgenic mice demonstrate that MHC-restricted hTERT-specific CTL can be activated following vaccination with I540, low-affinity epitopes, or DNA vaccines (Gross et al. 2004, Hernandez et al. 2002, Minev et al. 2000, Scardino et al. 2002). Nair and co-workers also tested the ability of mTert mRNA pulsed DC to elicit antitumor responses in C57BL/6 mice. Vaccination inhibited tumor growth in murine lung metastasis as well as bladder and breast tumor models with no evidence of autoimmunity against normal tissues despite the less stringent regulation of telomerase in mouse tissues as compared with human (Nair et al. 2000). Importantly, several studies have concluded that, while hTERT-specific CTL effectively target telomerase overexpressing cells such as cancer cells but not telomerase-negative normal cells, normal cell populations with low level hTERT expression, such as CD34+ progenitors, are also poor targets (Hernandez et al. 2002, Minev et al. 2000, Vonderheide et al. 1999, Vonderheide et al. 2001).

2.3 *Clinical Studies*

Stimulation of hTERT-specific CTL from PBMC of both healthy donors and cancer patients has been convincingly shown in preclinical models. Naturally occurring hTERT-specific CTL are also present at higher frequency in PBMC of some cancer patients compared with normal donors, providing important immunological for use in gene therapy since both promoters are active in the vast majority of cancer cells tested (but not normal cells), although their activities are quite different, with the hTERC promoter usually substantially stronger than hTERT (Abdul-Ghani et al. 2000, Bilsland et al. 2005, Bilsland et al. 2003, Boyd et al. 2004, Dufes et al. 2005, Groot-Wassink et al. 2004, Plumb et al. 2001). Since optimal cellular concentrations of therapeutic protein will differ between transgenes, it is useful that a wide range of expression levels can be achieved in telomerase gene therapy by taking advantage of the basic toolkit of both promoters. Transcriptional activity of the hTERC promoter is usually low in normal cells but substantially upregulated in cancer cells, while the hTERT promoter is essentially off in most normal cells but is turned on at low levels in cancer cells. Most telomerase gene therapy strategies that have been tested can be broadly categorized as cytotoxic gene therapy approaches or oncolytic virotherapy approaches, both of which aim directly to kill cells expressing telomerase while sparing normal counterparts, thereby circumventing the issue of the phenotypic lag (Keith et al. 2004).

3.2 Preclinical Studies

Cytotoxic gene therapy approaches tested to date can further be subdivided into strategies based on induction of apoptosis, enzyme prodrug therapy, targeted radiotherapy, immune-modulation, or inhibition of angiogenesis (Fig. 13.3). At least 20 individual therapeutic transgenes have so far been tested for their ability to selectively

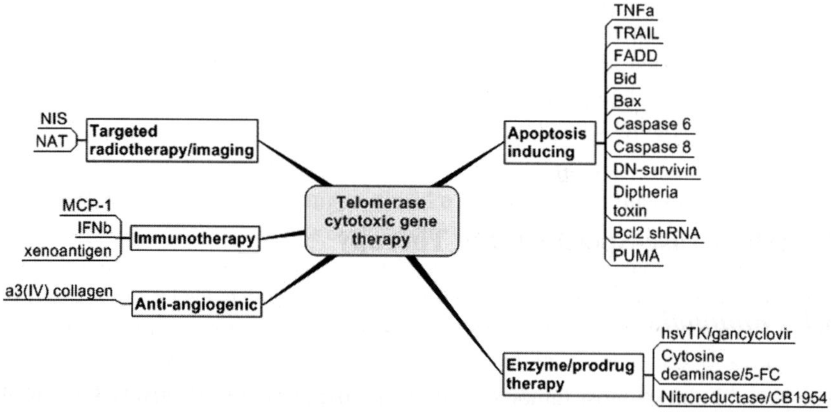

Fig. 13.3 Genes expressed under control of hTERC or hTERT promoters for preclinical models of cytotoxic gene therapy and their antitumor mechanisms of action. Abbreviations: **TNFα**, tumor necrosis factor α; **TRAIL**, TNFα-related apoptosis-inducing ligand; **FADD**, Fas-associated protein with death domain; **DN-survivin**, dominant negative survivin; **PUMA**, p53 upregulated mediator of apoptosis; **hsvTK**, herpes simplex viral thymidine kinase; **NIS**, sodium iodide symporter; **NAT**, noradrenaline transporter; **MCP-1**, monocyte chemoattractant protein 1; **IFNβ**, interferon β

induce cell death under the control of the hTERC and hTERT promoters (reviewed in Keith et al. 2004), with additional references below). As each approach elicits cell killing by a unique mechanism, with unique markers of activity and specific advantages and limitations, and as many of the constructs cited here have now also been tested in multiple reports using, for example, different delivery systems, cell models, or combining chemotherapeutics with telomerase-specific targeting, we will not give detailed descriptions of the primary data from each approach or all references for each construct. Apoptosis-inducing strategies include hTERT-restricted expression of the genes for Bax, Bid, TRAIL, TNFα, FADD, caspases 6 and 8, PUMA, dominant-negative survivin, diphtheria toxin, or downregulation of Bcl-2 with hTERT-specific expression of shRNA. Enzyme/pro-drug approaches to date have included the hsvTK/ gancyclovir, cytosine deaminase/ 5-FC, and bacterial nitroreductase/ CB1954 systems. Telomerase-specific expression of the radionuclide transport proteins NAT and NIS is reported to permit efficient targeted radiotherapy. TERT-specific expression of immune-modulator proteins IFNβ, “xeno-antigen,” and MCP-1 has also been reported. Finally, hTERT-specific expression of the noncollagenous domain of α3(IV) collagen is reported to inhibit angiogenesis. Delivery of cytotoxic gene therapy constructs in vitro or in vivo has been tested using naked DNA, adenovirus, adeno-associated virus, and lipid or dendrimer based nonviral vehicles (Abdul-Ghani et al. 2000, Bilsland et al. 2003, Boyd et al. 2004, Dufes et al. 2005, Groot-Wassink et al. 2004, Gu et al. 2000, Huang et al. 2006, Ito et al. 2005, Kazhdan et al. 2006, Koga et al. 2000, Komata et al. 2001a, Komata et al. 2001b, Li et al. 2006, Lin et al. 2002, Liu et al. 2002, Majumdar et al. 2001, Miyoshi et al. 2006, Nakamura et al. 2004, Sawada et al. 2002, Wang et al. 2004b).

However, most delivery systems suffer from poor biodistribution, which is a major limitation of cancer gene therapy, even in tumor models (Bilsland et al. 2003, Grace et al. 1999, Grill et al. 2002). Recently, attempts have been made to improve the limited tumor tropism observed for some wild-type viruses through genetic engineering (Liu and Kirn 2007). It has been proposed that these replicating virotherapy approaches, such as conditionally replicating adenoviruses (CRAds), may provide a solution by targeting virus replication and cytolysis in tumor cells, allowing viral progeny to be released into the tumor milieu and theoretically spread throughout target tumors. Most commonly, CRAds are constructed by harnessing expression of the adenovirus E1A gene, which is essential for the virus transcriptional program and for replication (Everts and van der Poel 2005, Frisch and Mymryk 2002). Several groups have constructed CRAds using the hTERT promoter to drive E1A expression (Huang et al. 2004, Huang et al. 2003, Irving et al. 2004, Kawashima et al. 2004, Kim et al. 2003a, Lanson et al. 2003, Wirth et al. 2003, Zou et al. 2004). One recent report compared both hTERC-E1A and hTERT-E1A specific CRAds (Bilsland et al. 2007). There is some conflict in the literature regarding the specificity of telomerase-specific CRAds. Whereas most reports indicate that hTERT-directed CRAds selectively replicate in and lyse cancer cells, but not normal cells, four studies have noted off-target E1 gene expression and replication in normal cells. Bilsland and colleagues provided direct evidence for feedback regulation of both hTERC and hTERT promoters by E1A in the CRAd system, suggesting that nonreplicating telomerase-specific vectors may be more predictable (Bilsland et al. 2007, Bortolanza et al. 2005, Hernandez-Alcoceba et al. 2002, Kim et al. 2003a).

Taking the gene therapy literature as a whole, cell lines covering most of the major common malignancies have been targeted effectively in vitro. Critically, almost all of these studies have shown specificity using normal cell strains. Efficacy against xenograft models has also been repeatedly shown across multiple tumor cell types. Although there is a concern that the activities of the core telomerase promoters in normal mouse tissues may not reflect their activities in humans, several groups have shown that systemic delivery of telomerase-specific gene therapy constructs does not result in significant off-target liver toxicity, and the biodistribution of systemically delivered hTERC and hTERT-specific transgene expression has been investigated by imaging and reporter assay; activity appears to be low or absent in normal tissues (for a review of efficacy and selectivity data, the reader is referred to tables in (Keith et al. 2004). These models must also be taken in context with the large body of data from normal and malignant human tissues showing tumor-specific hTERC and hTERT expression (Downey et al. 2001, Keith et al. 2002, Sarvesvaran et al. 1999, Shay and Bacchetti 1997, Soder et al. 1998, Soder et al. 1997, Wisman et al. 2000). Therefore, proof-of-principle broad-spectrum, selective targeting by the telomerase promoters has been convincingly shown in preclinical models.

3.3 Clinical Studies

The preclinical data in support of the predictability and specificity of telomerase-specific gene therapy is extremely compelling. However, no approach has entered clinical trials. Systemic cancer gene therapy is hampered by the generalized problem of inefficient delivery and biodistribution. Replicating virotherapy may provide a solution by allowing spread of vector throughout target tumors; however, no CRAd vector has yet demonstrated clinical efficacy (Liu and Kirn 2007, Liu et al. 2007). Nevertheless, Cell Genesys recently obtained license to add the Geron Corporation's hTERT-specific CRAd vector to their oncolytic virus development portfolio, which already includes a phase I trial in bladder cancer patients of CG0070, a GM-CSF armed, E2F-1 specific CRAd (Ramesh et al. 2006). Therefore, it seems reasonable to speculate that this hTERT-specific CRAd (now named CG5757) will be tested in humans in the near future.

3.4 Future Developments

Intense efforts are underway to develop more efficient tumor targeting vectors using both viral and nonviral building blocks (De Laporte et al. 2006, Glasgow et al. 2006). Until vectors with proven efficiency are developed, efficacious systemic delivery of cancer gene therapy will remain elusive. Consequently, there are few development routes for telomerase-specific gene therapy that are likely to be successful in the near future in terms of full clinical development leading to a drug approval, although intraperitoneal delivery may hold some promise for certain tumor types (Evans and Keith 2004). Several preclinical studies have evaluated the effects of combination therapies, which are likely to produce the best results in the clinical setting in the absence of efficient delivery vehicles (Chang et al. 2006, Lin et al. 2003, Shieh et al. 2006, Takeuchi et al. 2004). Nevertheless, well-designed phase I trials incorporating appropriate biomarkers will yield important information about the behavior of the promoters in real tissues and may provide early clinical evidence in support of continuing to develop these strategies. As more efficient future delivery systems emerge, telomerase-specific gene therapy may fulfill its promise. It should also be noted that production of the amounts of vector required to perform trials in large patient groups is likely to be a considerable hurdle for commercialization of viral vectors. In the longer term, development of more efficient nonviral vectors may be preferable in terms of manufacturing cost and reproducibility.

4 Oligonucleotide Targeting of Telomerase

4.1 Rationale

Oligonucleotide targeting approaches have the common objective of inhibiting telomerase activity using modified or unmodified DNA or RNA delivered to cells naked or in the context of an expression vector. Telomerase inhibition is achieved either by silencing hTERC or hTERT expression by transcript cleavage or by competitive inhibition using oligomers that hybridize to the hTERC template region. It should be noted that, as a mechanism for inhibition, downregulation of telomerase gene expression is not necessarily equivalent with active-site antagonism and may give rise to a different spectrum of biological effects. Thus, although inhibition of telomerase activity is the goal of all the agents listed in this section, and it would therefore be expected that the phenotypic lag should be observed in most cases, there is evidence that some methods induce rapid cytotoxicity independently of telomere length.

4.2 Preclinical Studies

Several studies have investigated the use of unmodified DNA oligomers or phosphorothioate backbone-modified oligodeoxynucleotides (PS-ODN) directed

to the phenotypic lag, which must be taken into account when planning preclinical models or trial endpoints. Additionally, signal transduction inhibitors may not exhibit their primary anti-neoplastic effects via telomerase, but rather through other targets modulated by the pathway that is inhibited. Therefore, it is important to consider the most relevant molecular markers of activity.

5.2 Preclinical Studies

Signaling upstream of the hTERC and hTERT promoters provides a multitude of targets for telomerase inhibition. Examples include estrogen and retinoid nuclear receptors, which contribute positively and negatively, respectively, to hTERT transcription. Treatment with retinoids in leukemia cells or the ERα antagonist tamoxifen in breast cancer cells inhibits hTERT expression and telomerase activity (Aldous et al. 1999, Pendino et al. 2001). Drugs targeting EGFR/MAPK pathways such as Iressa and PD98059 also inhibit telomerase expression in some cells (Budiyanto et al. 2003, Inui et al. 2002, Suenaga et al. 2006). An orally bioavailable derivative of PD98059, PD184352 (CL1040) has been tested in clinical trials, though its potential effects on telomerase are unknown (Kohno and Pouyssegur 2006). A small molecule c-Myc inhibitor, 10058-F4, also inhibits hTERT transcription, telomerase activity, and growth of hepatocellular carcinoma cells (Lin et al. 2007). Since the pathways leading to telomerase transcription are not fully understood, not all possible targets are currently known. However, this fact should provide an impetus for development of screening approaches, as discussed below.

The nucleoside analogue reverse transcriptase inhibitor AZT (3′-azido-3′-deoxythymidine) has been shown to inhibit telomerase activity and cell growth in a variety of cancer cell lines (Brown et al. 2003, Datta et al. 2006, Gan et al. 2002, Melana et al. 1998, Murakami et al. 1999). However, T cell proliferation was also inhibited by AZT, though the effect was transient and reversible (Beltz et al. 1999). Several other pleiotropic agents, such as the natural dietary compounds resveratrol and the green tea polyphenol ECGC (epigallocatechin gallate) are reported to inhibit telomerase, which may well constitute part of their overall antineoplastic effects (Lanzilli et al. 2006, Naasani et al. 1998). Notably, long-term culture in ECGC inhibited telomerase activity and resulted in delayed onset growth inhibition of cancer cells with discernable telomere shortening. ECGC did not affect telomere length in normal cells, and its plasma degradation products are more potent inhibitors than the parent molecule. Further, oral administration of ECGC inhibits growth of established xenografts (Naasani et al. 2003). The structure of ECGC provided the scaffold for design of a series of novel synthetic telomerase inhibitors with improved potency, including the compound MST-312, which inhibits telomerase activity in vitro at submicromolar concentrations and induces telomere shortening and delayed onset growth inhibition in several cell types (Seimiya et al. 2005, Seimiya et al. 2002). Naasani and colleagues also identified telomerase inhibitory properties of the rhodacyanine FJ5002 using a bioinformatic screen based on the weak activity of the plant alkaloid berberine (Naasani et al. 1999). No further reports of the use of FJ5002 have been published.

The non-nucleoside telomerase inhibitor BIBR1532 was identified by Boehringer Ingelheim and their academic partners (Damm et al. 2001). Long-term culture in BIBR1532 reversibly inhibited telomerase activity and decreased telomere lengths in lung, breast, prostate, and fibrosarcoma cancer cells. No short-term effect on cell proliferation was observed, but delayed onset growth retardation was shown, consistent with the predicted phenotypic lag expected of classic telomerase inhibitors. Growth inhibition was independent of p53 status, and the drug also inhibited growth of xenografts after oral administration. The authors stated that no effects on growth of normal lung fibroblasts were observed. BIBR1532 inhibits recombinant telomerase in vitro by a mixed-type noncompetitive mechanism, reducing the affinity of telomerase for deoxyribonucleotides and inhibiting formation of long reaction products without significantly affecting binding to the DNA primer (Pascolo et al. 2002). High concentrations of BIBR1532 have been shown to induce rapid direct cytotoxicity in leukemia cell lines and patient-derived primary AML and CLL cells. Surprisingly, this effect was not dependent on telomerase catalytic activity but was nonetheless mediated through telomere dysfunction, suggesting that the drug also mediates an uncapping effect at telomeres (El-Daly et al. 2005). Inhibition of telomerase activity with lower BIBR1532 concentrations may induce chemosensitivity to some agents in breast cancer cells (Ward and Autexier 2005). However, the effect of low-dose BIBR1532 on normal control cells should be investigated further. Notably, high doses caused delayed telomere-dependent toxicity in normal telomerase negative fibroblasts (El-Daly et al. 2005).

The guanine-rich 3′ telomeric single-stranded overhang also provides a potential target for drug development because of its ability to adopt a G-quadruplex secondary structure. One possible conformation arising from intramolecular folding involves Hoogsteen bonding between guanine bases of four telomere repeats on the same strand, resulting in a basket-like structure composed of stacked planar G-tetrads stabilized by cationic coordination. This conformation strongly inhibits the ability of telomerase to elongate its DNA primer and provides a unique structure for ligand development (Zahler et al. 1991). Agents that stabilize the quadruplex have therefore been seen as attractive candidate therapeutics, and a variety of diverse small molecules have been reported to act in this manner (for review, the reader is referred to (Kelland 2005)).

A 2,6-diamidoanthraquinone was the first reported agent of this type (Sun et al. 1997). Extensive structure activity investigations of disubstituted-amidoanthraquinones and disubstituted–acridines was performed, eventually leading to development of the 3,6,9-trisubstituted acridine BRACO19 (Neidle et al. 2000). In contrast to a number of early analogues with very unfavourable nonspecific cytotoxic profiles, BRACO19 inhibited telomerase in vitro with acceptable nonspecific cytotoxicity and repressed growth of breast, prostate, and uterine cancer cells. Recently, it was reported that BRACO19 repressed the growth of breast and uterine carcinoma xenografts (Burger et al. 2005, Gowan et al. 2002, Incles et al. 2003, Read et al. 2001). The pentacyclic acridine RHPS4 was also reported to inhibit telomerase activity at nontoxic concentrations in multiple cell lines (Gowan et al. 2001, Leonetti et al. 2004). Identification of a naturally occurring G-quadruplex stabilizing agent, termed telomestatin, was also reported. Telomestatin inhibits telomerase activity and reduces growth in a variety of cancer cell types and was recently reported to inhibit growth of lymphoma cell xenografts (Binz et al. 2005, Gomez et al. 2004, Gomez et al. 2006, Nakajima et al. 2003, Shammas et al. 2004, Tauchi et al. 2006).

G-quadruplex stabilizing agents were originally expected to behave as telomerase inhibitors, and most compounds do indeed inhibit telomerase activity in treated cells and cell extracts. Telomere shortening has also been observed in some studies (Burger et al. 2005, Shammas et al. 2004, Tahara et al. 2006, Zhou et al. 2006). However, the extent of telomere shortening is often not commensurate with effects on cell growth, and high concentrations induce rapid cytotoxicity prior to the onset of telomere shortening (Gowan et al. 2001, Gowan et al. 2002, Incles et al. 2004, Leonetti et al. 2004, Tahara et al. 2006). Recent data suggest that rapid cytotoxicity of G-quadruplex stabilizers results from telomere uncapping events that may include degradation of the 3′ overhang and loss of POT1 and/or TRF2 binding (Gomez et al. 2004, Gomez et al. 2006, Leonetti et al. 2004). Agents with high specificity for intrastrand G-quadruplex structures over other secondary structures appear more selective in terms of their tumor- and/or telomerase-specificity, whereas some the compound TMPyP4 is reported to be toxic to normal and cancer cells, including telomerase-negative ALT cells (Kelland 2005, Kim et al. 2003b, Rha et al. 2000, Shammas et al. 2004). Tumor specificity may be an issue for these agents, since uncapped telomeres could also be reasonably expected to induce senescence in normal cells. It is possible that the shorter or more dysfunctional telomeres of cancer cells as compared with normal cells might provide a therapeutic window for telomere targeting agents; however it is also possible that telomere maintenance mechanisms in cancer cells could render them more resistant. Normal control cells have been infrequently used in studies of G-quadruplex stabilizing agents, though concentrations of telomestatin that induced toxicity in cancer cells did not affect normal cells in at least two studies (Tahara et al. 2006, Tauchi et al. 2006).

5.3 Clinical Studies

Although several signal transduction inhibitors that may inhibit telomerase expression have been approved, or are undergoing clinical trials for various indications, their effects on telomerase are not considered primary mechanisms of action, though they may constitute some part of the overall therapeutic effect in some cases. AZT and tamoxifen are known to exhibit antitelomerase activity and are already widely used in clinical practice. In the case of tamoxifen, telomerase inhibition likely represents only a small part of the therapeutic effect, and there is little rationale for testing tamoxifen in tumors where estrogen receptor signaling does not contribute to the etiology. AZT has been used to treat some virus-associated human cancers (Matutes et al. 2001). Interestingly, there is some clinical evidence that telomerase activity is suppressed and telomere lengths are reduced by AZT therapy in HTLV-infected cells of adult T cell lymphoma patients (Datta et al. 2006).

5.4 Future Developments

Given the relative safety of AZT and the fact that it has been used in some cancer patients, although its mechanism of action is not specific for telomerase, this aspect of its activity may warrant investigation in trials of other tumor types. BIBR1532 seems to be a good candidate for progression to clinical trials, given its oral bioavailability, the consistency of its cellular effects with those of a classic telomerase inhibitor and, indeed, the fact of its discovery in collaboration with large pharma. However, no significant new studies evaluating BIBR1532 have emerged recently. MST-312 may also be a candidate, given that several clinical trials of related polyphenols have been reported (Bettuzzi et al. 2006, Chow et al. 2001, Chow et al. 2003). However, although telomerase inhibition, telomere shortening, and cell growth inhibition have been demonstrated for both compounds, telomerase-negative normal control cells have not been used widely, and their activity has been demonstrated in relatively few cancer cells and very few xenograft models. The expected phenotypic lag will need to be considered when designing trials of these compounds. In contrast, telomere targeting agents appear to induce relatively rapid toxicity mediated through telomere uncapping. AS1410 is a derivative of BRACO19 under development by the UK-based Antisoma. Preclinical data relating to the activity of AS1410 were reported at the 2005 AACR annual meeting (Kelland et al. 2005). In 2005, Antisoma announced an agreement with Heraeus to manufacture AS1410 for a phase I trial, though recruitment appears not to have begun to date.

In general, given the diversity of potential targets and leads, it is disappointing that relatively few selective small molecule telomerase inhibitors have been developed and no clinical trials have been reported so far. There would appear to be a compelling argument that new small molecule inhibitors of telomerase need to be identified with some urgency. Since most lead attrition occurs during the clinical phases of development, in which the cost of failure is at its highest, it is essential to employ adequate preclinical lead validation studies based on well-designed models so that only the best candidates progress to clinical testing. Substantial attention must now be paid to ensuring that both existing and newly identified leads are more thoroughly tested in the context of robust models demonstrating all or most of the expected preclinical characteristics of an efficient telomerase inhibitor: target inhibition, lack of short-term toxicity in vitro, effects on telomeres and delayed onset senescence or apoptosis, specificity for cancer cells over normal cells, broad-spectrum targeting of cancer cell lines, and efficacy with minimal toxicity in vivo.

In order to identify more new leads with these desirable properties, we should now consider the best assay development strategies. Cell-based assays offer potential to identify partially validated hits with improved lead-like qualities at the earliest stages of discovery and offer a number of critical advantages over conventional biochemical screening assays. Firstly, purification of the target protein in a functional conformation is not required. Instead, the key requirement is a measurable marker of endogenous target inhibition such as a change in reporter activity, protein phosphorylation, or cell morphology. Cell-based assays thus provide a better format for early development of biomarkers in parallel with validation of the screening assay. Importantly, cell-based assays may also discriminate between different drug effects, such as antagonism versus agonism and identify hits that interact with different target conformations that may be present in a physiological setting. Furthermore, membrane-impermeable compounds are discarded at the earliest stage of screening, and nonspecific cytotoxicity is easily identified. There is increasing industry enthusiasm for cell-based assays and, in particular, for high content phenotypic assays in which multiple outputs may be measured in the same well and correlated with complex phenotypes such as apoptosis, mitosis, or morphological change (Fox et al. 2006, Mitchison 2005). It is therefore likely that the use of cell-based screening could accelerate discovery of new small molecules targeting telomerase.

6 Applying Cell-Based Screening to Telomerase

Telomerase presents a number of opportunities for cell-based assay development. At the level of transcription, signaling pathways feeding either the hTERC or hTERT promoters, or the transcription factors that bind them, are excellent candidates for inhibition. Reporter assays of promoter activity are the most frequently used method to rapidly test candidate factors that influence gene promoter regions and are readily scalable (Kiss-Toth et al. 2000, Kiss-Toth et al. 2006). In a reporter screen for de-repressors of hTERT promoter activity in normal cells, Won and co-workers identified the molecule CGK1026 from a library of 20,000 compounds (Won et al. 2004). Promoter deletion and mutagenesis identified a CGK1026-responsive non-canonical E2F binding site in the hTERT promoter. CGK1026 was shown by chromatin immunoprecipitation (ChIP) to inhibit recruitment of histone deacetylases 1 and 2 to E2F-pocket protein complexes at the hTERT promoter. In the case of hTERC, the JNK inhibitor SP600125 was shown to increase hTERC-reporter activity and endogenous hTERC levels in several cell lines. Mutation of Sp1 sites in the hTERC promoter attenuated induction of reporter activity by the drug, whereas co-transfection of MEKK1 with Sp3 synergistically repressed promoter activity. It was shown by ChIP that SP600125 caused a switch in the ratio of Sp1/Sp3 binding to the endogenous promoter (Bilsland et al. 2006).

Such data validate the reporter assay approach to cell-based screening for novel small molecules with potential to regulate telomerase. Benefits of this approach are the speed and ease of the assay systems, as well as the ability to dissect pathways at the level of transcription factor binding via promoter mutagenesis. It may also be possible to tentatively identify the candidate target pathways and place novel inhibitors within them using complementary approaches such as RNAi or overexpression if some upstream regulatory factors are known. Importantly, however, no prior knowledge of the exact target pathway is required, since hits scored by this approach will target pathways that are intrinsically druggable (Fig. 13.4). Further, given the close relationship between telomerase expression and prosurvival signaling, the telomerase promoters can be regarded as probes for both known and unknown cell-survival and cell-death pathways, in the sense that it may not even be necessary to use leads identified in this way as classic telomerase inhibitors since they may have more rapidly acting anticancer properties mediated through inhibition of key survival pathways (Horikawa and Barrett 2003). However, target identification is not necessarily easy, and to demonstrate specificity it will be essential to perform adequate secondary assays, ideally employing normal control cells.

Although core telomerase activity can be reconstituted in vitro with the hTERC and hTERT components alone, telomerase is a multiprotein complex in

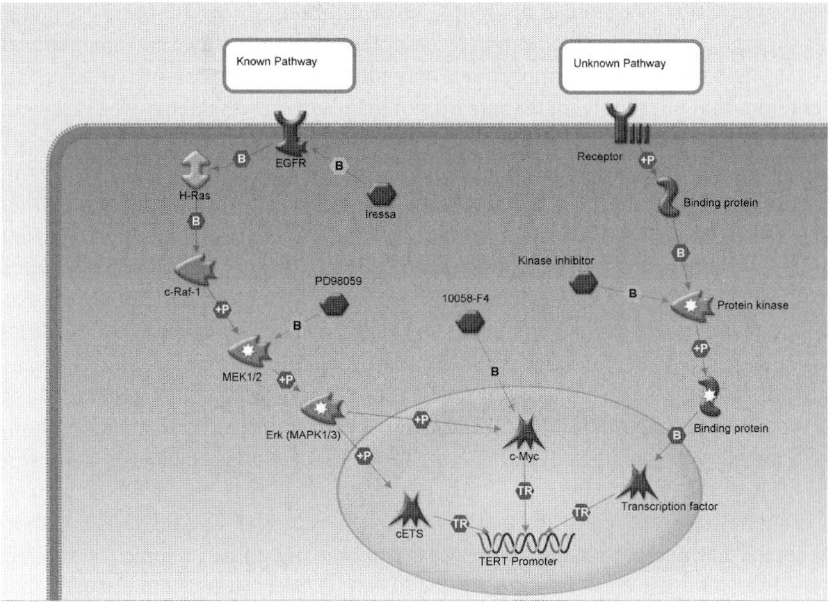

Fig. 13.4 Screening for novel inhibitors of telomerase gene expression. Several small molecule inhibitors of EGFR/MAPK signaling at various stages of development have been demonstrated to inhibit hTERT expression in cell lines. Although several pro- and anti-survival pathways are known to regulate the hTERC and hTERT promoters, the mechanisms underlying transcriptional regulation of the genes is incomplete. Therefore, a cell-based assay approach using the cloned promoters as probes could identify novel small molecule inhibitors of both known and unknown pathways

vivo, and its activity is modulated by interaction partners such as the chaperone protein hsp90 (Akalin et al. 2001, Forsythe et al. 2001, Holt et al. 1999). It is possible that hsp90 may be required for telomerase stability or for mediating conformational changes allowing the holoenzyme to translocate after synthesis of each individual telomere repeat. Interfering with hsp90 activity using the inhibitors geldanamycin, 17-AAG, and novobiocin inhibits telomerase activity and increases chemosensitivity in melanoma and head and neck cancer cells, providing experimental support for the functional importance of the interaction (Chang et al. 2006, Villa et al. 2003). 17-AAG has been tested in several phase I clinical trials in patients with advanced cancers refractory to standard therapies (Goetz et al. 2005, Grem et al. 2005, Nowakowski et al. 2006). However, hsp90 has multiple client proteins in both normal and cancer cells, and blanket inhibition of hsp90 may give rise to toxicity in normal cells. In contrast, selectively targeting the hTERT-hsp90 interaction or other functionally important hTERT interactions would be predicted to have a highly specific effect. Although protein-protein interactions have conventionally been considered difficult to drug with small molecules, there are noteworthy exceptions such as the nutlins, which inhibit the p53-MDM2 interaction, or FJ9, which inhibits the interaction between the PDZ domain of dishevelled and the Frizzled-7 receptor on the canonical Wnt pathway (Fujii et al. 2007, Vassilev et al. 2004). Protein-protein interaction assays such as the mammalian-2-hybrid or reporter complementation assays are amenable to up-scaling for screening purposes and may present a useful platform to screen for molecules that interfere with the integrity of the telomerase holoenzyme (Park et al. 2007, Paulmurugan and Gambhir 2003, Zhao et al. 2004).

Subcellular targeting has also recently emerged as a mechanism regulating telomerase activity (Akiyama et al. 2003b, Seimiya et al. 2000, Wong et al. 2002). Fluorescence tagging, either with antibodies, small molecules, or creation of fusion proteins, is central in studies of subcellular localization. Assay development is therefore dependent on the availability of tagging reagents that will not disrupt natural behavior of the target protein. In an elegant study employing a GFP-hTERT fusion protein, Wong and co-workers demonstrated transformation and DNA damage induced subnuclear shuttling of hTERT. In primary cells, ectopically expressed nonsaturating concentrations of GFP-hTERT localized primarily in nucleoli during G1/S but were excluded from nucleoli in late S/G2. In contrast, cancer cells showed constitutive nucleolar exclusion of the fusion, except in response to DNA damage, whereupon transient relocalization was observed (Wong et al. 2002). The authors speculate that nucleolar localization may sequester functional telomerase holoenzyme from telomere ends. Importantly, the fusion was demonstrated to be functional, as shown by reconstitution of telomerase activity and telomere maintenance in hTERT-negative cells, providing a potential screening platform to identify small molecules that interfere with pathways involved in hTERT nucleolar exclusion in cancer cells.

Screens based on fluorescent imaging of subcellular relocalization have previously been successful (Kau et al. 2003). Fluorescent imaging screens have the advantage of being intrinsically "high-content" in that multiple phenotypic parameters can simultaneously be assessed (Eggert and Mitchison 2006). However, visual interpretation of data has sometimes been necessary, and automated quantitative image analysis and data extraction have conventionally been important bottlenecks for high-throughput application of imaging screens. Although a number of commercial image processing packages have been used, availability of versatile and open-source image cytometry platforms such as CellProfiler and ImageJ is an important development (Carpenter et al. 2006, Rajwa et al. 2004). The ability to rapidly extract multiple diverse quantitative measurements such as cell number, shape, size, cell-cycle status, and the levels or location of fluorescent tag, is essential to support a variety of phenotypic screens.

Tankyrase also presents an intriguing target for inhibitor development. Tankyrase is a poly(ADP-ribose) polymerase (PARP) present in the nucleus and at other sites in the cell which associates with the telomere-binding protein TRF1 and mediates its poly(ADP-ribosyl)ation. TRF1 is a specific negative regulator of telomere length and its poly(ADP-ribosyl)ation by tankyrase leads to loss of telomere association and subsequent degradation by the ubiquitin/proteasome pathway (Chang et al. 2003, Smith et al. 1998). Therefore, tankyrase in the nucleus acts as a positive regulator of telomere length. Overexpression of nuclear localization signal-tagged tankyrase is able to confer resistance to telomerase inhibitors by blocking telomere shortening in several cancer cell lines (Seimiya et al. 2005). Following long-term treatment with the synthetic telomerase inhibitor MST-312, cells overexpressing tankyrase showed stabilized shortened telomere lengths, whereas cells treated with both the telomerase inhibitor and a conventional PARP inhibitor overcame the block and reached telomere-dependent crisis at significantly earlier time points than those treated with the telomerase inhibitor alone. Interestingly, fluorescent immunostaining showed that TRF1 expression was not observed in NLS-tankyrase expressing cells, but co-localization was observed on administration of a PARP inhibitor (Seimiya et al. 2005). Since cells expressing NLS-tagged tankyrase show significantly reduced TRF1 expression, a screening assay to identify selective tankyrase inhibitors based on TRF1 levels using the "in-cell western" technique can be readily envisaged (Wong 2004). Alternatively, co-localization of tankyrase and TRF1 in the presence of a PARP inhibitor suggests that a two-color cell imaging or FRET approach could be applied.

7 Conclusions

The paucity of effective anticancer treatment options and our improved understanding of the molecular bases of cancer phenotypes provide compelling rationales for target-based drug development. To generate cost-effective and efficacious treatments, it is essential to exploit the best targets by the most efficient development routes using evidence-based decision making to prioritize targets and leads at an early stage and to determine appropriate clinical endpoints to identify relevant beneficial biological effects which may sometimes occur in the absence of an objective reduction of tumor volume. A large amount of evidence supports the clinical development of strategies to target telomerase, and we are now seeing the fruits of intense target validation efforts manifest in several active clinical trials. However, while some approaches are relatively advanced, development of small molecule strategies has been slow. The development paths for telomerase-directed therapeutics are diverse and in some cases will require issues such as the phenotypic lag to be addressed. Telomerase consists of a number of targets that can theoretically be exploited by the small molecule route, suggesting that there is still substantial scope for development of leads with alternative mechanisms of action. Cell-based screening could be applied to investigate most of the telomerase-related targets, hopefully improving the quality of early leads. In some cases, such as cell-based screening using the telomerase promoters, it may be possible to identify novel anticancer signal transduction inhibitors whose principle mode of action need not be telomerase-specific. Secondary assays and target identification approaches will need to be defined on a screen-by-screen basis, but preclinical models should include demonstration of selectivity for cancer cells and efficacy in relevant in vivo models. Concerted efforts are now required to translate laboratory findings into viable therapeutic candidates.

References

Abdul-Ghani R, Ohana P, Matouk I, Ayesh S, Ayesh B, Laster M, Bibi O, Giladi H, Molnar-Kimber K, Sughayer MA, de Groot N, Hochberg A (2000) Use of transcriptional regulatory sequences of telomerase (hTER and hTERT) for selective killing of cancer cells. Mol Ther 2: 539–44

Adotevi O, Mollier K, Neuveut C, Cardinaud S, Boulanger E, Mignen B, Fridman WH, Zanetti M, Charneau P, Tartour E, Lemonnier F, Langlade-Demoyen P (2006) Immunogenic HLA-B*0702-restricted epitopes derived from human telomerase reverse transcriptase that elicit antitumor cytotoxic T-cell responses. Clin Cancer Res 12: 3158–67

Akalin A, Elmore LW, Forsythe HL, Amaker BA, McCollum ED, Nelson PS, Ware JL Holt SE (2001) A novel mechanism for chaperone-mediated telomerase regulation during prostate cancer progression. Cancer Res 61: 4791–96

Akiyama M, Hideshima T, Shammas MA, Hayashi T, Hamasaki M, Tai YT, Richardson P, Gryaznov S, Munshi NC, Anderson KC (2003a) Effects of oligonucleotide N3′→P5′ thiophosphoramidate (GRN163) targeting telomerase RNA in human multiple myeloma cells. Cancer Res 63: 6187–94

Akiyama M, Hideshima T, Hayashi T, Tai YT, Mitsiades CS, Mitsiades N, Chauhan D, Richardson P, Munshi NC, Anderson KC (2003b) Nuclear factor-kappaB p65 mediates tumor necrosis factor alpha-induced nuclear translocation of telomerase reverse transcriptase protein. Cancer Res 63: 18–21

Aldous WK, Marean AJ, DeHart MJ, Matej LA, Moore KH (1999) Effects of tamoxifen on telomerase activity in breast carcinoma cell lines. Cancer 85: 1523–29

Amarnath SM, Dyer CE, Ramesh A, Iwuagwu O, Drew PJ, Greenman J (2004) In vitro quantification of the cytotoxic T lymphocyte response against human telomerase reverse transcriptase in breast cancer. Int J Oncol 25: 211–17

Arai J, Yasukawa M, Ohminami H, Kakimoto M, Hasegawa A, Fujita S (2001) Identification of human telomerase reverse transcriptase-derived peptides that induce HLA-A24-restricted antileukemia cytotoxic T lymphocytes. Blood 97: 2903–07

Asai A, Oshima Y, Yamamoto Y, Uochi TA, Kusaka H, Akinaga S, Yamashita Y, Pongracz K, Pruzan R, Wunder E, Piatyszek M, Li S, Chin AC, Harley CB, Gryaznov S (2003) A novel telomerase template antagonist (GRN163) as a potential anticancer agent. Cancer Res 63: 3931–39

Ayyoub M, Migliaccio M, Guillaume P, Lienard D, Cerottini JC, Romero P, Levy F, Speiser DE, Valmori D (2001) Lack of tumor recognition by hTERT peptide 540–548-specific CD8(+) T cells from melanoma patients reveals inefficient antigen processing. Eur J Immunol 31: 2642–51

Bajpai AK, Park JH, Moon IJ, Kang H, Lee YH, Doh KO, Suh SI, Chang BC, Park JG (2005) Rapid blockade of telomerase activity and tumor cell growth by the DPL lipofection of ribbon antisense to hTERC. Oncogene 24: 6492–6501

Beltz L, Moran R, Elsawy O, Sadler J, Jurgenson J (1999) The effects of telomerase inhibitors on lymphocyte function. Anticancer Res 19: 3205–11

Berzofsky JA, Terabe M, Oh S, Belyakov IM, Ahlers JD, Janik JE, Morris JC (2004) Progress on new vaccine strategies for the immunotherapy and prevention of cancer. J Clin Invest 113: 1515–25

Bettuzzi S, Brausi M, Rizzi F, Castagnetti G, Peracchia G, Corti A (2006) Chemoprevention of human prostate cancer by oral administration of green tea catechins in volunteers with high-grade prostate intraepithelial neoplasia: a preliminary report from a one-year proof-of-principle study. Cancer Res 66: 1234–40

Bilsland AE, Fletcher-Monaghan A and Keith WN (2005) Properties of a telomerase-specific Cre/Lox switch for transcriptionally targeted cancer gene therapy. Neoplasia 7: 1020–29

Bilsland AE, Merron A, Vassaux G, Keith WN (2007) Modulation of telomerase promoter tumor selectivity in the context of oncolytic adenoviruses. Cancer Res 67: 1299–1307

Bilsland AE, Stevenson K, Atkinson S, Kolch W, Keith WN (2006) Transcriptional repression of telomerase RNA gene expression by c-Jun-NH2-kinase and Sp1/Sp3. Cancer Res 66: 1363–70

Bilsland AE, Anderson CJ, Fletcher-Monaghan AJ, McGregor F, Jeffry Evans TR, Ganly I, Knox RJ, Plumb JA, Nicol Keith W (2003) Selective ablation of human cancer cells by telomerase-specific adenoviral suicide gene therapy vectors expressing bacterial nitroreductase. Oncogene 22: 370–80

Binz N, Shalaby T, Rivera P, Shin-ya K, Grotzer MA (2005) Telomerase inhibition, telomere shortening, cell growth suppression and induction of apoptosis by telomestatin in childhood neuroblastoma cells. Eur J Cancer 41: 2873–81

Bortolanza S, Qian C, Kramer MG, Gomar C, Prieto J, Farinati F, Hernandez-Alcoceba R (2005) An oncolytic adenovirus controlled by a modified telomerase promoter is attenuated in telomerase-negative cells, but shows reduced activity in cancer cells. J Mol Med 83: 736–47

Boyd M, Mairs RJ, Keith WN, Ross SC, Welsh P, Akabani G, Owens J, Vaidyanathan G, Carruthers R, Dorrens J, Zalutsky MR (2004) An efficient targeted radiotherapy/gene therapy strategy utilising human telomerase promoters and radioastatine and harnessing radiation-mediated bystander effects. J Gene Med 6:937–47

Brown T, Sigurdson E, Rogatko A, Broccoli D (2003) Telomerase inhibition using azidothymidine in the HT-29 colon cancer cell line. Ann Surg Oncol 10: 910–15

Brunsvig PF, Aamdal S, Gjertsen MK, Kvalheim G, Markowski-Grimsrud CJ, Sve I, Dyrhaug M, Trachsel S, Moller M, Eriksen JA, Gaudernack G (2006) Telomerase peptide vaccination: a phase I/II study in patients with non-small cell lung cancer. Cancer Immunol Immunother 55: 1553–64

Budiyanto A, Bito T, Kunisada M, Ashida M, Ichihashi M, Ueda M (2003) Inhibition of the epidermal growth factor receptor suppresses telomerase activity in HSC-1 human cutaneous squamous cell carcinoma cells. J Invest Dermatol 121: 1088–94

Burger AM, Dai F, Schultes CM, Reszka AP, Moore MJ, Double JA, Neidle S (2005) The G-quadruplex-interactive molecule BRACO-19 inhibits tumor growth, consistent with telomere targeting and interference with telomerase function. Cancer Res 65: 1489–96

Canales BK, Li Y, Thompson MG, Gleason JM, Chen Z, Malaeb B, Corey DR, Herbert BS, Shay JW, Koeneman KS (2006) Small molecule, oligonucleotide-based telomerase template inhibition in combination with cytolytic therapy in an in vitro androgen-independent prostate cancer model. Urol Oncol 24: 141–51

Carpenter AE, Jones TR, Lamprecht MR, Clarke C, Kang IH, Friman O, Guertin DA, Chang JH, Lindquist RA, Moffat J, Golland P, Sabatini DM (2006) CellProfiler: image analysis software for identifying and quantifying cell phenotypes. Genome Biol 7: R100

Carpenter EL, Vonderheide RH (2006) Telomerase-based immunotherapy of cancer. Expert Opin Biol Ther 6: 1031–39

Chamuleau ME, Ossenkoppele GJ, van de Loosdrecht AA (2006) MHC class II molecules in tumor immunology: prognostic marker and target for immune modulation. Immunobiology 211: 619–25

Chang JT, Lu YC, Chen YJ, Tseng CP, Chen YL, Fang CW, Cheng AJ (2006) hTERT phosphorylation by PKC is essential for telomerase holoprotein integrity and enzyme activity in head neck cancer cells. Br J Cancer 94: 870–78

Chang JY, Zhang X, Komaki R, Cheung R, Fang B (2006) Tumor-specific apoptotic gene targeting overcomes radiation resistance in esophageal adenocarcinoma. Int J Radiat Oncol Biol Phys 64: 1482–94

Chang W, Dynek JN, Smith S (2003) TRF1 is degraded by ubiquitin-mediated proteolysis after release from telomeres. Genes Dev 17: 1328–33

Chen Z, Monia BP, Corey DR (2002) Telomerase inhibition, telomere shortening, and decreased cell proliferation by cell permeable 2′-O-methoxyethyl oligonucleotides. J Med Chem 45: 5423–25

Chen Z, Koeneman KS, Corey DR (2003) Consequences of telomerase inhibition and combination treatments for the proliferation of cancer cells. Cancer Res 63: 5917–25

Chow HH, Cai Y, Alberts DS, Hakim I, Dorr R, Shahi F, Crowell JA, Yang CS, Hara Y (2001) Phase I pharmacokinetic study of tea polyphenols following single-dose administration of epigallocatechin gallate and polyphenon E. Cancer Epidemiol Biomarkers Prev 10: 53–58

Chow HH, Cai Y, Hakim IA, Crowell JA, Shahi F, Brooks CA, Dorr RT, Hara Y, Alberts DS (2003) Pharmacokinetics and safety of green tea polyphenols after multiple-dose administration of epigallocatechin gallate and polyphenon E in healthy individuals. Clin Cancer Res 9: 3312–19

d'Adda di Fagagna F, Reaper PM, Clay-Farrace L, Fiegler H, Carr P, Von Zglinicki T, Saretzki G, Carter NP, Jackson SP (2003) A DNA damage checkpoint response in telomere-initiated senescence. Nature 426: 194–98

Damm K, Hemmann U, Garin-Chesa P, Hauel N, Kauffmann I, Priepke H, Niestroj C, Daiber C, Enenkel B, Guilliard B, Lauritsch I, Muller E, Pascolo E, Sauter G, Pantic M, Martens UM, Wenz C, Lingner J, Kraut N, Rettig WJ, Schnapp A (2001) A highly selective telomerase inhibitor limiting human cancer cell proliferation. Embo J 20: 6958–68

Datta A, Bellon M, Sinha-Datta U, Bazarbachi A, Lepelletier Y, Canioni D, Waldmann TA, Hermine O, Nicot C (2006) Persistent inhibition of telomerase reprograms adult T-cell leukemia to p53-dependent senescence. Blood 108: 1021–29

de Chastonay J (2005) The outlook for peptide drugs and the intricate relationship between the Active Pharmaceutical Ingredients manufacturer and their sponsors. J Pept Sci 11: 754–5

De Laporte L, Cruz Rea J, Shea LD (2006) Design of modular non-viral gene therapy vectors. Biomaterials 27: 947–54

Dikmen ZG, Gellert GC, Jackson S, Gryaznov S, Tressler R, Dogan P, Wright WE, Shay JW (2005) In vivo inhibition of lung cancer by GRN163L: a novel human telomerase inhibitor. Cancer Res 65: 7866–73

DiMasi JA, Grabowski HG (2007) Economics of new oncology drug development. J Clin Oncol 25: 209–16

DiMasi JA, Hansen RW, Grabowski HG (2003) The price of innovation: new estimates of drug development costs. J Health Econ 22: 151–85

Djojosubroto MW, Chin AC, Go N, Schaetzlein S, Manns MP, Gryaznov S, Harley CB, Rudolph KL (2005) Telomerase antagonists GRN163 and GRN163L inhibit tumor growth and increase chemosensitivity of human hepatoma. Hepatology 42: 1127–36

Downey MG, Going JJ, Stuart RC, Keith WN (2001) Expression of telomerase RNA in oesophageal and oral cancer. J Oral Pathol Med 30: 577–81

Dufes C, Keith WN, Bilsland A, Proutski I, Uchegbu IF, Schatzlein AG (2005) Synthetic anticancer gene medicine exploits intrinsic antitumor activity of cationic vector to cure established tumors. Cancer Res 65: 8079–84

Duncan VE, Ajmani PS, Hughes JA (1998) Cellular delivery is a major obstacle for oligodeoxynucleotide inhibition of telomerase activity. Anticancer Res 18: 4105–08

Eggert US, Mitchison TJ (2006) Small molecule screening by imaging. Curr Opin Chem Biol 10: 232–37

Elayadi AN, Demieville A, Wancewicz EV, Monia BP, Corey DR (2001) Inhibition of telomerase by 2′-O-(2-methoxyethyl) RNA oligomers: effect of length, phosphorothioate substitution and time inside cells. Nucleic Acids Res 29: 1683–89

El-Daly H, Kull M, Zimmermann S, Pantic M, Waller CF, Martens UM (2005) Selective cytotoxicity and telomere damage in leukemia cells using the telomerase inhibitor BIBR1532. Blood 105: 1742–49

Emens LA (2006) Roadmap to a better therapeutic tumor vaccine. Int Rev Immunol 25: 415–43

Evans TR, Keith WN (2004) Intra-peritoneal administration of genetic therapies: promises and pitfalls. Minerva Ginecol 56: 529–38

Everts B, van der Poel HG (2005) Replication-selective oncolytic viruses in the treatment of cancer. Cancer Gene Ther 12: 141–61

Filaci G, Fravega M, Setti M, Traverso P, Millo E, Fenoglio D, Negrini S, Ferrera F, Romagnoli A, Basso M, Contini P, Rizzi M, Ghio M, Benatti U, Damonte G, Ravetti JL, Carmignani G, Zanetti M, Indiveri F (2006) Frequency of telomerase-specific CD8+ T lymphocytes in patients with cancer. Blood 107: 1505–12

Folini M, De Marco C, Orlandi L, Daidone MG, Zaffaroni N (2000a) Attenuation of telomerase activity does not increase sensitivity of human melanoma cells to anticancer agents. Eur J Cancer 36: 2137–45

Folini M, Colella G, Villa R, Lualdi S, Daidone MG, Zaffaroni N (2000b) Inhibition of telomerase activity by a hammerhead ribozyme targeting the RNA component of telomerase in human melanoma cells. J Invest Dermatol 114: 259–67

Folini M, Berg K, Millo E, Villa R, Prasmickaite L, Daidone MG, Benatti U, Zaffaroni N (2003) Photochemical internalization of a peptide nucleic acid targeting the catalytic subunit of human telomerase. Cancer Res 63: 3490–94

Forsythe HL, Jarvis JL, Turner JW, Elmore LW, Holt SE (2001) Stable association of hsp90 and p23, but Not hsp70, with active human telomerase. J Biol Chem 276: 15571–74

Fox S, Farr-Jones S, Sopchak L, Boggs A, Nicely HW, Khoury R, Biros M (2006) High-throughput screening: update on practices and success. J Biomol Screen 11: 864–69

Fratantoni JC, Dzekunov S, Singh V, Liu LN (2003) A non-viral gene delivery system designed for clinical use. Cytotherapy 5: 208–10

Frisch SM, Mymryk JS (2002) Adenovirus-5 E1A: paradox and paradigm. Nat Rev Mol Cell Biol 3: 441–52

Fujii N, You L, Xu Z, Uematsu K, Shan J, He B, Mikami I, Edmondson LR, Neale G, Zheng J, Guy RK, Jablons DM (2007) An antagonist of dishevelled protein-protein interaction suppresses beta-catenin-dependent tumor cell growth. Cancer Res 67: 573–79

Gan Y, Mo Y, Johnston J, Lu J, Wientjes MG, Au JL (2002) Telomere maintenance in telomerase-positive human ovarian SKOV-3 cells cannot be retarded by complete inhibition of telomerase. FEBS Lett 527: 10–14

Gannage M, Abel M, Michallet AS, Delluc S, Lambert M, Giraudier S, Kratzer R, Niedermann G, Saveanu L, Guilhot F, Camoin L, Varet B, Buzyn A, Caillat-Zucman S (2005) Ex vivo characterization of multiepitopic tumor-specific CD8 T cells in patients with chronic myeloid leukemia: implications for vaccine development and adoptive cellular immunotherapy. J Immunol 174: 8210–18

Glasgow JN, Everts M, Curiel DT (2006) Transductional targeting of adenovirus vectors for gene therapy. Cancer Gene Ther 13: 830–44

Glukhov AI, Zimnik OV, Gordeev SA, Severin SE (1998) Inhibition of telomerase activity of melanoma cells in vitro by antisense oligonucleotides. Biochem Biophys Res Commun 248: 368–71

Goetz MP, Toft D, Reid J, Ames M, Stensgard B, Safgren S, Adjei AA, Sloan J, Atherton P, Vasile V, Salazaar S, Adjei A, Croghan G, Erlichman C (2005) Phase I trial of 17-allylamino-17-demethoxygeldanamycin in patients with advanced cancer. J Clin Oncol 23: 1078–87

Goldkorn A, Blackburn EH (2006) Assembly of mutant-template telomerase RNA into catalytically active telomerase ribonucleoprotein that can act on telomeres is required for apoptosis and cell cycle arrest in human cancer cells. Cancer Res 66: 5763–71

Gomez D, Paterski R, Lemarteleur T, Shin-Ya K, Mergny JL, Riou JF (2004) Interaction of telomestatin with the telomeric single-strand overhang. J Biol Chem 279: 41487–94

Gomez D, Wenner T, Brassart B, Douarre C, O'Donohue MF, El Khoury V, Shin-Ya K, Morjani H, Trentesaux C, Riou JF (2006) Telomestatin-induced telomere uncapping is modulated by POT1 through G-overhang extension in HT1080 human tumor cells. J Biol Chem 281: 38721–29

Gowan SM, Heald R, Stevens MF, Kelland LR (2001) Potent inhibition of telomerase by small-molecule pentacyclic acridines capable of interacting with G-quadruplexes. Mol Pharmacol 60: 981–88

Gowan SM, Harrison JR, Patterson L, Valenti M, Read MA, Neidle S, Kelland LR (2002) A G-quadruplex-interactive potent small-molecule inhibitor of telomerase exhibiting in vitro and in vivo antitumor activity. Mol Pharmacol 61: 1154–62

Grace MJ, Xie L, Musco ML, Cui S, Gurnani M, DiGiacomo R, Chang A, Indelicato S, Syed J, Johnson R, Nielsen LL (1999) The use of laser scanning cytometry to assess depth of penetration of adenovirus p53 gene therapy in human xenograft biopsies. Am J Pathol 155: 1869–78

Grem JL, Morrison G, Guo XD, Agnew E, Takimoto CH, Thomas R, Szabo E, Grochow L, Grollman F, Hamilton JM, Neckers L, Wilson RH (2005) Phase I and pharmacologic study of 17-(allylamino)-17-demethoxygeldanamycin in adult patients with solid tumors. J Clin Oncol 23: 1885–93

Grill J, Lamfers ML, van Beusechem VW, Dirven CM, Pherai DS, Kater M, Van der Valk P, Vogels R, Vandertop WP, Pinedo HM, Curiel DT, Gerritsen WR (2002) The organotypic multicellular spheroid is a relevant three-dimensional model to study adenovirus replication and penetration in human tumors in vitro. Mol Ther 6: 609–14

Groot-Wassink T, Aboagye EO, Wang Y, Lemoine NR, Keith WN, Vassaux G (2004) Noninvasive imaging of the transcriptional activities of human telomerase promoter fragments in mice. Cancer Res 64: 4906–11

Gross DA, Graff-Dubois S, Opolon P, Cornet S, Alves P, Bennaceur-Griscelli A, Faure O, Guillaume P, Firat H, Chouaib S, Lemonnier FA, Davoust J, Miconnet I, Vonderheide RH, Kosmatopoulos K (2004) High vaccination efficiency of low-affinity epitopes in antitumor immunotherapy. J Clin Invest 113: 425–33

Gryaznov S, Pongracz K, Matray T, Schultz R, Pruzan R, Aimi J, Chin A, Harley C, Shea-Herbert B, Shay J, Oshima Y, Asai A, Yamashita Y (2001) Telomerase inhibitors–oligonucleotide phosphoramidates as potential therapeutic agents. Nucleosides Nucleotides Nucleic Acids 20: 401–10

Gryaznov S, Asai A, Oshima Y, Yamamoto Y, Pongracz K, Pruzan R, Wunder E, Piatyszek M, Li S, Chin A, Harley C, Akinaga S, Yamashita Y (2003) Oligonucleotide N3′ → P5′ thiophosphoramidate telomerase template antagonists as potential anticancer agents. Nucleosides Nucleotides Nucleic Acids 22: 577–81

Gu J, Kagawa S, Takakura M, Kyo S, Inoue M, Roth JA, Fang B (2000) Tumor-specific transgene expression from the human telomerase reverse transcriptase promoter enables targeting of the therapeutic effects of the Bax gene to cancers. Cancer Res 60: 5359–64

Hamilton SE, Simmons CG, Kathiriya IS, Corey DR (1999) Cellular delivery of peptide nucleic acids and inhibition of human telomerase. Chem Biol 6: 343–51
Hamilton SE, Pitts AE, Katipally RR, Jia X, Rutter JP, Davies BA, Shay JW, Wright WE, Corey DR (1997) Identification of determinants for inhibitor binding within the RNA active site of human telomerase using PNA scanning. Biochemistry 36: 11873–80
Hao ZM, Luo JY, Cheng J, Li L, He D, Wang QY and Yang GX (2005) Intensive inhibition of hTERT expression by a ribozyme induces rapid apoptosis of cancer cells through a telomere length-independent pathway. Cancer Biol Ther 4:1098–1103
Herbert BS, Pongracz K, Shay JW, Gryaznov SM (2002) Oligonucleotide N3′→P5′ phosphoramidates as efficient telomerase inhibitors. Oncogene 21: 638–42
Herbert BS, Gellert GC, Hochreiter A, Pongracz K, Wright WE, Zielinska D, Chin AC, Harley CB, Shay JW, Gryaznov SM (2005) Lipid modification of GRN163, an N3′→P5′ thio-phosphoramidate oligonucleotide, enhances the potency of telomerase inhibition. Oncogene 24: 5262–68
Hermiston TW, Kirn DH (2005) Genetically based therapeutics for cancer: similarities and contrasts with traditional drug discovery and development. Mol Ther 11: 496–507
Hernandez J, Garcia-Pons F, Lone YC, Firat H, Schmidt JD, Langlade-Demoyen P, Zanetti M (2002) Identification of a human telomerase reverse transcriptase peptide of low affinity for HLA A2.1 that induces cytotoxic T lymphocytes and mediates lysis of tumor cells. Proc Natl Acad Sci USA 99: 12275–80
Hernandez-Alcoceba R, Pihalja M, Qian D, Clarke MF (2002) New oncolytic adenoviruses with hypoxia- and estrogen receptor-regulated replication. Hum Gene Ther 13: 1737–50
Hinz T, Buchholz CJ, van der Stappen T, Cichutek K, Kalinke U (2006) Manufacturing and quality control of cell-based tumor vaccines: a scientific and a regulatory perspective. J Immunother 29: 472–76
Hirano N, Butler MO, Xia Z, Berezovskaya A, Murray AP, Ansen S, Kojima S, Nadler LM (2006) Identification of an immunogenic CD8+ T-cell epitope derived from gamma-globin, a putative tumor-associated antigen for juvenile myelomonocytic leukemia. Blood 108: 2662–68
Hochreiter AE, Xiao H, Goldblatt EM, Gryaznov SM, Miller KD, Badve S, Sledge GW, Herbert BS (2006) Telomerase template antagonist GRN163L disrupts telomere maintenance, tumor growth, and metastasis of breast cancer. Clin Cancer Res 12: 3184–92
Holt SE, Aisner DL, Baur J, Tesmer VM, Dy M, Ouellette M, Trager JB, Morin GB, Toft DO, Shay JW, Wright WE, White MA (1999) Functional requirement of p23 and Hsp90 in telomerase complexes. Genes Dev 13: 817–26
Hoos A, Parmiani G, Hege K, Sznol M, Loibner H, Eggermont A, Urba W, Blumenstein B, Sacks N, Keilholz U, Nichol G (2007) A clinical development paradigm for cancer vaccines and related biologics. J Immunother 30: 1–15
Horikawa I, Barrett JC (2003) Transcriptional regulation of the telomerase hTERT gene as a target for cellular and viral oncogenic mechanisms. Carcinogenesis 24: 1167–76
Huang Q, Zhang X, Wang H, Yan B, Kirkpatrick J, Dewhrist MW, Li CY (2004) A novel conditionally replicative adenovirus vector targeting telomerase-positive tumor cells. Clin Cancer Res 10: 1439–45
Huang SL, Wu Y, Yu H, Zhang P, Zhang XQ, Ying L, Zhao HF (2006) Inhibition of Bcl-2 expression by a novel tumor-specific RNA interference system increases chemosensitivity to 5-fluorouracil in Hela cells. Acta Pharmacol Sin 27: 242–48
Huang TG, Savontaus MJ, Shinozaki K, Sauter BV, Woo SL (2003) Telomerase-dependent oncolytic adenovirus for cancer treatment. Gene Ther 10: 1241–47
Humphreys RE, Hillman GG, von Hofe E, Xu M (2004) Forcing tumor cells to present their own tumor antigens to the immune system: a necessary design for an efficient tumor immunotherapy. Cell Mol Immunol 1: 180–85
Incles CM, Schultes CM, Kelland LR, Neidle S (2003) Acquired cellular resistance to flavopiridol in a human colon carcinoma cell line involves up-regulation of the telomerase catalytic subunit and telomere elongation. Sensitivity of resistant cells to combination treatment with a telomerase inhibitor. Mol Pharmacol 64: 1101–08

Incles CM, Schultes CM, Kempski H, Koehler H, Kelland LR, Neidle S (2004) A G-quadruplex telomere targeting agent produces p16-associated senescence and chromosomal fusions in human prostate cancer cells. Mol Cancer Ther 3: 1201–06

Inui T, Shinomiya N, Fukasawa M, Kobayashi M, Kuranaga N, Ohkura S, Seki S (2002) Growth-related signaling regulates activation of telomerase in regenerating hepatocytes. Exp Cell Res 273: 147–56

Irving J, Wang Z, Powell S, O'Sullivan C, Mok M, Murphy B, Cardoza L, Lebkowski JS, Majumdar AS (2004) Conditionally replicative adenovirus driven by the human telomerase promoter provides broad-spectrum antitumor activity without liver toxicity. Cancer Gene Ther 11: 174–85

Ito H, Kanzawa T, Miyoshi T, Hirohata S, Kyo S, Iwamaru A, Aoki H, Kondo Y, Kondo S (2005) Therapeutic efficacy of PUMA for malignant glioma cells regardless of p53 status. Hum Gene Ther 16: 685–98

Ji XM, Xie CH, Fang MH, Zhou FX, Zhang WJ, Zhang MS, Zhou YF (2006) Efficient inhibition of human telomerase activity by antisense oligonucleotides sensitizes cancer cells to radiotherapy. Acta Pharmacol Sin 27: 1185–91

Kanazawa Y, Ohkawa K, Ueda K, Mita E, Takehara T, Sasaki Y, Kasahara A, Hayashi N (1996) Hammerhead ribozyme-mediated inhibition of telomerase activity in extracts of human hepatocellular carcinoma cells. Biochem Biophys Res Commun 225: 570–76

Kau TR, Schroeder F, Ramaswamy S, Wojciechowski CL, Zhao JJ, Roberts TM, Clardy J, Sellers WR, Silver PA (2003) A chemical genetic screen identifies inhibitors of regulated nuclear export of a Forkhead transcription factor in PTEN-deficient tumor cells. Cancer Cell 4: 463–76

Kawashima T, Kagawa S, Kobayashi N, Shirakiya Y, Umeoka T, Teraishi F, Taki M, Kyo S, Tanaka N, Fujiwara T (2004) Telomerase-specific replication-selective virotherapy for human cancer. Clin Cancer Res 10: 285–92

Kazhdan I, Long L, Montellano R, Cavazos DA, Marciniak RA (2006) Targeted gene therapy for breast cancer with truncated Bid. Cancer Gene Ther 13: 141–49

Keith WN, Sarvesvaran J, Downey M (2002) Analysis of telomerase RNA gene expression by in situ hybridization. Methods Mol Biol 191: 65–81

Keith WN, Bilsland A, Evans TR, Glasspool RM (2002) Telomerase-directed molecular therapeutics. Expert Rev Mol Med 2002: 1–25

Keith WN, Bilsland A, Hardie M, Evans TR (2004) Drug insight: Cancer cell immortality-telomerase as a target for novel cancer gene therapies. Nat Clin Pract Oncol 1: 88–96

Kelland LR (2005) Overcoming the immortality of tumor cells by telomere and telomerase based cancer therapeutics–current status and future prospects. Eur J Cancer 41: 971–79

Kelland LR, Green C, Clark S, Griffiths-Johnson D, Dunmore K, Robson M, Martins C, Gunaratnam M, Reszka T, Neidle S (2005) A small molecule trisubstituted acridine-based telomere targeting agent (TTAs) possessing preclinical in vivo antitumor activity. In: Proceedings of the American Association for Cancer Research, Volume 46, 2005. Abstract no. 2837

Kim E, Kim JH, Shin HY, Lee H, Yang JM, Kim J, Sohn JH, Kim H, Yun CO (2003a) Ad-mTERT-delta19, a conditional replication-competent adenovirus driven by the human telomerase promoter, selectively replicates in and elicits cytopathic effect in a cancer cell-specific manner. Hum Gene Ther 14: 1415–28

Kim MY, Gleason-Guzman M, Izbicka E, Nishioka D, Hurley LH (2003b) The different biological effects of telomestatin and TMPyP4 can be attributed to their selectivity for interaction with intramolecular or intermolecular G-quadruplex structures. Cancer Res 63: 3247–56

Kiss-Toth E, Guesdon FM, Wyllie DH, Qwarnstrom EE, Dower SK (2000) A novel mammalian expression screen exploiting green fluorescent protein-based transcription detection in single cells. J Immunol Methods 239: 125–35

Kiss-Toth E, Wyllie DH, Holland K, Marsden L, Jozsa V, Oxley KM, Polgar T, Qwarnstrom EE, Dower SK (2006) Functional mapping and identification of novel regulators for the Toll/Interleukin-1 signalling network by transcription expression cloning. Cell Signal 18: 202–14

Koga S, Kondo Y, Komata T, Kondo S (2001) Treatment of bladder cancer cells in vitro and in vivo with 2-5A antisense telomerase RNA. Gene Ther 8: 654–58

Koga S, Hirohata S, Kondo Y, Komata T, Takakura M, Inoue M, Kyo S, Kondo S (2000) A novel telomerase-specific gene therapy: gene transfer of caspase-8 utilizing the human telomerase catalytic subunit gene promoter. Hum Gene Ther 11: 1397–1406

Kohno M, Pouyssegur J (2006) Targeting the ERK signaling pathway in cancer therapy. Ann Med 38: 200–11

Komata T, Kondo Y, Koga S, Ko SC, Chung LW, Kondo S (2000) Combination therapy of malignant glioma cells with 2-5A-antisense telomerase RNA and recombinant adenovirus p53. Gene Ther 7: 2071–79

Komata T, Koga S, Hirohata S, Takakura M, Germano IM, Inoue M, Kyo S, Kondo S, Kondo Y (2001a) A novel treatment of human malignant gliomas in vitro and in vivo: FADD gene transfer under the control of the human telomerase reverse transcriptase gene promoter. Int J Oncol 19: 1015–20

Komata T, Kondo Y, Kanzawa T, Hirohata S, Koga S, Sumiyoshi H, Srinivasula SM, Barna BP, Germano IM, Takakura M, Inoue M, Alnemri ES, Shay JW, Kyo S, Kondo S (2001b) Treatment of malignant glioma cells with the transfer of constitutively active caspase-6 using the human telomerase catalytic subunit (human telomerase reverse transcriptase) gene promoter. Cancer Res 61: 5796–5802

Kondo S, Kondo Y, Li G, Silverman RH, Cowell JK (1998) Targeted therapy of human malignant glioma in a mouse model by 2-5A antisense directed against telomerase RNA. Oncogene 16: 3323–30

Kondo Y, Koga S, Komata T, Kondo S (2000) Treatment of prostate cancer in vitro and in vivo with 2-5A-anti-telomerase RNA component. Oncogene 19: 2205–11

Kosciolek BA, Kalantidis K, Tabler M, Rowley PT (2003) Inhibition of telomerase activity in human cancer cells by RNA interference. Mol Cancer Ther 2: 209–16

Kraemer K, Fuessel S, Schmidt U, Kotzsch M, Schwenzer B, Wirth MP, Meye A (2003) Antisense-mediated hTERT inhibition specifically reduces the growth of human bladder cancer cells. Clin Cancer Res 9: 3794–3800

Kummar S, Gutierrez M, Doroshow JH, Murgo AJ (2006) Drug development in oncology: classical cytotoxics and molecularly targeted agents. Br J Clin Pharmacol 62: 15–26

Kummar S, Kinders R, Rubinstein L, Parchment RE, Murgo AJ, Collins J, Pickeral O, Low J, Steinberg SM, Gutierrez M, Yang S, Helman L, Wiltrout R, Tomaszewski JE, Doroshow JH (2007) Compressing drug development timelines in oncology using phase '0' trials. Nat Rev Cancer 7: 131–39

Kushner DM, Paranjape JM, Bandyopadhyay B, Cramer H, Leaman DW, Kennedy AW, Silverman RH, Cowell JK (2000) 2-5A antisense directed against telomerase RNA produces apoptosis in ovarian cancer cells. Gynecol Oncol 76: 183–92

Lanson NA, Jr., Friedlander PL, Schwarzenberger P, Kolls JK, Wang G (2003) Replication of an adenoviral vector controlled by the human telomerase reverse transcriptase promoter causes tumor-selective tumor lysis. Cancer Res 63: 7936–41

Lanzilli G, Fuggetta MP, Tricarico M, Cottarelli A, Serafino A, Falchetti R, Ravagnan G, Turriziani M, Adamo R, Franzese O, Bonmassar E (2006) Resveratrol down-regulates the growth and telomerase activity of breast cancer cells in vitro. Int J Oncol 28: 641–48

Lavelle F, Riou JF, Laoui A, Mailliet P (2000) Telomerase: a therapeutic target for the third millennium? Crit Rev Oncol Hematol 34: 111–26

Lee SH, Kim JW, Lee HW, Cho YS, Oh SH, Kim YJ, Jung CH, Zhang W, Lee JH (2003) Interferon regulatory factor-1 (IRF-1) is a mediator for interferon-gamma induced attenuation of telomerase activity and human telomerase reverse transcriptase (hTERT) expression. Oncogene 22: 381–91

Leonetti C, Amodei S, D'Angelo C, Rizzo A, Benassi B, Antonelli A, Elli R, Stevens MF, D'Incalci M, Zupi G, Biroccio A (2004) Biological activity of the G-quadruplex ligand RHPS4 (3,11-difluoro-6,8,13-trimethyl-8H-quino[4,3,2-kl]acridinium methosulfate) is associated with telomere capping alteration. Mol Pharmacol 66: 1138–46

Lesko LJ (2007) Paving the critical path: how can clinical pharmacology help achieve the vision? Clin Pharmacol Ther 81: 170–77

Lev A, Denkberg G, Cohen CJ, Tzukerman M, Skorecki KL, Chames P, Hoogenboom HR, Reiter Y (2002) Isolation and characterization of human recombinant antibodies endowed with the antigen-specific, major histocompatibility complex-restricted specificity of T cells directed toward the widely expressed tumor T-cell epitopes of the telomerase catalytic subunit. Cancer Res 62: 3184–94
Li B, Fan J, Liu X, Qi R, Bo L, Gu J, Qian C (2006) Suppression of colorectal tumor growth by regulated survivin targeting. J Mol Med 84: 1077–86
Li S, Crothers J, Haqq CM, Blackburn EH (2005) Cellular and gene expression responses involved in the rapid growth inhibition of human cancer cells by RNA interference-mediated depletion of telomerase RNA. J Biol Chem 280: 23709–17
Lin CP, Liu JD, Chow JM, Liu CR, Liu HE (2007) Small-molecule c-Myc inhibitor, 10058-F4, inhibits proliferation, downregulates human telomerase reverse transcriptase and enhances chemosensitivity in human hepatocellular carcinoma cells. Anticancer Drugs 18: 161–70
Lin T, Gu J, Zhang L, Huang X, Stephens LC, Curley SA, Fang B (2002) Targeted expression of green fluorescent protein/tumor necrosis factor-related apoptosis-inducing ligand fusion protein from human telomerase reverse transcriptase promoter elicits antitumor activity without toxic effects on primary human hepatocytes. Cancer Res 62: 3620–25
Lin T, Zhang L, Davis J, Gu J, Nishizaki M, Ji L, Roth JA, Xiong M, Fang B (2003) Combination of TRAIL gene therapy and chemotherapy enhances antitumor and antimetastasis effects in chemosensitive and chemoresistant breast cancers. Mol Ther 8: 441–48
Liu J, Zou WG, Lang MF, Luo J, Sun LY, Wang XN, Qian QJ, Liu XY (2002) Cancer-specific killing by the CD suicide gene using the human telomerase reverse transcriptase promoter. Int J Oncol 21: 661–66
Liu TC, Kirn D (2007) Systemic efficacy with oncolytic virus therapeutics: clinical proof-of-concept and future directions. Cancer Res 67: 429–32
Liu TC, Galanis E, Kirn D (2007) Clinical trial results with oncolytic virotherapy: a century of promise, a decade of progress. Nat Clin Pract Oncol 4: 101–17
Ludwig A, Saretzki G, Holm PS, Tiemann F, Lorenz M, Emrich T, Harley CB, von Zglinicki T (2001) Ribozyme cleavage of telomerase mRNA sensitizes breast epithelial cells to inhibitors of topoisomerase. Cancer Res 61: 3053–61
Maecker B, von Bergwelt-Baildon MS, Anderson KS, Vonderheide RH, Anderson KC, Nadler LM, Schultze JL (2005) Rare naturally occurring immune responses to three epitopes from the widely expressed tumor antigens hTERT and CYP1B1 in multiple myeloma patients. Clin Exp Immunol 141: 558–62
Majumdar AS, Hughes DE, Lichtsteiner SP, Wang Z, Lebkowski JS, Vasserot AP (2001) The telomerase reverse transcriptase promoter drives efficacious tumor suicide gene therapy while preventing hepatotoxicity encountered with constitutive promoters. Gene Ther 8: 568–78
Masutomi K, Yu EY, Khurts S, Ben-Porath I, Currier JL, Metz GB, Brooks MW, Kaneko S, Murakami S, DeCaprio JA, Weinberg RA, Stewart SA, Hahn WC (2003) Telomerase maintains telomere structure in normal human cells. Cell 114: 241–53
Matthes E, Lehmann C (1999) Telomerase protein rather than its RNA is the target of phosphorothioate-modified oligonucleotides. Nucleic Acids Res 27: 1152–58
Matutes E, Taylor GP, Cavenagh J, Pagliuca A, Bareford D, Domingo A, Hamblin M, Kelsey S, Mir N, Reilly JT (2001) Interferon alpha and zidovudine therapy in adult T-cell leukaemia lymphoma: response and outcome in 15 patients. Br J Haematol 113: 779–84
Melana SM, Holland JF, Pogo BG (1998) Inhibition of cell growth and telomerase activity of breast cancer cells in vitro by 3′-azido-3′-deoxythymidine. Clin Cancer Res 4: 693–96
Michaelis LC, Ratain MJ (2006) Measuring response in a post-RECIST world: from black and white to shades of grey. Nat Rev Cancer 6: 409–14
Minev B, Hipp J, Firat H, Schmidt JD, Langlade-Demoyen P, Zanetti M (2000) Cytotoxic T cell immunity against telomerase reverse transcriptase in humans. Proc Natl Acad Sci USA 97: 4796–4801
Mitchison TJ (2005) Small-molecule screening and profiling by using automated microscopy. Chembiochem 6: 33–39

Miyoshi T, Hirohata S, Ogawa H, Doi M, Obika M, Yonezawa T, Sado Y, Kusachi S, Kyo S, Kondo S, Shiratori Y, Hudson BG, Ninomiya Y (2006) Tumor-specific expression of the RGD-alpha3(IV)NC1 domain suppresses endothelial tube formation and tumor growth in mice. Faseb J 20: 1904–06

Mukai S, Kondo Y, Koga S, Komata T, Barna BP, Kondo S (2000) 2-5A antisense telomerase RNA therapy for intracranial malignant gliomas. Cancer Res 60: 4461–67

Murakami J, Nagai N, Shigemasa K, Ohama K (1999) Inhibition of telomerase activity and cell proliferation by a reverse transcriptase inhibitor in gynaecological cancer cell lines. Eur J Cancer 35: 1027–34

Naasani I, Seimiya H, Tsuruo T (1998) Telomerase inhibition, telomere shortening, and senescence of cancer cells by tea catechins. Biochem Biophys Res Commun 249: 391–96

Naasani I, Seimiya H, Yamori T, Tsuruo T (1999) FJ5002: a potent telomerase inhibitor identified by exploiting the disease-oriented screening program with COMPARE analysis. Cancer Res 59: 4004–11

Naasani I, Oh-Hashi F, Oh-Hara T, Feng WY, Johnston J, Chan K, Tsuruo T (2003) Blocking telomerase by dietary polyphenols is a major mechanism for limiting the growth of human cancer cells in vitro and in vivo. Cancer Res 63: 824–30

Nair S, Boczkowski D, Moeller B, Dewhirst M, Vieweg J, Gilboa E (2003) Synergy between tumor immunotherapy and antiangiogenic therapy. Blood 102: 964–71

Nair SK, Heiser A, Boczkowski D, Majumdar A, Naoe M, Lebkowski JS, Vieweg J, Gilboa E (2000) Induction of cytotoxic T cell responses and tumor immunity against unrelated tumors using telomerase reverse transcriptase RNA transfected dendritic cells. Nat Med 6: 1011–17

Nakajima A, Tauchi T, Sashida G, Sumi M, Abe K, Yamamoto K, Ohyashiki JH, Ohyashiki K (2003) Telomerase inhibition enhances apoptosis in human acute leukemia cells: possibility of antitelomerase therapy. Leukemia 17: 560–67

Nakamura M, Masutomi K, Kyo S, Hashimoto M, Maida Y, Kanaya T, Tanaka M, Hahn WC, Inoue M (2005) Efficient inhibition of human telomerase reverse transcriptase expression by RNA interference sensitizes cancer cells to ionizing radiation and chemotherapy. Hum Gene Ther 16: 859–68

Nakamura M, Kyo S, Kanaya T, Yatabe N, Maida Y, Tanaka M, Ishida Y, Fujii C, Kondo T, Inoue M, Mukaida N (2004) hTERT-promoter-based tumor-specific expression of MCP-1 effectively sensitizes cervical cancer cells to a low dose of cisplatin. Cancer Gene Ther 11: 1–7

Neidle S, Harrison RJ, Reszka AP, Read MA (2000) Structure-activity relationships among guanine-quadruplex telomerase inhibitors. Pharmacol Ther 85: 133–09

Nicol Keith W, Jeffry Evans TR, Glasspool RM (2001) Telomerase and cancer: time to move from a promising target to a clinical reality. J Pathol 195: 404–14

Norton JC, Piatyszek MA, Wright WE, Shay JW, Corey DR (1996) Inhibition of human telomerase activity by peptide nucleic acids. Nat Biotechnol 14: 615–19

Nosrati M, Li S, Bagheri S, Ginzinger D, Blackburn EH, Debs RJ, Kashani-Sabet M (2004) Antitumor activity of systemically delivered ribozymes targeting murine telomerase RNA. Clin Cancer Res 10: 4983–90

Nowakowski GS, McCollum AK, Ames MM, Mandrekar SJ, Reid JM, Adjei AA, Toft DO, Safgren SL, Erlichman C (2006) A phase I trial of twice-weekly 17-allylamino-demethoxygeldanamycin in patients with advanced cancer. Clin Cancer Res 12: 6087–93

Ozawa T, Gryaznov SM, Hu LJ, Pongracz K, Santos RA, Bollen AW, Lamborn KR, Deen DF (2004) Antitumor effects of specific telomerase inhibitor GRN163 in human glioblastoma xenografts. Neuro-oncol 6: 218–26

Pallini R, Sorrentino A, Pierconti F, Maggiano N, Faggi R, Montano N, Maira G, Larocca LM, Levi A, Falchetti ML (2006) Telomerase inhibition by stable RNA interference impairs tumor growth and angiogenesis in glioblastoma xenografts. Int J Cancer 118: 2158–67

Park K, Yi SY, Lee CS, Kim KE, Pai HS, Seol DW, Chung BH, Kim M (2007) A split enhanced green fluorescent protein-based reporter in yeast two-hybrid system. Protein J 26: 107–16

Parkhurst MR, Riley JP, Igarashi T, Li Y, Robbins PF, Rosenberg SA (2004) Immunization of patients with the hTERT:540–548 peptide induces peptide-reactive T lymphocytes that do not recognize tumors endogenously expressing telomerase. Clin Cancer Res 10: 4688–98

Pascolo E, Wenz C, Lingner J, Hauel N, Priepke H, Kauffmann I, Garin-Chesa P, Rettig WJ, Damm K, Schnapp A (2002) Mechanism of human telomerase inhibition by BIBR1532, a synthetic, non-nucleosidic drug candidate. J Biol Chem 277: 15566–72

Paulmurugan R, Gambhir SS (2003) Monitoring protein-protein interactions using split synthetic renilla luciferase protein-fragment-assisted complementation. Anal Chem 75: 1584–89

Pendino F, Flexor M, Delhommeau F, Buet D, Lanotte M, Segal-Bendirdjian E (2001) Retinoids down-regulate telomerase and telomere length in a pathway distinct from leukemia cell differentiation. Proc Natl Acad Sci USA 98: 6662–67

Pitts AE, Corey DR (1998) Inhibition of human telomerase by 2′-O-methyl-RNA. Proc Natl Acad Sci USA 95: 11549–54

Plumb JA, Bilsland A, Kakani R, Zhao J, Glasspool RM, Knox RJ, Evans TR, Keith WN (2001) Telomerase-specific suicide gene therapy vectors expressing bacterial nitroreductase sensitize human cancer cells to the pro-drug CB1954. Oncogene 20: 7797–7803

Rajwa B, McNally HA, Varadharajan P, Sturgis J, Robinson JP (2004) AFM/CLSM data visualization and comparison using an open-source toolkit. Microsc Res Tech 64: 176–84

Ramesh N, Ge Y, Ennist DL, Zhu M, Mina M, Ganesh S, Reddy PS, Yu DC (2006) CG0070, a conditionally replicating granulocyte macrophage colony-stimulating factor–armed oncolytic adenovirus for the treatment of bladder cancer. Clin Cancer Res 12: 305–13

Read M, Harrison RJ, Romagnoli B, Tanious FA, Gowan SH, Reszka AP, Wilson WD, Kelland LR, Neidle S (2001) Structure-based design of selective and potent G quadruplex-mediated telomerase inhibitors. Proc Natl Acad Sci USA 98: 4844–49

Rha SY, Izbicka E, Lawrence R, Davidson K, Sun D, Moyer MP, Roodman GD, Hurley L, Von Hoff D (2000) Effect of telomere and telomerase interactive agents on human tumor and normal cell lines. Clin Cancer Res 6: 987–93

Rock KL, Goldberg AL (1999) Degradation of cell proteins and the generation of MHC class I-presented peptides. Annu Rev Immunol 17: 739–79

Sams-Dodd F (2005) Target-based drug discovery: is something wrong? Drug Discov Today 10: 139–47

Saretzki G, Ludwig A, von Zglinicki T, Runnebaum IB (2001) Ribozyme-mediated telomerase inhibition induces immediate cell loss but not telomere shortening in ovarian cancer cells. Cancer Gene Ther 8: 827–34

Sarvesvaran J, Going JJ, Milroy R, Kaye SB, Keith WN (1999) Is small cell lung cancer the perfect target for anti-telomerase treatment? Carcinogenesis 20: 1649–51

Sawada T, Yamada O, Yoshimura N, Hatori K, Fuchinoue S, Teraoka S (2002) Xenoantigen, an alphaGal epitope-expression construct driven by the hTERT-promoter, specifically kills human pancreatic cancer cell line. Cancer Cell Int 2: 14

Scardino A, Gross DA, Alves P, Schultze JL, Graff-Dubois S, Faure O, Tourdot S, Chouaib S, Nadler LM, Lemonnier FA, Vonderheide RH, Cardoso AA, Kosmatopoulos K (2002) HER-2/neu and hTERT cryptic epitopes as novel targets for broad spectrum tumor immunotherapy. J Immunol 168: 5900–06

Schreurs MW, Kueter EW, Scholten KB, Kramer D, Meijer CJ, Hooijberg E (2005) Identification of a potential human telomerase reverse transcriptase-derived, HLA-A1-restricted cytotoxic T-lymphocyte epitope. Cancer Immunol Immunother 54: 703–12

Schroers R, Huang XF, Hammer J, Zhang J, Chen SY (2002) Identification of HLA DR7-restricted epitopes from human telomerase reverse transcriptase recognized by CD4+ T-helper cells. Cancer Res 62: 2600–05

Schroers R, Shen L, Rollins L, Rooney CM, Slawin K, Sonderstrup G, Huang XF, Chen SY (2003) Human telomerase reverse transcriptase-specific T-helper responses induced by promiscuous major histocompatibility complex class II-restricted epitopes. Clin Cancer Res 9: 4743–55

Seimiya H, Muramatsu Y, Ohishi T, Tsuruo T (2005) Tankyrase 1 as a target for telomere-directed molecular cancer therapeutics. Cancer Cell 7: 25–37

Seimiya H, Sawada H, Muramatsu Y, Shimizu M, Ohko K, Yamane K, Tsuruo T (2000) Involvement of 14-3-3 proteins in nuclear localization of telomerase. Embo J 19: 2652–61

Seimiya H, Oh-hara T, Suzuki T, Naasani I, Shimazaki T, Tsuchiya K, Tsuruo T (2002) Telomere shortening and growth inhibition of human cancer cells by novel synthetic telomerase inhibitors MST-312, MST-295, and MST-1991. Mol Cancer Ther 1: 657–65

Shammas MA, Shmookler Reis RJ, Li C, Koley H, Hurley LH, Anderson KC, Munshi NC (2004) Telomerase inhibition and cell growth arrest after telomestatin treatment in multiple myeloma. Clin Cancer Res 10: 770–76

Shammas MA, Koley H, Batchu RB, Bertheau RC, Protopopov A, Munshi NC, Goyal RK (2005) Telomerase inhibition by siRNA causes senescence and apoptosis in Barrett's adenocarcinoma cells: mechanism and therapeutic potential. Mol Cancer 4: 24

Shay JW, Bacchetti S (1997) A survey of telomerase activity in human cancer. Eur J Cancer 33: 787–91

Shay JW, Wright WE (2006) Telomerase therapeutics for cancer: challenges and new directions. Nat Rev Drug Discov 5: 577–84

Shieh GS, Shiau AL, Yo YT, Lin PR, Chang CC, Tzai TS, Wu CL (2006) Low-dose etoposide enhances telomerase-dependent adenovirus-mediated cytosine deaminase gene therapy through augmentation of adenoviral infection and transgene expression in a syngeneic bladder tumor model. Cancer Res 66: 9957–66

Smith S, Giriat I, Schmitt A, de Lange T (1998) Tankyrase, a poly(ADP-ribose) polymerase at human telomeres. Science 282: 1484–87

Soder AI, Going JJ, Kaye SB, Keith WN (1998) Tumor specific regulation of telomerase RNA gene expression visualized by in situ hybridization. Oncogene 16: 979–83

Soder AI, Hoare SF, Muir S, Going JJ, Parkinson EK, Keith WN (1997) Amplification, increased dosage and in situ expression of the telomerase RNA gene in human cancer. Oncogene 14: 1013–21

Su Z, Vieweg J, Weizer AZ, Dahm P, Yancey D, Turaga V, Higgins J, Boczkowski D, Gilboa E, Dannull J (2002) Enhanced induction of telomerase-specific CD4(+) T cells using dendritic cells transfected with RNA encoding a chimeric gene product. Cancer Res 62: 5041–48

Su Z, Dannull J, Yang BK, Dahm P, Coleman D, Yancey D, Sichi S, Niedzwiecki D, Boczkowski D, Gilboa E, Vieweg J (2005) Telomerase mRNA-transfected dendritic cells stimulate antigen-specific CD8+ and CD4+ T cell responses in patients with metastatic prostate cancer. J Immunol 174: 3798–3807

Suenaga M, Yamaguchi A, Soda H, Orihara K, Tokito Y, Sakaki Y, Umehara M, Terashi K, Kawamata N, Oka M, Kohno S, Tei C (2006) Antiproliferative effects of gefitinib are associated with suppression of E2F-1 expression and telomerase activity. Anticancer Res 26: 3387–91

Sun D, Thompson B, Cathers BE, Salazar M, Kerwin SM, Trent JO, Jenkins TC, Neidle S, Hurley LH (1997) Inhibition of human telomerase by a G-quadruplex-interactive compound. J Med Chem 40: 2113–16

Tahara H, Shin-Ya K, Seimiya H, Yamada H, Tsuruo T, Ide T (2006) G-Quadruplex stabilization by telomestatin induces TRF2 protein dissociation from telomeres and anaphase bridge formation accompanied by loss of the 3′ telomeric overhang in cancer cells. Oncogene 25: 1955–66

Tajima K, Ito Y, Demachi A, Nishida K, Akatsuka Y, Tsujimura K, Hida T, Morishima Y, Kuwano H, Mitsudomi T, Takahashi T, Kuzushima K (2004) Interferon-gamma differentially regulates susceptibility of lung cancer cells to telomerase-specific cytotoxic T lymphocytes. Int J Cancer 110: 403–12

Takai H, Smogorzewska A, de Lange T (2003) DNA damage foci at dysfunctional telomeres. Curr Biol 13: 1549–56

Takeuchi H, Kanzawa T, Kondo Y, Komata T, Hirohata S, Kyo S, Kondo S (2004) Combination of caspase transfer using the human telomerase reverse transcriptase promoter and conventional therapies for malignant glioma cells. Int J Oncol 25: 57–63

Tauchi T, Shin-ya K, Sashida G, Sumi M, Okabe S, Ohyashiki JH, Ohyashiki K (2006) Telomerase inhibition with a novel G-quadruplex-interactive agent, telomestatin: in vitro and in vivo studies in acute leukemia. Oncogene 25: 5719–25

Tressler R, Kornbrust D, Chin A, Harley C, Behrs M, Elias L (2006) GRN163L, a telomerase inhibitor under development for cancer treatment: data guiding clinical trial design. European Journal of Cancer Supplements 4: 188

Vassilev LT, Vu BT, Graves B, Carvajal D, Podlaski F, Filipovic Z, Kong N, Kammlott U, Lukacs C, Klein C, Fotouhi N, Liu EA (2004) In vivo activation of the p53 pathway by small-molecule antagonists of MDM2. Science 303: 844–48

Villa R, Folini M, Lualdi S, Veronese S, Daidone MG, Zaffaroni N (2000) Inhibition of telomerase activity by a cell-penetrating peptide nucleic acid construct in human melanoma cells. FEBS Lett 473: 241–48

Villa R, Folini M, Porta CD, Valentini A, Pennati M, Daidone MG, Zaffaroni N (2003) Inhibition of telomerase activity by geldanamycin and 17-allylamino, 17-demethoxygeldanamycin in human melanoma cells. Carcinogenesis 24: 851–59

Vonderheide RH (2002) Telomerase as a universal tumor-associated antigen for cancer immunotherapy. Oncogene 21: 674–79

Vonderheide RH, Hahn WC, Schultze JL, Nadler LM (1999) The telomerase catalytic subunit is a widely expressed tumor-associated antigen recognized by cytotoxic T lymphocytes. Immunity 10: 673–79

Vonderheide RH, Anderson KS, Hahn WC, Butler MO, Schultze JL, Nadler LM (2001) Characterization of HLA-A3-restricted cytotoxic T lymphocytes reactive against the widely expressed tumor antigen telomerase. Clin Cancer Res 7: 3343–78

Vonderheide RH, Domchek SM, Schultze JL, George DJ, Hoar KM, Chen DY, Stephans KF, Masutomi K, Loda M, Xia Z, Anderson KS, Hahn WC, Nadler LM (2004) Vaccination of cancer patients against telomerase induces functional antitumor CD8+ T lymphocytes. Clin Cancer Res 10: 828–39

Wang ES, Wu K, Chin AC, Chen-Kiang S, Pongracz K, Gryaznov S, Moore MA (2004a) Telomerase inhibition with an oligonucleotide telomerase template antagonist: in vitro and in vivo studies in multiple myeloma and lymphoma. Blood 103: 258–66

Wang YG, Wang JH, Zhang YH, Gu Q, Liu XY (2004b) Antitumor effect of a novel adeno-associated virus vector targeting to telomerase activity in tumor cells. Acta Biochim Biophys Sin (Shanghai) 36: 492–500

Ward RJ, Autexier C (2005) Pharmacological telomerase inhibition can sensitize drug-resistant and drug-sensitive cells to chemotherapeutic treatment. Mol Pharmacol 68: 779–86

White LK, Wright WE, Shay JW (2001) Telomerase inhibitors. Trends Biotechnol 19: 114–20

Wirth T, Zender L, Schulte B, Mundt B, Plentz R, Rudolph KL, Manns M, Kubicka S, Kuhnel F (2003) A telomerase-dependent conditionally replicating adenovirus for selective treatment of cancer. Cancer Res 63: 3181–88

Wisman GB, De Jong S, Meersma GJ, Helder MN, Hollema H, de Vries EG, Keith WN, van der Zee AG (2000) Telomerase in (pre)neoplastic cervical disease. Hum Pathol 31: 1304–12

Won J, Chang S, Oh S, Kim TK (2004) Small-molecule-based identification of dynamic assembly of E2F-pocket protein-histone deacetylase complex for telomerase regulation in human cells. Proc Natl Acad Sci USA 101: 11328–33

Wong JM, Kusdra L, Collins K (2002) Subnuclear shuttling of human telomerase induced by transformation and DNA damage. Nat Cell Biol 4: 731–36

Wong SK (2004) A 384-well cell-based phospho-ERK assay for dopamine D2 and D3 receptors. Anal Biochem 333: 265–72

Yatabe N, Kyo S, Kondo S, Kanaya T, Wang Z, Maida Y, Takakura M, Nakamura M, Tanaka M, Inoue M (2002) 2-5A antisense therapy directed against human telomerase RNA inhibits telomerase activity and induces apoptosis without telomere impairment in cervical cancer cells. Cancer Gene Ther 9: 624–30

Yeo M, Rha SY, Jeung HC, Hu SX, Yang SH, Kim YS, An SW, Chung HC (2005) Attenuation of telomerase activity by hammerhead ribozyme targeting human telomerase RNA induces growth retardation and apoptosis in human breast tumor cells. Int J Cancer 114: 484–89

Yokoyama Y, Takahashi Y, Shinohara A, Lian Z, Wan X, Niwa K, Tamaya T (1998) Attenuation of telomerase activity by a hammerhead ribozyme targeting the template region of telomerase RNA in endometrial carcinoma cells. Cancer Res 58: 5406–10

Yokoyama Y, Takahashi Y, Shinohara A, Wan X, Takahashi S, Niwa K, Tamaya T (2000) The 5′-end of hTERT mRNA is a good target for hammerhead ribozyme to suppress telomerase activity. Biochem Biophys Res Commun 273: 316–21

Zahler AM, Williamson JR, Cech TR, Prescott DM (1991) Inhibition of telomerase by G-quartet DNA structures. Nature 350: 718–20

Zhang PH, Zou L, Tu ZG (2006a) RNAi-hTERT inhibition hepatocellular carcinoma cell proliferation via decreasing telomerase activity. J Surg Res 131: 143–49

Zhang Z, Yang X, Zhang Y, Zeng B, Wang S, Zhu T, Roden RB, Chen Y, Yang R (2006b) Delivery of telomerase reverse transcriptase small interfering RNA in complex with positively charged single-walled carbon nanotubes suppresses tumor growth. Clin Cancer Res 12: 4933–39

Zhao HF, Kiyota T, Chowdhury S, Purisima E, Banville D, Konishi Y, Shen SH (2004) A mammalian genetic system to screen for small molecules capable of disrupting protein-protein interactions. Anal Chem 76: 2922–27

Zheng JN, Sun YF, Pei DS, Liu JJ, Chen JC, Li W, Sun XQ, Shi QD, Han RF, Ma TX (2006) Inhibition of proliferation and induction of apoptosis in human renal carcinoma cells by anti-telomerase small interfering RNAs. Acta Biochim Biophys Sin (Shanghai) 38: 500–06

Zhou JM, Zhu XF, Lu YJ, Deng R, Huang ZS, Mei YP, Wang Y, Huang WL, Liu ZC, Gu LQ, Zeng YX (2006) Senescence and telomere shortening induced by novel potent G-quadruplex interactive agents, quindoline derivatives, in human cancer cell lines. Oncogene 25: 503–11

Zielinska D, Pongracz K, Gryaznov S (2005) A new approach to oligonucleotide N3′–>P5′ phosphoramidate building blocks. Nucleosides Nucleotides Nucleic Acids 24: 1063–67

Zou L, Zhang P, Luo C, Tu Z (2006) shRNA-targeted hTERT suppress cell proliferation of bladder cancer by inhibiting telomerase activity. Cancer Chemother Pharmacol 57: 328–34

Zou W, Luo C, Zhang Z, Liu J, Gu J, Pei Z, Qian C, Liu X (2004) A novel oncolytic adenovirus targeting to telomerase activity in tumor cells with potent. Oncogene 23: 457–64

Chapter 14
Werner Syndrome, Telomeres, and Stress Signaling: Implications for Future Therapies?

Terence Davis and David Kipling(✉)

Abstract Werner syndrome (WS) is a premature ageing disorder used as a model of normal human ageing. WS individuals have several characteristics of normal ageing, such as cataracts, hair greying, and skin ageing, but manifest these at an early age. Additionally, WS individuals have high levels of inflammatory diseases such as atherosclerosis and type II diabetes. The in vivo ageing in WS is associated with premature ageing of fibroblasts in culture, and it is thought that WS is a disease of accelerated cell ageing. Normal fibroblasts senesce as a result of telomere erosion and WRNp is known to play a role in telomere maintenance. However, telomeres in WS cells do not appear to show accelerated rates of erosion. Thus the cause of the accelerated senescence is not understood. Several features of WS cells suggest that they are growing under conditions of stress, and low oxygen conditions and antioxidant treatment revert some of the accelerated senescence phenotype. In addition, WS cells have significant levels of genomic instability. Oxidative stress and genomic instability result in the activation of stress kinases such as p38, and the p38-specific inhibitor SB203580 essentially prevents the accelerated senescence seen in WS fibroblasts. The recent development of p38 inhibitors with different binding properties, specificities, and oral bioavailability, and of new potent and selective inhibitors of other stress kinases such as JNK and MK2, will make it possible to dissect the roles of various kinase pathways in the accelerated senescence of WS cells. If this accelerated senescence is reflective of WS ageing in vivo, these kinase inhibitors may well form the basis of anti-ageing therapeutics for individuals with WS.

1 Introduction

Werner syndrome (WS) was first characterized by Dr Otto Werner at Kiel University (Werner 1904) and is an autosomal-recessive disorder that belongs to a category of diseases called premature ageing disorders (Martin et al. 1999). These diseases are also known as segmental progerias, meaning that these conditions are

David Kipling
Department of Pathology, School of Medicine, Cardiff University, Heath Park, Cardiff CF14 4XN, United Kingdom
e-mail: kiplingd@cardiff.ac.uk

K.L. Rudolph (ed.), *Telomeres and Telomerase in Ageing, Disease, and Cancer.*

associated with many, but not all, of the clinical characteristics seen in normal ageing processes (Brown et al. 1985). An increasing number of human disorders belong to this category, including ataxia telangiectasia, Cockayne syndrome, Rothmund-Thomson syndrome, and Hutchinson-Gilford progeria. There is considerable interest in researching the underlying molecular pathology of these diseases due to the possibility that this research may provide insight into aspects of the normal ageing process as they are tractable systems that can be traced to single-gene defects (Opresko et al. 2003, Kipling et al. 2004).

WS can be regarded as a model system for the study of normal ageing, as well as of age-associated diseases. Individuals with WS display the premature onset of many age-related phenotypes, including juvenile bilateral cataracts, skin atrophy, premature hair-greying, thymic atrophy, and early onset of age-related diseases such as type II diabetes mellitus, osteoporosis, and atherosclerosis (Martin et al. 1999) (Table 14.1). The premature ageing phenotypes usually manifest themselves in the second or third decade of life, and the median age of diagnosis of this syndrome is in the early 30s. Thus, WS has been referred to as an adult progeroid syndrome and can be distinguished from the childhood progerias, such as Hutchinson Gilford. However, the clinical picture of WS is more complex than a simple global acceleration of age-linked pathology. Individuals with WS may show symptoms that are characteristic of aged individuals, but manifest them much more severely. These include calcification of the cardiac valves and atrophy of the testicles and skin appendages. Not all tissues show symptoms of premature

Table 14.1 Clinical Characteristics for Werner Syndrome

Symptoms[a]	Occurs in Normal Ageing?
Cataracts	Yes
Gray hair	Yes
Alopecia	Yes
Sclerotic skin	Yes
Skin atrophy	Yes
Pigmentary alterations	Yes
Osteoporosis	Yes
Diabetes mellitus type II	Yes
Atherosclerosis	Yes
Arteriosclerosis	Yes
Elderly facial appearance	Yes
Lack of pubertal growth spurt	No
Poor musculature on limbs	No
Hypogonadism	No
Reduced fertility	No
Skin ulceration	No
Short stature	No
Soft tissue calcification	No
High pitched voice	No
Mesenchymal neoplasms	Rare

[a]Based upon data in Martin et al. 1999, Hofer et al. 2005.

senescence, and WS individuals appear to have essentially normal immune function and lack notable pathology of the central nervous system (Postiglione et al. 1996), although there are some reports of premature senile dementia (Hofer et al. 2005). WS is also associated with non-age-related symptoms such as short stature, soft tissue calcification, and reduced fertility. WS individuals also display poor muscular development of the limbs, and ulcerative lesions often develop over pressure points. An elevated incidence of cancer is observed in WS, although this is generally limited to nonepithelial and, in particular, mesenchymal neoplasms that are rare in the normal population, although there is an elevated frequency of thyroid tumors. In addition, the cancer incidence is unusual in that there are often several sites of primary neoplasia, e.g., multiple meningiomas (Goto et al. 1996). The two major causes of death of WS individuals are myocardial infarction and cancer at a median age of 47 (Goto 1997).

The study of biological systems benefits greatly from the availability of models that reflect, either wholly or partially, the particular system under study. Thus it is hoped that research into WS will aid in the understanding of not only genome instability (a hallmark of WS), but also some aspects of the underlying mechanisms leading to normal human ageing (Kipling et al. 2004).

2 The *WRN* Gene Encodes a Structure-Specific DNA Helicase

WS is caused by mutations in the *WRN* gene located on chromosome 8 that encodes a 1432 amino acid protein (Yu et al. 1996). Cloning of *WRN* 10 years ago revealed that WRNp has several functional domains, including a helicase domain with significant homology to the RecQ helicases. Apart from WRNp (RecQ2), the RecQ family in humans includes proteins such as Bloom (RecQ3) and Rothmund-Thomson (RecQ4), whose germ line mutations are responsible for the corresponding diseases, both of which are associated with genomic instability (Hickson 2003). There are two other members of the RecQ family, RecQ1 and RecQ5, which have not yet been associated with human genetic disease. Unique among the RecQ helicases, WRNp has a 3′ to 5′ exonuclease activity with significant homology to the nuclease domain of *E. coli* DNA polymerase I (Moser et al. 1997). At the carboxyl terminal end of the protein are located nuclear (NLS) and nucleolar localization signals (Suzuki et al. 2001).

The vast majority of known mutations in the *WRN* gene result in the creation of premature stop codons, either directly as nonsense mutations, or due to structural rearrangements and coding sequence frame-shifts that would produce truncated proteins lacking the NLS (Moser et al. 1999). Any WRNp produced would fail to translocate to the nucleus and be degraded in the cytoplasm (Moser et al. 2000). In addition the *WRN* mutant mRNA is subjected to degradation by the mRNA processing system (Hentze and Kulozik 1999). One exception to date is an individual who is a compound heterozygote for mis-sense mutations close to the exonuclease domain. These do not affect WRNp function, but destabilize the *WRN* mRNA and thus act as "functional-null" mutations (Huang et al. 2006). Thus in most, if not all,

WS cases, no WRNp is detectable, and the full-blown WS phenotype appears to be due to complete absence of WRNp. However, cells from heterozygous individuals have an increased sensitivity to genotoxic drugs (Ogburn et al. 1997, Okada et al. 1998, Poot et al. 1999) and an increased genome instability (Moser et al. 2000), which may be due to the reduced protein levels that are found in heterozygotes as a result of low levels of *WRN* expression (Yamabe et al. 1997, Muller et al. 2005). As the prevalence of heterozygotes is expected to be high, 0.6% in Japan for example (Satoh et al. 1999), it has been proposed that this may be a significant worldwide health issue (Martin et al. 1999).

The full-blown Werner phenotype may require the complete absence of WRNp; however, it is possible that *WRN* polymorphisms may have phenotypic manifestations. Indeed, some studies have shown that an increased risk of myocardial infarction is associated with individuals that are homozygous for cysteine at the C1367R polymorphic site (Castro et al. 2000). However, in a separate study no such association was found (Bohr et al. 2004), and in general there are no significant associations between the various *WRN* polymorphisms and health or longevity (Castro et al. 1999, Kuningas et al. 2006).

WRNp interacts with numerous proteins involved in DNA metabolism and processing, including replication protein A, p53, proliferating cell nuclear antigen, topoisomerase I, DNA polymerases β and δ, the Ku complex, and the Mre11/Rad50/Nbs1 complex (Shen and Loeb 2001), suggesting that WRNp plays a role in DNA replication, recombination, and repair processes. A number of studies have indicated that WRNp can unwind and/or hydrolyze a number of distinctive DNA structures, such as duplex DNA, branched DNA structures, D-loops, and four-way DNA junctions (Fry and Loeb 1999, Kamath-Loeb et al. 2001, Brosh et al. 2002, Opresko et al. 2002, Orren et al. 2002), although recent data suggest that WRNp is not necessary for D-loop processing, at least in vitro (Verdun and Karlseder 2006). Of particular interest is the ability of WRNp to efficiently resolve G-quadruplexes (otherwise called tetrad DNA) that are thermodynamically stable intramolecular structures containing four hydrogen-bonded guanine residues on G-rich telomeric sequences (Mohaghegh et al. 2001, Chang et al. 2004). Many of these structures may form during, and thus impair, processes of DNA metabolism. In addition, WRNp can drive branch migration of Holliday junctions and thus might accelerate and facilitate DNA recombination (Constantinou et al. 2000).There appears to be good evidence for involvement of WRNp in two major DNA repair pathways, DNA double-strand break (DSB) repair via the nonhomologous end joining reaction and through homologous recombination, and base-excision repair (Bohr 2005). However, despite this, there does not appear to be a major DNA repair deficit in WS.

Increasing evidence suggests that WRNp plays a role at telomeres. Telomeres in mammalian cells are protected and sequestered as T-loop structures and, consequently, the telomeric end is protected and prevented from being recognized as a DSB by the DNA repair machinery (Fig. 14.1). The formation of T-loops requires

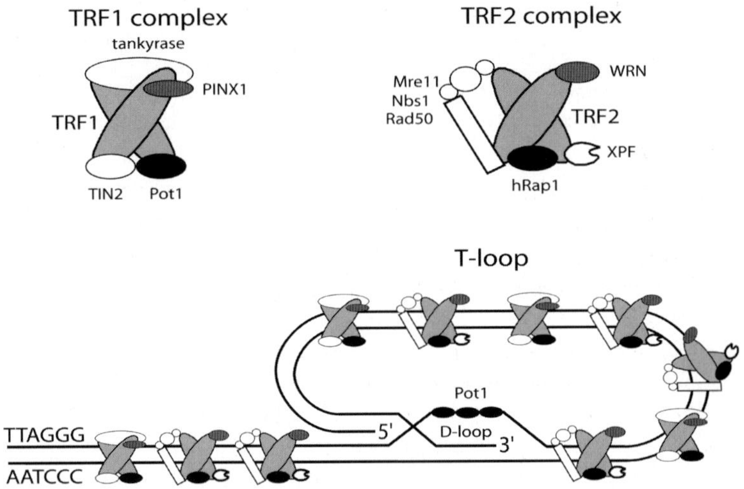

Fig. 14.1 Schematic depicting the telomere T-loop structure and bound protein complexes. At least two distinct complexes of proteins have been found, both of which bind to the double-stranded portion of telomeric DNA. The POT1 protein binds to both the TRF1 complex and to the G-rich single strand of the D-loop. Adapted from de Lange 2004

the telomeric binding proteins TRF1 and TRF2 (van Steensel et al. 1998, Stansel et al. 2001). In addition, telomeres have a single-stranded 3′ G-rich overhang known as the D-loop that is sequestered by POT1 into a deep binding pocket (de Lange 2005), and deletion of POT1 results in increased chromosomal aberrations (Hockemeyer et al. 2005). This suggests that POT1 functions to protect telomere ends. WRNp interacts with TRF1, TRF2, POT1, and with telomeric DNA in human cells (Johnson et al. 2001. Opresko et al. 2002, Opresko et al. 2004), and indeed TRF2 recruits WRNp exonuclease and replication protein A to facilitate the unwinding of long tracts of telomeric DNA (Opresko et al. 2002). Moreover, in the presence of POT1, WRNp helicase activity is stimulated and efficiently displaces a forked telomeric duplex releasing the invading tail of the D-loop (Opresko et al. 2005). These data, and the observation that WRNp is known to associate with telomeres during the S-phase of the cell cycle (Crabbe et al. 2004, Opresko et al. 2004), suggest that the WRNp unwinding function is involved in telomere replication and maintenance. This latter is supported by the observation that the *Saccharomyces cerevisiae* WRNp homolog Sgs1 is known to participate in telomere maintenance in telomerase null yeast (Johnson et al. 2001), and the observation that introduction of the dominant negative WRN^{K577M} construct into normal cells leads to a degree of telomere instability, manifested by the sporadic spontaneous truncation of whole telomeres (Bai and Murnane 2003, Crabbe et al. 2004). WRNp may therefore play a role in telomere replication by resolving aberrant secondary structures as the replication fork progresses through telomeres.

3 Cellular Aspects of Werner Syndrome

Many of the tissues showing premature ageing in WS individuals are those within the body that retain the ability to divide during life. This is clearly seen in the aged-skin pathology that includes sclerotic skin, skin atrophy, and pigmentary alterations (Table 14.1). The in vivo premature ageing is paralleled by accelerated cell ageing in vitro, with cultured WS dermal fibroblasts showing a shortened cellular life span, an aged appearance (Salk et al. 1981c, Tollefsbol and Cohen 1984), a transcriptional profile resembling that of aged cells (Vijg and Calder 2004), and a significant level of nuclear deformation (Adelfalk et al. 2005). This latter is also typical of Hutchinson Gilford dermal fibroblasts (De Sandre-Giovannoli 2003), and nuclear deformation often characterizes ageing cells (Lans and Hoeijmakers 2006). WS fibroblasts in culture manifest a poor rate of growth compared to normal cells (Salk et al. 1981c, Davis et al. 2005), partially due to a delayed passage through the S-phase of the cell cycle (Fujiwara et al. 1977, Rodriguez-Lopez et al. 2002). This slow growth is associated with a greatly increased rate of decline in the mitotic fraction of WS fibroblasts (Faragher et al. 1993, Kill et al. 1994) and is enhanced by the clonal succession, attenuation, and expansion observed in WS fibroblasts (Hoehn et al. 1975, Salk et al. 1981a, Schonberg et al. 1984, Benn 1985, Melcher et al. 2000, Baird et al. 2004), a phenotype displayed to a lesser degree by normal fibroblasts (Benn 1985). This clonal succession may result in the bulk of WS cell growth resulting from small numbers of dividing cells in the population (see Fig. 14.2). These phenomena acting together result in a more rapid turnover of the dividing cell fraction and a subsequent increased rate of production of senescent cells (Bird et al. 2003). Literature comparisons of the life spans of WS strains show that 90% of WS cultures have an in vitro life span of less than 20 PDs (Salk et al. 1981c, Tollefsbol and Cohen 1984). This compares to a life span for normal adult dermal fibroblasts that ranges from 20 to 85 PDs (Cristofalo et al. 1998). Not all division-competent tissues in WS show accelerated ageing, an example being the immune system, and it is interesting that WS T cells do not show premature senescence in culture (James et al. 2000). Whether cellular senescence plays a role in organismal ageing has been hotly disputed (Rubin 2002); however, recent observations clearly demonstrate that senescent fibroblasts accumulate in the skin of aged baboons (Herbig et al. 2006, Jeyapalan et al. 2007). These observations suggest the hypothesis that accelerated cellular senescence may underlie several of the features of accelerated ageing in WS, as cells from a dividing tissue that shows premature in vivo ageing also show accelerated in vitro ageing, whereas cells from a dividing tissue that does not manifest premature ageing in vivo do not show premature senescence in vitro (Ostler et al. 2002).

Cellular senescence has the potential to contribute to a large range of age-related degenerations in humans in two nonexclusive ways. The first is reduced proliferative capacity (i.e., fewer divisions "left on the clock") in organs where cell division is central to normal function or repair (e.g., small intestine, immune system, and skin) (see Allsopp, this volume). The second route reflects the observation that

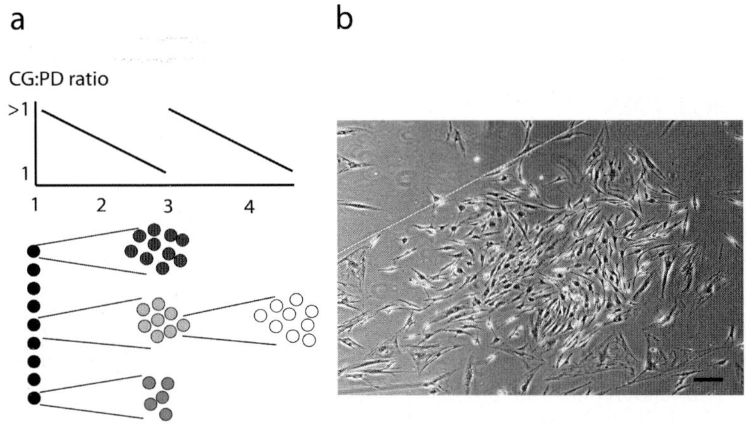

Fig. 14.2 **A)** Clonal growth model for WS cultures (adapted from Salk et al. 1981a, Melcher et al. 2000, Baird et al. 2004). Stage 1 is a culture where many of the cells are slow growing and/or growth arrested. At stage 2, some cells have a growth advantage and expand into clonal populations (stage 3). Most of the cells at late stage 3 are growth arrested and a second clonal expansion is seen in stage 4. The ratio of cell divisions (CD) to population doublings (PD) is high in stage 1 and in early stage 2, reduces to parity by late stage 2 and in stage 3, and increases again in stage 4. Thus, as the telomere erosion rate is calculated as [(bp telomeric DNA eroded) / PDs], the apparent rate of telomere erosion is high initially (CD > PD) and then reduces to a normal rate as the culture approaches stage 3 (CD ~ PD). **B)** Phase contrast photograph of a WS AG00780 fibroblast culture. The picture shows a clonal outgrowth of WS cells from a background of senescent cells. Many such clonal outgrowths were observed, and this culture achieved greater than 9 PDs despite the majority of cells being senescent, and had an initially high telomeric erosion rate that fell to a normal rate as the culture progressed (Baird et al. 2004). Bar = 200 μm

senescent cells display deleterious biochemical features resulting from patterns of gene expression that differ markedly from their dividing counterparts (see Gutierrez and Ju, this volume). Senescent cells express elevated levels of ICAM-1 on their plasma membrane, a feature that is associated with inflammatory conditions, being heavily expressed in Parkinson's disease, ulcerative colitis, osteoporosis, atherosclerosis, and rheumatoid arthritis (Lavigne et al. 2004, Gorgoulis et al. 2005, Vainer 2005, Miklossy et al. 2006). Senescent cells also excrete inflammatory cytokines such as IL-1, IL-15, and transforming growth factor β (TGFβ) (Maier 1990, Kumar et al. 1992, Garfinkel et al. 1994, Shelton et al. 1999, Krtolica and Campisi 2002, Schnabl et al. 2003, Vande Berg et al. 2005). In addition, aged normal fibroblasts and fibroblasts from WS individuals have upregulated levels of IL-1α and IL-1β (Kumar et al. 1993). These data suggest that the presence of senescent cells in human tissues may help to create a markedly proinflammatory microenvironment, and it is interesting that individuals with WS are prone to inflammatory conditions such as type II diabetes and atherosclerosis and have elevated plasma levels of tumor necrosis factor α (TNFα) and ICAM-1 (Murano et al. 1997, Martin et al. 1999, Yokote et al. 2004). Thus, premature cellular senescence may underlie some of the in vivo manifestations of WS (Kipling et al. 2004, Davis and Kipling 2006). However, the relationship of these cellular manifestations of the syndrome to the clinical presentation of the WS individual, remains unclear (Ostler et al. 2002).

4 The Role of Telomeres in Werner Syndrome

Due to the nature of the DNA replication process, it is not possible for the DNA replication machinery to efficiently replicate the 5′ ends of linear chromosomes, leading to the so-called "end replication problem" (Olovnikov 1973, Blackburn 1991). This end replication problem, together with processes such as C-strand resection and oxidative damage, results in the progressive loss of telomeric repeats at the end of the chromosomes (see Dillin and Karlseder, this volume). Telomere shortening is believed to underlie the replicative senescence of normal adult dermal fibroblasts (Bodnar et al. 1998, Vaziri and Benchimol 1998) (see Allsopp, this volume), although there are exceptions, including some strains of fetal and mammary fibroblasts (O'Hare et al. 2001, Forsyth et al. 2003). This has led to the telomeric loss hypothesis as a mitotic clock mechanism in fibroblasts (Harley et al. 1990, Hastie et al. 1990, Allsopp et al. 1992, Vaziri et al. 1994), a hypothesis that proposes that when telomeres shorten to a critical limit they lose their function, resulting in a signal to arrest the cell cycle that appears to be qualitatively similar to a DNA DSB response (d'Adda di Fagagna et al. 2003) (see Allsopp, this volume). This hypothesis is supported by the observation that telomerase activity is associated with the ability of cells to proliferate indefinitely, is upregulated during neoplastic progression (Campisi 2000; see also Rudolph, this volume), and can immortalize many human fibroblasts in culture (Bodnar et al. 1998, Vaziri and Benchimol 1998), leading to cell lines that no longer show telomere erosion. In normal cells the DNA damage-like signal from critically eroded telomeres is transduced via molecules such as the tumor suppressor p53, leading to activation of the cyclin-dependent kinase inhibitor $p21^{WAF1}$ and exit from the cell cycle (Vaziri and Benchimol 1996).

Since many fibroblasts senesce through telomeric erosion, it is possible that WS fibroblasts show accelerated replicative senesce as a result of accelerated telomeric erosion. This would be consistent with the reported association of WRNp with telomeric proteins such as TRF2 (Opresko et al. 2002; see also Blasco, this volume) and the telomere instability that results from introduction of the dominant negative WRN^{K577M} construct into normal cells (Bai and Murnane 2003, Crabbe et al. 2004).

Indeed, it has been reported that telomeric erosion was accelerated in four WS fibroblast strains (Schulz et al. 1996). However, the erosion rates overlapped those seen in normal strains and no increased erosion rate was seen in two further WS strains (Choi et al. 2001). Additionally, EBV-transformed WS T cells showed an altered telomere dynamics during cell division, in that they appear to undergo phases of rapid erosion interspersed with phases of lengthening (Tahara et al. 1997). These data suggest a direct role for WRNp in the regulation of telomere dynamics. Recent work, however, has shown that telomeric erosion in WS cells is not accelerated at the single cell level (Baird et al. 2004), suggesting that accelerated telomere erosion is not responsible for the shortened replicative life span of WS fibroblasts.

This seems to be at odds with the capability of ectopic hTERT expression to immortalize WS fibroblasts (Ouellette et al. 2000, Wyllie et al. 2000, Choi et al. 2001). However, despite being immortalized the telomerized WS cells still exhibit the slow growth and senescent-like morphology that are typical of primary WS cells (Wyllie et al. 2000, Davis et al. 2005, Davis et al. 2006). In addition, the telomerized cells still show hypersensitivity to DNA damaging agents such as 4-nitroquinoline-1-oxide (Hisama et al. 2000). Furthermore, the altered gene expression patterns seen in the young WS primary cells resemble those patterns seen in old cells (Vijg and Calder 2004) and are not fully corrected by telomerase expression (Choi et al. 2001). These lines of evidence suggest that the WS cellular phenotype cannot be fully corrected simply by forced expression of telomerase and the prevention of telomere erosion, and instead suggest the existence of roles for WRNp in other biological processes that cause a cellular phenotype when WRNp is absent, even in the presence of telomerase.

How then are we to reconcile these data? The apparent increase in telomeric erosion rates observed in some WS cultures is consistent with a change in the growth kinetics seen in the cultures, in that a population doubling of the culture represents more than one cell division for the fraction of the cells in the population that can actually divide, a process of so-called clonal succession (Hoehn et al. 1975, Salk et al. 1981a, Benn 1985, Schonberg et al. 1984, Melcher et al. 2000). This phenomenon can in result in the bulk of WS cell growth resulting from small numbers of dividing cells in the population (Schonberg et al. 1984, Benn 1985). A side effect of this process is that it then causes telomeric erosion to appear to be increased, since this crucially is measured as a function of PDs, and not actual cell divisions (Baird et al. 2004). The effects of clonal outgrowth on the apparent telomere erosion rate are illustrated in Fig. 14.2. A similar increased telomeric erosion rate has been reported for normal cultures that are growth arrested as a result of exposure to short periods of oxidative stress. After removal of the stress a minority of the cells resume growth and take over the culture, resulting in an apparent initial increase in telomere erosion (Dumont et al. 2001). This mechanism could also explain the altered dynamics seen in EBV-transformed WS T cells that express telomerase, where the telomeres appear to undergo phases of rapid erosion interspersed with phases of lengthening (Tahara et al. 1997). The rapid telomere erosion would occur during phases of clonal outgrowth, with telomerase re-extending the telomeres at phases of slower growth. These combined data suggest that premature senescence in WS cells involves a mechanism that is independent of telomere erosion, superimposed upon the usual telomere dependent senescence (Baird et al. 2004).

Although these data argue against a model whereby loss of WRNp leads to faster telomere erosion and thus accelerated replicative senescence, this does not rule out a role for WRNp at telomeres, nor changes in telomere behavior in WS cells. For example, recent data have indicated that replicative senescence may not be induced by telomere length per se, but due to alterations in telomeric structure (Chin et al. 1999, Blackburn 2000, Masutomi et al. 2003; see Blasco, this volume). Telomeres form terminal loops (T-loop) stabilized by proteins such as TRF2 (van Steensel et al. 1998, Griffith et al. 1999). This loop structure probably provides telomere capping functions, and its opening exposes the single-strand overhang triggering senescence (van Steensel et al. 1998, Chin et al. 1999, Saretzki et al. 1999, Blackburn 2000, von Zglinicki et al. 2001, Karlseder et al. 2002). Alternatively, it has been proposed that replicative senescence is due to the erosion of the telomeric single-strand overhang (Masutomi et al. 2003, Stewart et al. 2003), though this has been disputed (Keys et al. 2004). Thus, the observation that accelerated telomere erosion is not responsible for accelerated senescence of WS fibroblasts does not preclude a role for telomeres in this process. Rather, since telomere lengths are random on chromosomal ends, and a single (or a small subset of) dysfunctional telomere(s), are sufficient to limit cellular survival in vivo (Hemann et al. 2001, Zou et al. 2004), WS fibroblasts may be particularly sensitive to dysfunctional telomeres that arise stochastically during culture.

5 Genome Instability as a Hallmark of Werner Syndrome

WS cells show genomic instability that goes beyond telomeres and that might contribute significantly to the WS phenotype, both in patients and in cultured cells. WS cells exhibit slow DNA synthesis, although they initiate replication forks successfully (Fujiwara et al. 1977). In addition, WS cells show a significantly higher level of replication fork stalling compared to normal cells (Rodriguez-Lopez et al. 2002). WS cells are hypersensitive to DNA damaging agents (e.g., 4-nitroquinoline-1-oxide) and chemicals that induce interstrand cross-links, such as 8-methoxypsoralen (Prince et al. 1999, Hisama et al. 2000, Poot et al. 2002). Treatment of WS fibroblasts with these agents leads to reduced cell proliferation, but not p53-dependent apoptosis (Poot et al. 2002), as the latter is attenuated in WS cells (Spillare et al. 1999). The observation that cells from individuals with WS are not sensitive to a variety of agents that damage DNA suggests that the defect in WS is not a general defect in DNA repair. Instead, the evidence focuses on WRNp being a DNA structure-resolution enzyme that assists cells to resolve tangled DNA strands and thus facilitate DNA transactions. Interstrand cross-links may be repaired by homologous recombination (Hoeijmakers 2001), and it appears that recombinational repair is initiated in WS cells at replication fork stalls, but the lack of WRNp results in a failure to resolve the recombinational intermediate. Indeed, ectopic expression of a dominant negative version of the RAD51 protein corrects the growth defect of cisplatin-treated immortalized WS cells, presumably by preventing the initiation of recombinational repair of the replication fork stalls induced by the *cis*-Platin treatment (Saintigny et al. 2002). In addition, the ectopic expression of the bacterial resolvase RusA was effective in the resolution of stalled replication forks in these cells. This suggests that WS cells are deficient in undertaking the recombinational resolution of stalled replication forks (Rodriguez-Lopez et al. 2002), a deficiency that leads to a hyperrecombinant phenotype (Cheng et al. 1990, Elli et al. 1996).

The hyperrecombinant phenotype may explain the tendency for WS cells to exhibit variegated translocation mosaicism (VTM) (Hoehn et al. 1975), in that some cells show pseudodiploidy, with multiple, variable, and clonal chromosome rearrangements. VTM has been observed in a variety of WS cell types, including putative glial cells, fibroblasts, T cells, and EBV-transformed lymphoblastoid cell lines (Hoehn et al. 1975, Salk et al. 1981a, Salk 1982, Schonberg et al. 1984, Gebhart et al. 1988, Tahara et al. 1997). VTM probably reflects the workings of the mutator phenotype of WS (Fukuchi et al. 1989) that is most readily demonstrated by selection experiments designed to detect loss of function mutations at the HPRT locus by treatment with 6-thioguanine (6-TG). WS cultures produce a significantly higher fraction of 6-TG-resistant colonies than wild-type controls, most of which show large deletions at the HPRT locus. It is this mutator phenotype that presumably leads to the observed genomic instability that is the hallmark of WS. The potential consequences of this genome instability for the WS phenotype will be discussed below.

6 Premature Senescence in WS Cells May be a Stress Response

Several lines of evidence presented above argue that the premature senescence of WS cells cannot be readily explained by accelerated telomere erosion. This is consistent with the observation that in mice the phenotype of a mTERC$^{-/-}$ Wrn$^{-/-}$ mouse is not simply that of a mTERC$^{-/-}$ Wrn$^{+/+}$ with shorter telomeres, such as a later-generation mTERC$^{-/-}$ Wrn$^{+/+}$ mouse (Rudolph et al. 1999, Chang et al. 2004, Chang 2005) (see Chang, this volume). Although in the mouse short telomeres are needed for loss of WRNp to yield the WS-like phenotypes, the loss of WRNp is clearly having an effect that goes beyond simply accelerating telomere erosion (if indeed it does that). This parallels the data for human cells, where forced expression of telomerase can prevent premature senescence (of a bulk culture) and where both telomere-dependent and telomere-independent mechanisms of senescence are being postulated.

Since telomeres in WS fibroblasts do not have increased rates of telomere erosion per cell division (Baird et al. 2004), what then is the cause of the accelerated replicative decline? Several features of the growth of WS cells would suggest that the mechanism leading to their senescence is not the same as that seen in normal cells.

Even at low passages an appreciable proportion of cells display an enlarged morphology with prominent F-actin stress fibers resembling senescent cells (Davis et al. 2005). The characteristics of slow growth and enlarged morphology are reminiscent of cells growing under a condition that has been termed "replication stress" (Pichierri and Franchitto 2004), and many young WS cells resemble fibroblasts that have undergone oncogenic *ras*- or arsenite-induced premature senescence (Huot et al. 1997, Deng et al. 2004).

Premature senescence induced by ectopic expression of *ras* can result from activation of the stress-associated p38 MAPK (Wang et al. 2002) and the use of the p38 selective inhibitor SB203580 prevents *ras*-induced senescence in human BJ fibroblasts (Deng et al. 2004). Cells arrested by forced activation of the p38 pathway have many features reminiscent of senescence (Haq et al. 2002, Wang et al. 2002, Iwasa et al. 2003), and p38 is also believed to play a role in normal cellular senescence as a result of telomere erosion (Iwasa et al. 2003, Satyanarayana et al. 2004). Activation of p38 leads to the stabilization of p53 and the cyclin-dependent kinase inhibitor $p21^{WAF1}$ and subsequent cell cycle arrest (Bulavin et al. 1999, Kim et al. 2002). In addition, p38 activation contributes to the phosphorylation of the small heat shock protein HSP27 and the subsequent formation of F-actin stress fibers that leads to the altered morphology of senescent cells. This resemblance between normal fibroblasts that have undergone a stress-induced premature senescence and WS cells suggests that the latter may have activated a stress-induced cell cycle arrest involving p38 activation.

The possible involvement of p38 in the accelerated senescence of WS fibroblasts prompted us to test the effects of a known selective inhibitor of the p38 signaling pathway, SB203580. The results showed that SB203580 treatment increased the growth rate of primary WS cells to that seen for normal cells, rescued the senescent-like morphology, and extended the overall cellular life span to within the range seen for normal fibroblasts (Davis et al. 2005, Davis et al. 2006). In addition, it was shown that the p38 pathway was indeed activated in young WS cells, and that this activation was downregulated with drug treatment. Treatment of the cells with SB203580 reverted essentially all of the phenotypic characteristics of WS fibroblasts, strongly implicating a role for both p38 and stress signaling in WS. The cause of the stress-induced arrest is unknown, although it may be due to the increase in replication fork stalling and/or the generally increased level of genomic instability seen in WS fibroblasts (Davis and Kipling 2006).

7 Therapeutic Possibilities for Werner Syndrome

If the shortened in vitro life span of WS fibroblasts is reflective of events in vivo, this suggests that SB203580 and/or related drugs that prevent accelerated senescence may have therapeutic possibilities for WS. What makes this discovery of wider interest, however, is the fact that p38 is also stimulated in a wide range of inflammatory conditions (Beddy et al. 2004, Chabaud-Riou and Firestein 2004,

Hollenbach et al. 2004, Hollenbach et al. 2005). This raises the novel suggestion that although some features of WS may reflect a direct response to its genome instability phenotype (e.g., the increased cancer incidence), or result from the accelerated accumulation of senescent cells (e.g., the aged skin), there may be a third spectrum of clinical symptoms that are related to the consequences of the chronic p38 activation in vivo that these data would suggest is occurring (Davis and Kipling 2006). As p38 activation has been implicated in atherosclerosis, type II diabetes, and osteoporosis, the production of proinflammatory cytokines as a consequence of p38 activation could contribute to the high rates of these diseases seen in WS individuals (Fig. 14.3). This raises the exciting possibility that p38 inhibitors may form the basis of anti-ageing therapeutics for individuals with WS.

However, although SB203580 suppresses inflammation in model systems for ulcerative colitis and Crohn's disease by blocking p38 activation, it is not suitable for use in vivo due to toxicity, significant off-target effects, and a lack of oral bioavailability (Force et al. 2004, Hollenbach et al. 2004, Hollenbach et al. 2005). In addition, many of the current p38 inhibitors have been associated with hepatotoxicity, cardiotoxicity, CNS problems, and gastrointestinal tract symptoms when in clinical use (Goldstein and Gabriel 2005). There is therefore a pressing need for new classes of p38 inhibitors to address these issues, and significant efforts are being made to this end. Several inhibitors (e.g., BIRB796, VX702, VX745, R1487, and SCIO-469) have been in Phase II and III clinical trials for chronic inflammatory conditions such as Crohn's disease, rheumatoid arthritis, ulcerative colitis, and psoriasis, with mixed success (Force et al. 2004). Thus, toxicity data and efficacy profiles for these drugs are already available, and to date they are proving to have low toxicity, although elevated liver enzyme levels have been reported (Goldstein and Gabriel 2005). These new compounds have different modes of binding to that of SB203580, and have a different range of target specificities. Indeed some, such as VX702 and VX745, are exquisitely selective in their activities and have IC_{50} values for the p38 pathway in cell systems in the nM range (Goldstein and Gabriel 2005). Additionally, the selectivity problem may be countered by the use of inhibitors that can dissect the pathways downstream of p38, and several selective inhibitors of the HSP27 kinase MK2 are being developed (Anderson et al. 2005). New classes of p38 inhibitors that are orally bioavailable are also in development, e.g., RO3201195 (Goldstein et al. 2006).

A further problem is that the activation of other stress-related kinase pathways may play a role in the accelerated replicative senescence. Werner syndrome cells have an elevated oxidative state (Pagano et al. 2005a, Pagano et al. 2005b) and the treatment of Werner syndrome cells with antioxidants and reduced oxygen tension can extend the replicative life span (Salk et al. 1981b, Saito et al. 1995, Kashino et al. 2003, Davis et al. 2007). Although it is suppressed by SB203580 treatment, this oxidative stress does not appear to be transduced by the p38 kinase pathway, as p38 activity is still high in cells grown in reduced oxygen and ascorbic acid (Davis et al. 2007). One candidate for an oxidative stress-activated pathway that is known to be inhibited by SB203580 is JNK (Godl and Daub 2004), and it is interesting that JNK is activated in young WS cells (unpublished data). To address this problem the

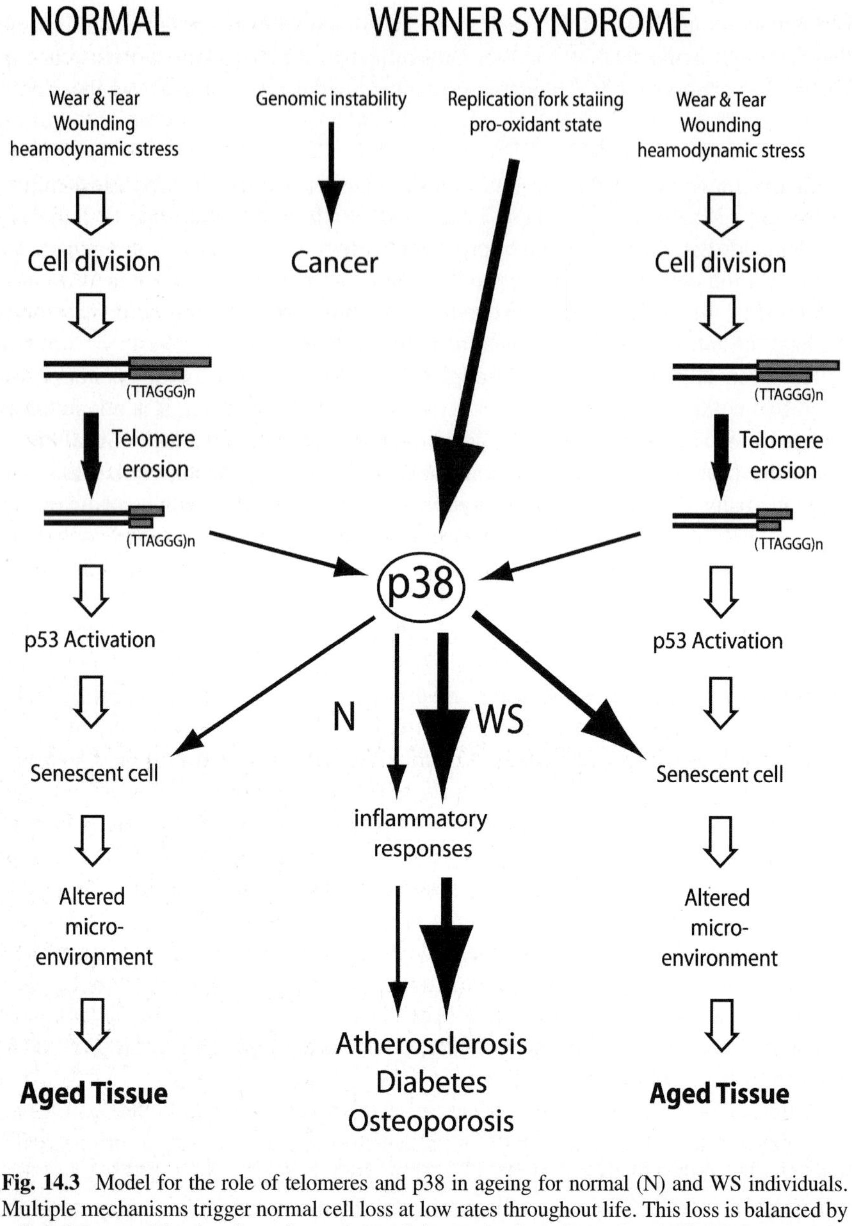

Fig. 14.3 Model for the role of telomeres and p38 in ageing for normal (N) and WS individuals. Multiple mechanisms trigger normal cell loss at low rates throughout life. This loss is balanced by cell division resulting in telomeric erosion and the production of senescent cells. Senescent cells have altered patterns of gene expression with the potential to produce deleterious changes in tissue microenvironment (Funk et al. 2000). These changes may result in tissue and organismal ageing. In addition, telomere erosion will upregulate p38 leading to inflammatory processes. In WS individuals, the same process of cell turnover leading to telomere erosion is occurring. However, the loss of WRN RecQ helicase results in an elevated genome instability, pro-oxidant state, and fork stalling during DNA replication. The genome instability may lead directly to increased cancer incidence. Genome instability, pro-oxidant state, and replication fork stall would lead to an

development of new potent and selective JNK inhibitors is needed and efforts toward these ends are being made by several groups (Gaillard et al. 2005, Liu et al. 2006, Szczepankiewicz et al. 2006). With these new tools in hand, a determined effort can be made toward dissecting the various pathways that are involved in the rapid senescence of WS cells. Finally, the development of a fully reflective mouse model of WS (Chang et al. 2004) makes future "proof of principle" studies possible as to whether inhibitors of the stress-related kinase pathways will be effective as therapeutic agents in a premature ageing disease.

8 Concluding Comments

The available data suggest a role for WRNp in telomere biology and a requirement for telomere erosion to occur to allow the cellular (human) or organismal (mouse) phenotype of loss of WRNp to be observed. Paradoxically, the data also argue against loss of WRNp having a major effect on telomere erosion rates and therefore contributing by this mechanism to the observed accelerated replicative senescence that is believed to be a key cellular phenotype underlying the disease. We believe that the key to resolving this is the genome instability throughout the genome that is associated with loss of WRNp, such as the increased frequency of stalled replication forks. It seems likely that such nuclear aberrations will be recognized by the cell as some form of "stress" signal, potentially sufficiently strong so as to elicit cell cycle arrest (which would be manifested as a stochastic, telomere-independent senescence). This telomere-independent senescence would then act alongside telomere-dependent senescence (or even in synergy with it), to produce the reduced replicative capacity of WS cells (Fig. 14.3). Provided the rate of telomere-independent senescence is not so high as to block completely the growth of a culture, this model then provides an explanation for ability of telomerase to immortalize primary WS cells (and yet produce cell lines that remain slow-growing, in this model because of the ongoing process of telomere-independent senescence) and the requirement for short telomeres in the mouse WS model.

Whatever the mechanistic details linking loss of WRNp to accelerated replicative senescence, the recent observations of a role for the proinflammatory p38 pathway in WS together with genome instability such as stalled replication forks provides a hitherto unexpected new insight into disease mechanism. We would like

Fig. 14.3 (continued) increase in intracellular stress that would activate the p38-signaling pathway. This would lead to telomere-independent cell senescence that would synergize with the ongoing telomere-dependent cell senescence, thus increasing tissue ageing. The much elevated p38 activity would lead to increased production of pro-inflammatory responses and increased incidence of atherosclerosis, diabetes, and osteoporosis. In this model, drugs that block the activation of p38 may alleviate many of these clinical manifestations and possibly lead to an increased life span. (Scheme adapted from Davis and Kipling 2005, Davis and Kipling 2006.)

to propose that genome instability, particularly in dividing cells (in the form of replication stress due to stalled replication forks) leads not only to accelerated replicative senescence of WS cultures, but also a series of clinical symptoms that are the direct consequence of chronic activation of p38 and the subsequent production of circulating proinflammatory cytokines such as TNFα. These in turn predispose toward diseases such as atherosclerosis, type II diabetes, and osteoporosis through well-known mechanisms (Fig. 14.3). In many respects, the clinical presentation of WS has many similarities to metabolic syndrome, with the obvious exception of morbid obesity. This may reflect a similar chronic production of cytokines in both instances, albeit from different sources. With Western society facing a growing problem of obesity and subsequent disease, it is perhaps of concern that many of the cytokines produced in response to p38 also act on cells to stimulate just that pathway. As part of a wider concept of "inflamm-ageing" (Franceschi et al. 2000, Davis and Kipling 2006) it is possible that long-term proinflammatory cytokine production in individuals will not only increase their disease risk but will also drive the ageing process in general, as we speculate is occurring in WS.

Acknowledgements Work in our laboratory is supported by the BBSRC, MRC and CR-UK. We thank our colleagues for critical reading of this manuscript.

References

Adelfalk C, Scherthan H, Hirsch-Kauffmann M, Schweiger M (2005) Nuclear deformation characterises Werner syndrome cells. Cell Biol Int 29: 1032–37

Allsopp RC, Vaziri H, Patterson C, Goldstein S, Younglai EV, Futcher AB, Greider CW, Harley CB (1992) Telomere length predicts replicative capacity of human fibroblasts. Proc Natl Acad Sci USA 89: 10114–18

Anderson DR, Hegde S, Reinhard E, Gomez L, Vernier WF, Lee L, Liu S, Sambandam A, Snider PA Masih L (2005) Aminocyanopyridine inhibitors of mitogen activated protein kinase-activated protein kinase 2 (MK-2). Bioorg Med Chem Lett 15: 1587–90

Bai Y, Murnane JP (2003) Telomere instability in a human tumor cell line expressing a dominant-negative WRN protein. Hum Genet 113: 337–47

Baird DM, Davis T, Rowson J, Jones CJ, Kipling D (2004) Normal telomere erosion rates at the single cell level in Werner syndrome fibroblast cells. Hum Mol Genet 13: 1515–24

Beddy DJ, Watson WR, Fitzpatrick JM, O'Connell PR (2004) Critical involvement of stress-activated mitogen-activated protein kinases in the regulation of intracellular adhesion molecule-1 in serosal fibroblasts isolated from patients with Crohn's disease. J Am Coll Surg 199: 234–42

Benn PA (1985) Chromosome translocations in fibroblast cultures derived from patients with Werner's syndrome. Am J Hum Genet 37: 221–23

Bird J, Ostler EL, Faragher RG (2003) Can we say that senescent cells cause ageing? Exp Gerontol 38: 1319–26

Blackburn EH (1991) Structure and function of telomeres. Nature 350: 569–73

Blackburn EH (2000) Telomere states and cell fates. Nature 408: 53–56

Bodnar AG, Ouellette M, Frolkis M, Holt SE, Chiu CP, Morin GB, Harley CB, Shay JW, Lichtsteiner S,Wright WE (1998) Extension of life-span by introduction of telomerase into normal human cells. Science 279: 349–52

Bohr VA (2005) Deficient DNA repair in the human progeroid disorder, Werner syndrome. Mutat Res 577: 252–59

Bohr VA, Metter EJ, Harrigan JA, von Kobbe C, Liu JL, Gray MD, Majumdar A, Wilson DM, Seidman MM (2004) Werner syndrome protein 1367 variants and disposition towards coronary artery disease in Caucasian patients. Mech Ageing Dev 125: 491–96

Brosh RM, Waheed J, Sommers JA (2002) Biochemical characterization of the DNA substrate specificity of Werner syndrome helicase. J Biol Chem 277: 23236–45

Brown WT, Kieras FJ, Houck GE, Dutkowski R, Jenkins EC (1985) A comparison of adult and childhood progerias: Werner syndrome and Hutchinson-Gilford progeria syndrome. Adv Exp Med Biol 190: 229–44

Bulavin DV, Saito S, Hollander MC, Sakaguchi K, Anderson CW, Appella E, Fornace AJ (1999) Phosphorylation of human p53 by p38 kinase coordinates N-terminal phosphorylation and apoptosis in response to UV radiation. EMBO J 18: 6845–54

Campisi J (2000) Cancer, aging and cellular senescence. In Vivo 14: 183–88

Castro E, Ogburn CE, Hunt KE, Tilvis R, Louhija J, Penttinen R, Erkkola R, Panduro A, Riestra R, Piussan C, Deeb SS, Wang L, Edland SD, Martin GM, Oshima J (1999) Polymorphisms at the Werner locus: I. Newly identified polymorphisms, ethnic variability of 1367Cys/Arg, and its stability in a population of Finnish centenarians. Am J Med Genet 82: 399–403

Castro E, Edland SD, Lee L, Ogburn CE, Deeb SS, Brown G, Panduro A, Riestra R, Tilvis R, Louhija J, Penttinen R, Erkkola R, Wang L, Martin GM, Oshima J (2000) Polymorphisms at the Werner locus: II. 1074Leu/Phe, 1367Cys/Arg, longevity, and atherosclerosis. Am J Med Genet 95: 374–80

Chabaud-Riou M, Firestein GS (2004) Expression and activation of mitogen-activated protein kinase kinases-3 and -6 in rheumatoid arthritis. Am J Pathol 164: 177–84

Chang S (2005) A mouse model of Werner Syndrome: what can it tell us about aging and cancer? Int J Biochem Cell Biol 37: 991–99

Chang S, Multani AS, Cabrera NG, Naylor ML, Laud P, Lombard D, Pathak S, Guarente L, DePinho RA (2004) Essential role of limiting telomeres in the pathogenesis of Werner syndrome. Nat Genet 36: 877–82

Cheng RZ, Murano S, Kurz B, Shmookler Reis RJ (1990) Homologous recombination is elevated in some Werner-like syndromes but not during normal in vitro or in vivo senescence of mammalian cells. Mutat Res 237: 259–69

Chin L, Artandi SE, Shen Q, Tam A, Lee SL, Gottlieb GJ, Greider CW, DePinho RA (1999) p53 deficiency rescues the adverse effects of telomere loss and cooperates with telomere dysfunction to accelerate carcinogenesis. Cell 97: 527–38

Choi D, Whittier PS, Oshima J, Funk WD (2001) Telomerase expression prevents replicative senescence but does not fully reset mRNA expression patterns in Werner syndrome cell strains. FASEB J 15: 1014–20

Constantinou A, Tarsounas M, Karow JK, Brosh RM, Bohr VA, Hickson ID West SC (2000) Werner's syndrome protein (WRN) migrates Holliday junctions and co-localizes with RPA upon replication arrest. EMBO Rep 1: 80–84

Crabbe L, Verdun RE, Haggblom CI, Karlseder J (2004) Defective telomere lagging strand synthesis in cells lacking WRN helicase activity. Science 306: 1951–53

Cristofalo VJ, Allen RG, Pignolo RJ, Martin BG, Beck JC (1998) Relationship between donor age and the replicative lifespan of human cells in culture: a reevaluation. Proc Natl Acad Sci USA 95: 10614–19

d'Adda di Fagagna F, Reaper PM, Clay-Farrace L, Fiegler H, Carr P, Von Zglinicki T, Saretzki G, Carter NP, Jackson SP (2003) A DNA damage checkpoint response in telomere-initiated senescence. Nature 426: 194–98

Davis T, Kipling D (2005) Telomeres and telomerase biology in vertebrates: progress towards a non-human model for replicative senescence and ageing. Biogerontology 6: 371–85

Davis T, Kipling D (2006) Werner Syndrome as an example of inflamm-ageing: possible therapeutic opportunities for a progeroid syndrome? Rejuv Res 9: 402–07

Davis T, Baird DM, Haughton MF, Jones CJ, Kipling D (2005) Prevention of accelerated cell aging in Werner Syndrome using a p38 mitogen-activated protein kinase inhibitor. J Gerontol A Biol Sci Med Sci 60: 1386–93

Davis T, Haughton MF, Jones CJ, Kipling D (2006) Prevention of accelerated cell ageing in Werner Syndrome. Ann NY Acad Sci 1067: 243–47

Davis T, Wyllie FS, Rokicki MJ, Bagley MC, Kipling D (2007) The role of cellular senescence in Werner syndrome: towards therapeutic intervention in human premature ageing. Ann NY Acad Sci 1100: 455–69

de Lange T (2004) T-loops and the origin of telomeres. Nat Rev Mol Cell Biol 5: 323–29

de Lange T (2005) Shelterin: the protein complex that shapes and safeguards human telomeres. Genes Dev 19: 2100–10

Deng Q, Liao R, Wu BL, Sun P (2004) High intensity ras signaling induces premature senescence by activating p38 pathway in primary human fibroblasts. J Biol Chem 279: 1050–59

De Sandre-Giovannoli A, Bernard R, Cau P, Navarro C, Amiel J, Boccaccio I, Lyonnet S, Stewart CL, Munnich A, Le Merrer M, Levy N (2003) Lamin A truncation in Hutchinson-Gilford progeria. Science 300: 2055

Dumont P, Royer V, Pascal T, Dierick JF, Chainiaux F, Frippiat C, de Magalhaes JP, Eliaers F, Remacle J, Toussaint O (2001) Growth kinetics rather than stress accelerate telomere shortening in cultures of human diploid fibroblasts in oxidative stress-induced premature senescence. FEBS Lett 502: 109–12

Elli R, Chessa L, Antonelli A, Petrinelli P, Ambra R, Marcucci L (1996) Effects of topoisomerase II inhibition in lymphoblasts from patients with progeroid and "chromosome instability" syndromes. Cancer Genet Cytogenet 87: 112–16

Faragher RGA, Kill IR, Hunter JA, Pope FM, Tannock C, Shall S (1993) The gene responsible for Werner syndrome may be a cell division "counting" gene. Proc Natl Acad Sci USA 90: 12030–34

Force T, Kuida K, Namchuk M, Parang K, Kyriakis JM (2004) Inhibitors of protein kinase signaling pathways: emerging therapies for cardiovascular disease. Circulation 109: 1196–1205

Forsyth NR, Evans AP, Shay JW, Wright WE (2003) Developmental differences in the immortalization of lung fibroblasts by telomerase. Aging Cell 2: 235–43

Franceschi C, Bonafe M, Valensin S, Olivieri F, De Luca M, Ottaviani E, De Benedictis G (2000) Inflamm-aging. An evolutionary perspective on immunosenescence. Ann NY Acad Sci 908: 244–54

Fry M, Loeb LA (1999) Human Werner Syndrome DNA helicase unwinds tetrahelical structures of the fragile X syndrome repeat sequence d(CGG)n. J Biol Chem 274: 12797–802

Fujiwara Y, Higashikawa T, Tatsumi M (1977) A retarded rate of DNA replication and normal level of DNA repair in Werner's syndrome fibroblasts in culture. J Cell Physiol 92: 365–74

Fukuchi K, Martin GM, Monnat RJ (1989) Mutator phenotype of Werner syndrome is characterized by extensive deletions. Proc Natl Acad Sci USA 86: 5893–97

Funk WD, Wang CK, Shelton DN, Harley CB, Pagon GD, Hoeffler WK (2000) Telomerase expression restores dermal integrity to in vitro-aged fibroblasts in a reconstituted skin model. Exp Cell Res 258: 270–78

Gaillard P, Jeanclaude-Etter I, Ardissone V, Arkinstall S, Cambet Y, Camps M, Chabert C, Church D, Cirillo R, Gretener D, Halazy S, Nichols A, Szyndralewiez C, Vitte PA, Gotteland JP (2005) Design and synthesis of the first generation of novel potent, selective, and in vivo active (benzothiazol-2-yl)acetonitrile inhibitors of the c-Jun N-terminal kinase. J Med Chem 48: 4596–607

Garfinkel S, Brown S, Wessendorf JHM, Maciag T (1994) Post-transcriptional regulation of interleukin 1 in various strains of young and senescent human umbilical vein endothelial cells. Proc Natl Acad Sci USA 91: 1559–63

Gebhart E, Bauer R, Raub U, Schinzel M, Ruprecht KW, Jonas JB (1988) Spontaneous and induced chromosomal instability in Werner syndrome. Hum Genet 80: 135–39

Godl K, Daub H (2004) Proteomic analysis of kinase inhibitor selectivity and function. Cell Cycle 3: 393–95

Goldstein DM, Gabriel T (2005) Pathway to development of p38 MAP kinase. A review of ten chemotypes selected for development. Curr Top Med Chem 5: 1017–29

Goldstein DM, Alfredson T, Bertrand J, Browner MF, Clifford K, Dalrymple SA, Dunn J, Freire-Moar J, Harris S, Labadie SS, La Fargue J, Lapierr, JM, Larrabee S, Li F, Papp E, McWeeney D, Ramesha C, Roberts R, Rotstein D, San Pablo B, Sjogren EB, So OY, Talamas FX, Tao W, Trejo A, Villasenor A, Welch M, Welch T, Weller P, Whiteley PE, Young K, Zipfel S (2006) Discovery of S-[5-amino-1-(4-fluorophenyl)-1H-pyrazol-4-yl]-[3-(2,3-dihydroxypropoxy)phenyl]methanone (RO3201195), an orally bioavailable and highly selective inhibitor of p38 MAP kinase. J Med Chem 49: 1562–75

Gorgoulis VG, Pratsinis H, Zacharatos P, Demoliou C, Sigala F, Asimacopoulos PJ, Papavassiliou AG, Kletsas D (2005) p53-dependent ICAM-1 overexpression in senescent human cells identified in atherosclerotic lesions. Lab Invest 85: 502–11

Goto M, Miller RW, Ishikawa Y, Sugano H (1996) Excess of rare cancers in Werner syndrome (adult progeria). Cancer Epidemiol Biomarkers Prev 5: 236–46

Goto M (1997) Hierarchical deterioration of body systems in Werner's syndrome: implications for normal ageing. Mech Ageing Dev 98: 239–54

Griffith JD, Comeau L, Rosenfield S, Stansel RM, Bianchi A, Moss H, de Lange T (1999) Mammalian telomeres end in a large duplex loop. Cell 97: 503–14

Haq R, Brenton JD, Takahashi M, Finan D, Finkielsztein A, Damaraju S, Rottapel R, Zanke B (2002) Constitutive p38HOG mitogen-activated protein kinase activation induces permanent cell cycle arrest and senescence. Cancer Res 62: 5076–82

Harley CB, Futcher AB, Greider CW (1990) Telomeres shorten during ageing of human fibroblasts. Nature 345: 458–60

Hastie ND, Dempster M, Dunlop MG, Thompson AM, Green DK, Allshire RC (1990) Telomere reduction in human colorectal carcinoma and with ageing. Nature 346: 866–68

Hemann MT, Strong MA, Hao LY, Greider CW (2001) The shortest telomere, not average telomere length, is critical for cell viability and chromosome stability. Cell 107: 67–77

Hentze MW, Kulozik AE (1999) A perfect message: RNA surveillance and nonsense-mediated decay. Cell 96: 307–10

Herbig U, Ferreira M, Condel L, Carey D, Sedivy JM (2006) Cellular senescence in aging primates. Science 311: 1257

Hickson ID (2003) RecQ helicases: caretakers of the genome. Nat Rev Cancer 3: 169–78

Hisama FM, Chen YH, Meyn MS, Oshima J, Weissman SM (2000) WRN or telomerase constructs reverse 4-nitroquinoline 1-oxide sensitivity in transformed Werner syndrome fibroblasts. Cancer Res 60: 2372–76

Hockemeyer D, Sfeir AJ, Shay JW, Wright WE, de Lange T (2005) POT1 protects telomeres from a transient DNA damage response and determines how human chromosomes end. EMBO J 24: 2667–78

Hoehn H, Bryant EM, Au K, Norwood TH, Boman H, Martin GM (1975) Variegated translocation mosaicism in human skin fibroblast cultures. Cytogenet Cell Genet 15: 282–98

Hoeijmakers JH (2001) DNA repair mechanisms. Maturitas 38: 17–22

Hofer AC, Tran RT, Aziz OZ, Wright W, Novelli G, Shay J, Lewis M (2005) Shared phenotypes among segmental progeroid syndromes suggest underlying pathways of aging. J Gerontol A Biol Sci Med Sci 60: 10–20

Hollenbach E, Neumann M, Vieth M, Roessner A, Malfertheiner P, Naumann M (2004) Inhibition of p38 MAP kinase- and RICK/NF-kappaB-signaling suppresses inflammatory bowel disease. FASEB J 18: 1550–52

Hollenbach E, Vieth M, Roessner A, Neumann M, Malfertheiner P, Naumann M (2005) Inhibition of RICK/nuclear factor-kappaB and p38 signaling attenuates the inflammatory response in a murine model of Crohn disease. J Biol Chem 280: 14981–88

Huang S, Lee L, Hanson NB, Lenaerts C, Hoehn H, Poot M, Rubin CD, Chen DF, Yang CC, Juch H, Dorn, T, Spiegel R, Oral EA, Abid M, Battisti C, Lucci-Cordisco E, Neri G, Steed EH, Kidd A, Isley W, Showalter D, Vittone JL, Konstantinow A, Ring J, Meyer P, Wenger SL, von Herbay A, Wollina U, Schuelke M, Huizenga CR, Leistritz DF, Martin GM, Mian IS, Oshima J (2006) The spectrum of WRN mutations in Werner syndrome patients. Hum Mutat 27: 558–67

Huot J, Houle F, Marceau F, Landry J (1997) Oxidative stress-induced actin reorganization mediated by the p38 mitogen-activated protein kinase/heat shock protein 27 pathway in vascular endothelial cells. Circ Res 80: 383–92

Iwasa H, Han J, Ishikawa F (2003) Mitogen-activated protein kinase p38 defines the common senescence-signalling pathway. Genes Cells 8: 131–44

James SE, Faragher RGA, Burke JF, Shall S, Mayne LV (2000) Werner's syndrome T lymphocytes display a normal in vitro life-span. Mech Ageing Dev 121: 139–49

Jeyapalan JC, Ferreira M, Sedivy JM, Herbig U (2007) Accumulation of senescent cells in mitotic tissue of aging primates. Mech Ageing Dev 12: 36–44

Johnson FB, Marciniak RA, McVey M, Stewart SA, Hahn WC, Guarente L (2001) The Saccharomyces cerevisiae WRN homolog Sgs1p participates in telomere maintenance in cells lacking telomerase. EMBO J 20: 905–13

Kamath-Loeb AS, Loeb LA, Johansson E, Burgers PM, Fry M (2001) Interactions between the Werner syndrome helicase and DNA polymerase delta specifically facilitate copying of tetraplex and hairpin structures of the d(CGG)n trinucleotide repeat sequence. J Biol Chem 276: 16439–46

Karlseder J, Smogorzewska A, de Lange T (2002) Senescence induced by altered telomere state, not telomere loss. Science 295: 2446–49

Kashino G, Kodama S, Nakayama Y, Suzuki K, Fukase K, Goto M, Watanabe M (2003) Relief of oxidative stress by ascorbic acid delays cellular senescence of normal human and Werner syndrome fibroblast cells. Free Radic Biol Med 35: 438–43

Keys B, Serra V, Saretzki G, Von Zglinicki T (2004) Telomere shortening in human fibroblasts is not dependent on the size of the telomeric-3′-overhang. Aging Cell 3: 103–09

Kill IR, Faragher RGA, Lawrence K, Shall S (1994) The expression of proliferation-dependent antigens during the lifespan of normal and progeroid human fibroblasts in culture. J Cell Sci 107: 571–79

Kim GY, Mercer SE, Ewton DZ, Yan Z, Jin K, Friedman E (2002) The stress-activated protein kinases p38 alpha and JNK1 stabilize p21(Cip1) by phosphorylation. J Biol Chem 277: 29792–802

Kipling D, Davis T, Ostler EL, Faragher RG (2004) What can progeroid syndromes tell us about human aging? Science 305: 1426–31

Krtolica AR, Campisi J (2002) Cancer and aging: a model for the cancer promoting effects of the aging stroma. Int J Biochem Cell Biol 34: 1401–14

Kumar S, Millis AJ, Baglioni C (1992) Expression of interleukin 1-inducible genes and production of interleukin 1 by aging human fibroblasts. Proc Natl Acad Sci USA 89: 4683–87

Kumar S, Vinci JM, Millis AJ, Baglioni C (1993) Expression of interleukin-1 alpha and beta in early passage fibroblasts from aging individuals. Exp Gerontol 28: 505–13

Kuningas M, Slagboom PE, Westendorp RG, van Heemst D (2006). Impact of genetic variations in the WRN gene on age related pathologies and mortality. Mech Ageing Dev 127: 307–13

Lans H, Hoeijmakers JHJ (2006) Ageing nucleus gets out of shape. Nature 440: 32–34

Lavigne P, Benderdour M, Lajeunesse D, Shi Q, Fernandes JC (2004) Expression of ICAM-1 by osteoblasts in healthy individuals and in patients suffering from osteoarthritis and osteoporosis. Bone 35: 463–70

Liu M, Xin Z, Clampit JE, Wang,S, Gum RJ, Haasch DL, Trevillyan JM, Abad-Zapatero C, Fry EH, Sham HL, Liu G (2006) Synthesis and SAR of 1,9-dihydro-9-hydroxypyrazolo[3,4-b] quinolin-4-ones as novel, selective c-Jun N-terminal kinase inhibitors. Bioorg Med Chem Lett 16: 2590–94

Maier JA, Voulalas P, Roeder D, Maciag T (1990) Extension of the life-span of human endothelial cells by an interleukin-1 alpha antisense oligomer. Science 249: 1570–74

Martin GM, Oshima J, Gray MD, Poot M (1999) What geriatricians should know about the Werner syndrome. J Am Geriatr Soc 47: 1136–44

Masutomi K, Yu EY, Khurts S, Ben-Porath I, Currier JL, Metz GB, Brooks MW, Kaneko S, Murakami S, DeCaprio JA, Weinberg RA, Stewart SA, Hahn WC (2003) Telomerase maintains telomere structure in normal human cells. Cell 114: 241–53

Melcher R, von Golitschek R, Steinlein C, Schindler D, Neitzel H, Kainer K, Schmid M, Hoehn H (2000) Spectral karyotyping of Werner syndrome fibroblast cultures. Cytogenet Cell Genet 91: 180–85

Miklossy J, Doudet DD, Schwab C, Yu S, McGeer EG, McGeer PL (2006) Role of ICAM-1 in persisting inflammation in Parkinson disease and MPTP monkeys. Exp Neurol 197: 275–83

Mohaghegh P, Karow JK, Brosh RM, Bohr VA, Hickson ID (2001) The Bloom's and Werner's syndrome proteins are DNA structure-specific helicases. Nucleic Acids Res 29: 2843–49

Moser MJ, Holley WR, Chatterjee A, Mian IS (1997) The proofreading domain of Escherichia coli DNA polymerase I and other DNA and/or RNA exonuclease domains. Nucleic Acids Res 25: 5110–18

Moser MJ, Oshima J, Monnat RJ (1999) WRN mutations in Werner syndrome. Hum Mutat 13: 271–79

Moser MJ, Kamath-Loeb AS, Jacob JE, Bennett SE, Oshima J, Monnat RJ (2000) WRN helicase expression in Werner syndrome cell lines. Nucleic Acids Res 28: 648–54

Muller FB, Tsianakas A, Kuwert C, Korge BP, Hunzelmann N (2005) A novel compound heterozygous mutation in Werner syndrome results in WRN transcript decay. Br J Dermatol 152: 1030–32

Murano S, Nakazawa A, Saito I, Masuda M, Morisaki N, Akikusa B, Tsuboyama T, Saito Y (1997) Increased blood plasminogen activator inhibitor-1 and intercellular adhesion molecule-1 as possible risk factors of atherosclerosis in Werner syndrome. Gerontology 43 Suppl 1: 43–52

O'Hare MJ, Bond J, Clarke C, Takeuchi Y, Atherton AJ, Berry C, Moody J, Silver AR, Davies DC, Alsop AE, Neville AM, Jat PS (2001) Conditional immortalization of freshly isolated human mammary fibroblasts and endothelial cells. Proc Natl Acad Sci USA 98, 646–51

Ogburn CE, Oshima J, Poot M, Chen R, Hunt KE, Gollahon KA, Rabinovitch PS, Martin GM (1997) An apoptosis-inducing genotoxin differentiates heterozygotic carriers for Werner helicase mutations from wild-type and homozygous mutants. Hum Genet 101: 121–25

Okada M, Goto M, Furuichi Y, Sugimoto M (1998) Differential effects of cytotoxic drugs on mortal and immortalized B-lymphoblastoid cell lines from normal and Werner's syndrome patients. Biol Pharm Bull 21: 235–39

Olovnikov AM (1973) A theory of marginotomy. The incomplete copying of template margin in enzymic synthesis of polynucleotides and biological significance of the phenomenon. J Theor Biol 41: 181–90

Opresko PL, von Kobbe C, Laine JP, Harrigan J, Hickson ID, Bohr VA (2002) Telomere-binding protein TRF2 binds to and stimulates the Werner and Bloom syndrome helicases. J Biol Chem 277: 41110–19

Opresko PL, Cheng WH, von Kobbe C, Harrigan JA, Bohr VA (2003) Werner syndrome and the function of the Werner protein; what they can teach us about the molecular aging process. Carcinogenesis 24: 791–802

Opresko PL, Otterlei M, Graakjaer J, Bruheim P, Dawut L, Kolvraa S, May A, Seidman MM, Bohr VA (2004) The Werner syndrome helicase and exonuclease cooperate to resolve telomeric D loops in a manner regulated by TRF1 and TRF2. Mol Cell 14: 763–74

Opresko PL, Mason PA, Podell ER, Lei M, Hickson ID, Cech TR, Bohr VA (2005) POT1 stimulates RecQ helicases WRN and BLM to unwind telomeric DNA substrates. J Biol Chem 280: 32069–80

Orren DK, Theodore S, Machwe A (2002) The Werner syndrome helicase/exonuclease (WRN) disrupts and degrades D-loops in vitro. Biochemistry 41: 13483–88

Ostler EL, Wallis CV, Sheerin AN, Faragher RGA (2002) A model for the phenotypic presentation of Werner's syndrome. Exp Gerontol 37: 285–92

Ouellette MM, McDaniel LD, Wright WE, Shay JW, Schultz RA (2000) The establishment of telomerase-immortalized cell lines representing human chromosome instability syndromes. Hum Mol Genet 9: 403–11

Pagano G, Zatterale A, Degan P, d'Ischia M, Kelly FJ, Pallardo FV, Calzone R, Castello G, Dunster C, Giudice A, Kilinc Y, Lloret A, Manini P, Masella R, Vuttariello E, Warnau M (2005a) In vivo prooxidant state in Werner syndrome (WS): results from three WS patients and two WS heterozygotes. Free Radic Res 39: 529–33

Pagano G, Zatterale A, Degan P, d'Ischia M, Kelly FJ, Pallardo FV, Kodama S (2005b) Multiple involvement of oxidative stress in Werner syndrome phenotype. Biogerontology 6: 233–43

Pichierri P, Franchitto A (2004) Werner syndrome protein, the MRE11 complex and ATR: menage-a-trois in guarding genome stability during DNA replication? Bioessays 26: 306–13

Postiglione A, Soricelli A, Covelli EM, Iazzetta N, Ruocco A, Milan G, Santoro L, Alfano B, Brunett A (1996) Premature aging in Werner's syndrome spares the central nervous system. Neurobiol Aging 17: 325–30

Poot M, Gollahon KA, Rabinovitch PS (1999) Werner syndrome lymphoblastoid cells are sensitive to camptothecin-induced apoptosis in S-phase. Hum Genet 104: 10–14

Poot M, Gollahon KA, Emond MJ, Silber JR, Rabinovitch PS (2002) Werner syndrome diploid fibroblasts are sensitive to 4-nitroquinoline-N-oxide and 8-methoxypsoralen: implications for the disease phenotype. FASEB J 16: 757–58

Prince PR, Ogburn CE, Moser MJ, Emond MJ, Martin GM, Monnat RJ (1999) Cell fusion corrects the 4-nitroquinoline 1-oxide sensitivity of Werner syndrome fibroblast cell lines. Hum Genet 105: 132–38

Rodriguez-Lopez AM, Jackson DA, Iborra F, Cox LS (2002) Asymmetry of DNA replication fork progression in Werner's syndrome. Aging Cell 1: 30–39

Rubin H (2002) The disparity between human cell senescence in vitro and lifelong replication in vivo. Nat Biotechnol 20: 675–81

Rudolph KL, Chan S, Lee HW, Blasco M, Gottlieb GJ, Greider C, DePinho RA (1999) Longevity, stress response, and cancer in aging telomerase-deficient mice. Cell 96: 701–12

Saintigny Y, Makienko K, Swanson C, Emond MJ, Monnat RJ (2002) Homologous recombination resolution defect in werner syndrome. Mol Cell Biol 22: 6971–78

Saito H, Hammond AT, Moses RE (1995) The effect of low oxygen tension on the in vitro-replicative life span of human diploid fibroblast cells and their transformed derivatives. Exp Cell Res 217: 272–79

Salk D (1982) Werner's syndrome: a review of recent research with an analysis of connective tissue metabolism, growth control of cultured cells, and chromosomal aberrations. Hum Genet 62: 1–5

Salk D, Au K, Hoehn H, Martin GM (1981a) Cytogenetics of Werner's syndrome cultured skin fibroblasts: variegated translocation mosaicism. Cytogenet Cell Genet 30: 92–107

Salk D, Au K, Hoehn H, Martin GM (1981b) Effects of radical-scavenging enzymes and reduced oxygen exposure on growth and chromosome abnormalities of Werner syndrome cultured skin fibroblasts. Hum Genet 57: 269–75

Salk D, Bryant E, Au K, Hoehn H, Martin GM (1981c) Systematic growth studies, cocultivation, and cell hybridization studies of Werner syndrome cultured skin fibroblasts. Hum Genet 58: 310–16

Saretzki G, Sitte N, Merkel U, Wurm RE, von Zglinicki T (1999) Telomere shortening triggers a p53-dependent cell cycle arrest via accumulation of G-rich single stranded DNA fragments. Oncogene 18: 5148–58

Satoh M, Imai M, Sugimoto M, Goto M, Furuichi Y (1999) Prevalence of Werner's syndrome heterozygotes in Japan. Lancet 353: 1766

Satyanarayana A, Greenberg RA, Schaetzlein S, Buer J, Masutomi K, Hahn WC, Zimmermann S, Martens U, Manns MP, Rudolph KL (2004) Mitogen stimulation cooperates with telomere shortening to activate DNA damage responses and senescence signaling. Mol Cell Biol 24: 5459–547

Schnabl B, Purbeck CA, Choi YH, Hagedorn CH, Brenner D (2003) Replicative senescence of activated human hepatic stellate cells is accompanied by a pronounced inflammatory but less fibrogenic phenotype. Hepatology 37: 653–64

Schonberg S, Niermeijer MF, Bootsma D, Henderson E, German J (1984) Werner's syndrome: proliferation in vitro of clones of cells bearing chromosome translocations. Am J Hum Genet 36: 387–97

Schulz VP, Zakian VA, Ogburn CE, McKay J, Jarzebowicz AA, Edland SD, Martin GM (1996) Accelerated loss of telomeric repeats may not explain accelerated replicative decline of Werner syndrome cells. Hum Genet 97: 750–54

Shelton DN, Chang E, Whittier PS, Choi D, Funk WD (1999) Microarray analysis of replicative senescence. Curr Biol 9: 939–45

Shen J, Loeb LA (2001) Unwinding the molecular basis of the Werner syndrome. Mech Ageing Dev 122: 921–44

Spillare EA, Robles AI, Wang XW, Shen JC, Yu CE, Schellenberg GD, Harris CC (1999) p53-mediated apoptosis is attenuated in Werner syndrome cells. Genes Dev 13: 1355–60

Stansel RM, de Lange T, Griffith JD (2001) T-loop assembly in vitro involves binding of TRF2 near the 3′ telomeric overhang. EMBO J 20: 5532–40

Stewart SA, Ben-Porath I, Carey VJ, O'Connor BF, Hahn WC, Weinberg RA (2003) Erosion of the telomeric single-strand overhang at replicative senescence. Nat Genet 33: 492–96

Suzuki T, Shiratori M, Furuichi Y, Matsumoto T (2001) Diverged nuclear localization of Werner helicase in human and mouse cells. Oncogene 20: 2551–58

Szczepankiewicz BG, Kosogof C, Nelson LT, Liu G, Liu B, Zhao H, Serby MD, Xin Z, Liu M, Gum R, Haasch DL, Wang S, Clampit JE, Johnson EF, Lubben TH, Stashko MA, Olejniczak ET, Sun C, Dorwin SA, Haskins K, Abad-Zapatero C, Fry EH, Hutchins CW, Sham HL, Rondinone CM, Trevillyan JM (2006) Aminopyridine-based c-Jun N-terminal kinase inhibitors with cellular activity and minimal cross-kinase activity. J Med Chem 49: 3563–80

Tahara H, Tokutake Y, Maeda S, Kataoka H, Watanabe T, Satoh M, Matsumoto T, Sugawara M, Ide T, Goto M, Furuichi Y,Sugimoto M (1997) Abnormal telomere dynamics of B-lymphoblastoid cell strains from Werner's syndrome patients transformed by Epstein-Barr virus. Oncogene 15: 1911–20

Tollefsbol TO, Cohen HJ (1984) Werner's syndrome: An underdiagnosed disorder resembling premature aging. Age 7: 75–88

van Steensel B, Smogorzewska A, de Lange T (1998) TRF2 protects human telomeres from end-to-end fusions. Cell 92: 401–13

Vaziri H, Benchimol S (1996) From telomere loss to p53 induction and activation of a DNA-damage pathway at senescence: the telomere loss/DNA damage model of cell aging. Exp Gerontol 31: 295–301

Vaziri H, Benchimol S (1998) Reconstitution of telomerase activity in normal human cells leads to elongation of telomeres and extended replicative life span. Curr Biol 8: 279–82

Vaziri H, Dragowska W, Allsopp RC, Thomas TE, Harley CB, Lansdorp PM (1994) Evidence for a mitotic clock in human hematopoietic stem cells: loss of telomeric DNA with age. Proc Natl Acad Sci USA 91: 9857–60

Vainer B (2005) Intercellular adhesion molecule-1 (ICAM-1) in ulcerative colitis: presence, visualization, and significance. Inflamm Res 54: 313–27

Vande Berg JS, Rose MA, Haywood-Reid PL, Rudolph R, Payne WG, Robson MC (2005) Cultured pressure ulcer fibroblasts show replicative senescence with elevated production of plasmin, plasminogen activator inhibitor-1, and transforming growth factor-beta1. Wound Repair Regen. 13: 76–83

Verdun RE, Karlseder J (2006) The DNA damage machinery and homologous recombination pathway act consecutively to protect human telomeres. Cell 127: 709–20

Vijg J, Calder RB (2004) Transcripts of aging. Trends Genet 20: 221–24

von Zglinicki T, Burkle A, Kirkwood TB (2001) Stress, DNA damage and ageing – an integrative approach. Exp Gerontol 36: 1049–62

Wang W, Chen JX, Liao R, Deng Q, Zhou JJ, Huang S, Sun P (2002) Sequential activation of the MEK-extracellular signal-regulated kinase and MKK3/6-p38 mitogen-activated protein kinase pathways mediates oncogenic ras-induced premature senescence. Mol Cell Biol 22: 3389–3403

Werner O (1904). On cataract associated in conjunction with scleroderma. Doctoral Dissertation, Kiel University, Schmidt and Klaunig, Kiel

Wyllie FS, Jones CJ, Skinner JW, Haughton MF, Wallis C, Wynford-Thomas D, Faragher RGA, Kipling D (2000) Telomerase prevents the accelerated cell ageing of Werner syndrome fibroblasts. Nat Genet 24: 16–17

Yamabe Y, Sugimoto M, Satoh M, Suzuki N, Sugawara M, Goto M, Furuichi Y (1997) Down-regulation of the defective transcripts of the Werner's syndrome gene in the cells of patients. Biochem Biophys Res Commun 236: 151–54

Yokote K, Honjo S, Kobayashi K, Fujimoto M, Kawamura H, Mori S, Saito Y (2004) Metabolic improvement and abdominal fat redistribution in Werner syndrome by pioglitazone. J Am Geriatr Soc 52: 1582–83

Yu CE, Oshima J, Fu YH, Wijsman EM, Hisama F, Alisch R, Matthews S, Nakura J, Miki T, Ouais S, Martin GM, Mulligan J, Schellenberg GD (1996) Positional cloning of the Werner's syndrome gene. Science 272: 258–62

Zou Y, Sfeir A, Gryaznov SM, Shay JW, Wright WE (2004) Does a sentinel or a subset of short telomeres determine replicative senescence? Mol Biol Cell 15: 3709–18

Index

Color Plates

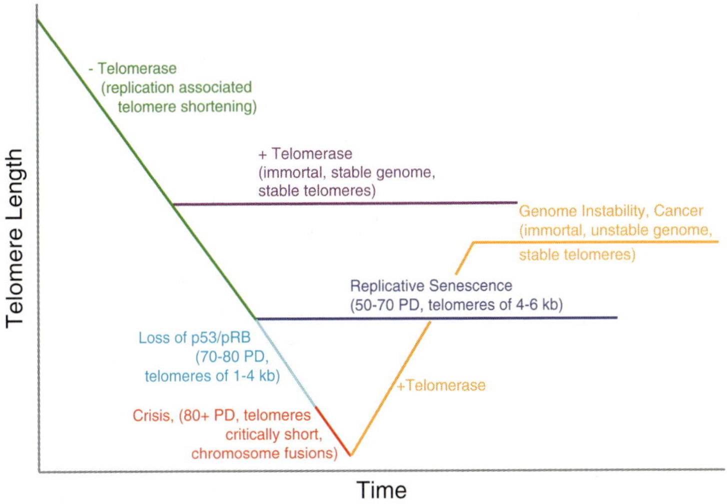

Fig. 1.1 Schematic of telomere dynamics during aging. Telomeres in telomerase-negative cells shorten over time (green line). When telomerase is introduced into such cells, telomere length is stabilized and the cells become immortal without accumulating chromosomal aberrations (purple line). When telomeres become short, cells arrest in senescence (blue line). Suppression of DNA damage pathways allows further telomere shortening (turquoise line), leading to dysfunctional and critically short telomeres, genome fragmentation and crisis (red line). Upregulation of telomerase allows escape from crisis with longer telomeres, and establishment of immortal clones with unstable genomes (orange line)

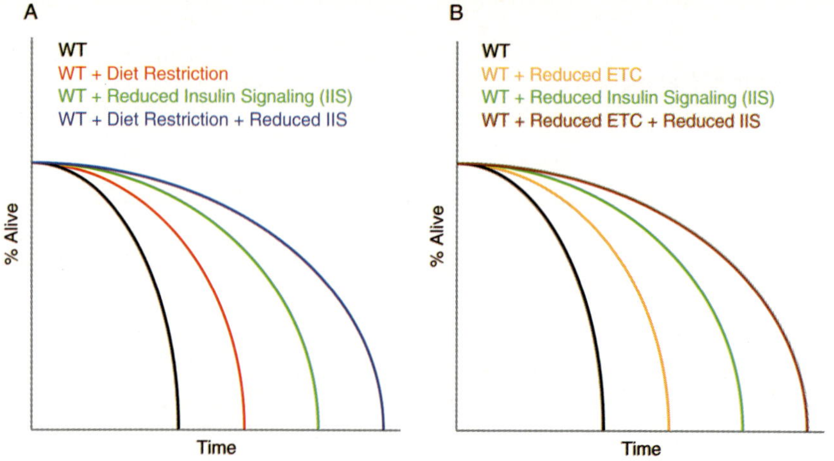

Fig. 1.2 Reduced insulin/IGF-1 signaling (IIS) synergizes with perturbations in other longevity pathways. Representative life span analysis of worms with combinatorial inactivation of IIS, diet restriction, or reduced mitochondrial electron transport chain (ETC) activity is depicted. A) Combination of diet restriction (depicted as red line) with reduced IIS (green line) results in animals that live longer than either single perturbation (blue line). B) Combination of reduced ETC (depicted as yellow line) with reduced IIS (green line) results in animals that live longer than either single perturbation (cayenne line). For more details of separation of these three pathways, please see text

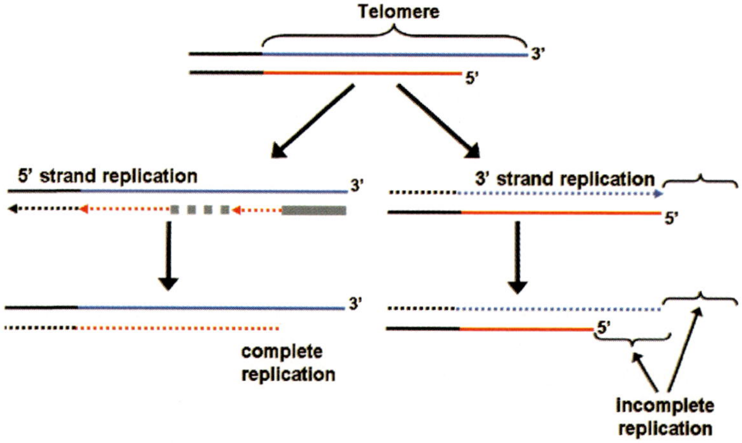

Fig. 2.1 The end replication problem. The replication of the 5′ strand (red) and 3′ strand (blue) of a single telomere are shown, as well as adjacent nontelomeric DNA (black). Newly synthesized DNA (i.e., the daughter strand) is indicated by the dashed lines. Replication of the 5′strand occurs by lagging strand synthesis, where Okazaki fragments are shown in grey. Replication of the 3′ strand occurs by leading strand synthesis. All telomeres end in a 3′overhang, which the normal DNA replication machinery cannot replicate. Degradation of the most 5′ Okazaki fragment presumably allows regeneration of the overhang during replication of the 5′ strand, allowing complete replication of the telomere on this daughter chromosome. However, the overhang cannot be regenerated by *de novo* DNA synthesis during leading strand synthesis, and therefore degradation of the 5′ strand is presumably required, which is hypothesized to account for the telomere shortening observed during cell division in cells that do not express telomerase

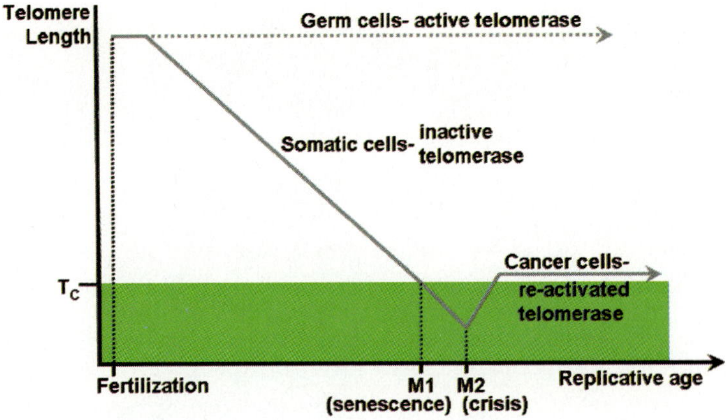

Fig. 2.2 The telomere hypothesis of cell aging and immortalization. The changes in telomere length during replicative aging of cells *in vivo*, beginning at the time of fertilization, is depicted. The change in telomere length during replicative aging for somatic cells and germ cells are indicated by the solid and dashed gray arrows respectively. The green bar shows the telomere size range for which there is insufficient telomeric DNA to allow a stable telomere complex to form. The upper size limit of this region is the critical telomere length (marked as T_C), which is the minimum length of telomeric DNA required to allow formation of a stable telomere complex at the end of a chromosome. The status of telomerase activity is also indicated. See text for further details

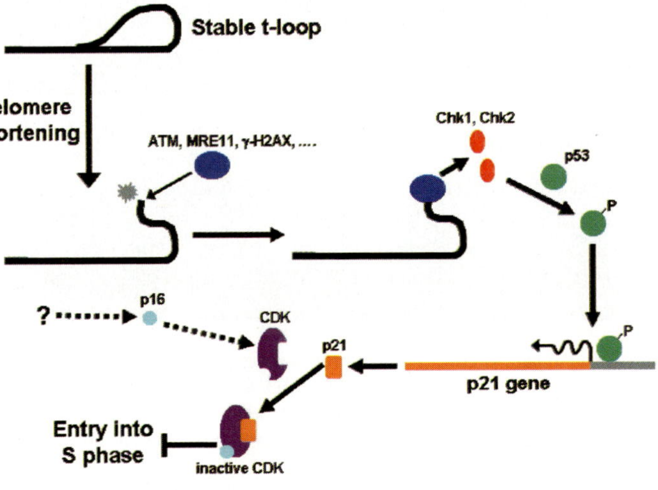

Fig. 2.3 The mechanism of telomere-induced replicative senescence. A stable telomere is shown in a t-loop confirmation, in which the end of the chromosome is protected. After a finite number of replication events, the telomere can no longer maintain a stable structure, represented as unfolding of the loop, and consequently the end of the telomere is recognized by DNA damage response sensors. See text for further details

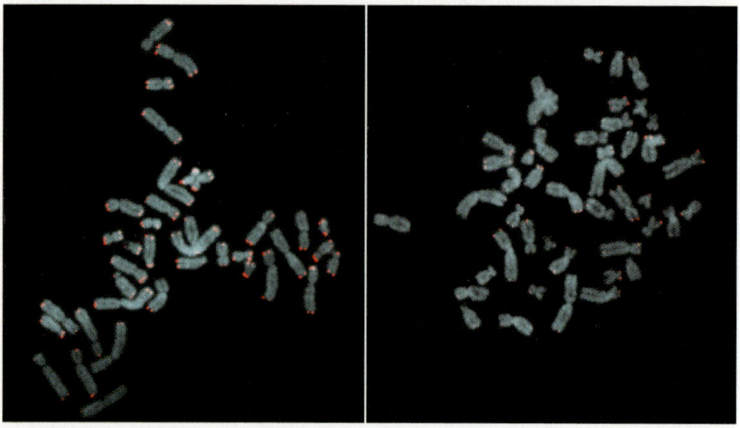

Fig. 3.1 Heterogeneity in telomere Q-FISH signals between metaphases from young proliferating fibroblast cultures (PD13). Most metaphases show strong telomeric signals (left panel), but some display signals as weak as in near-senescent cells (right panel)

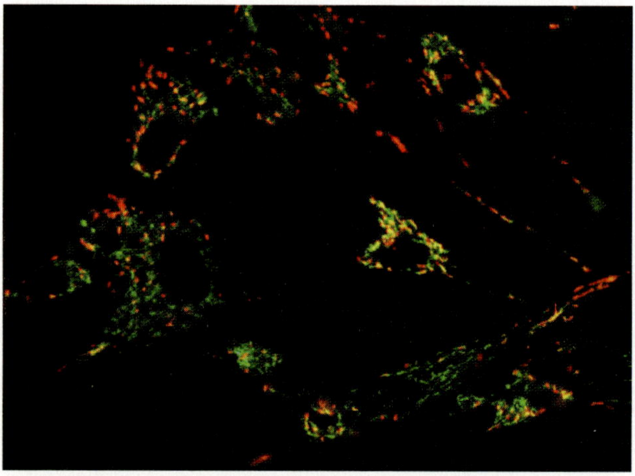

Fig. 3.2 Confocal image of live young fibroblasts staining with JC-1. The dye emits green fluorescence at low concentrations and red fluorescence at high density. It is transported into mitochondria in a membrane potential (MMP)-dependent fashion, so that green indicates low MMP, while mitochondria stained red have high MMP. Note the cell-to-cell heterogeneity in JC-1 ratios, as well as intracellular heterogeneity

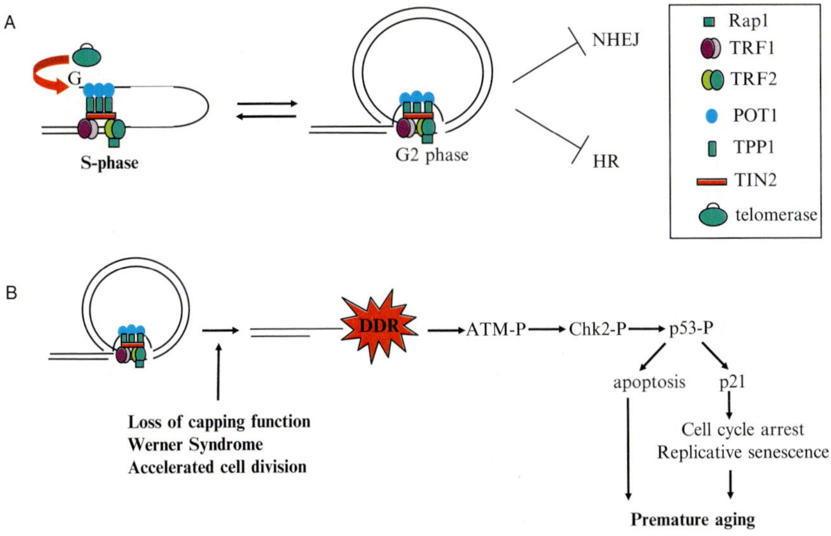

Fig. 4.1 (A) Telomere structure. The telomere folds back onto itself to form a double-stranded t-loop and a single-stranded d-loop. This complex protects telomeres at the G2 phase of the cell cycle from inappropriate nonhomologous end joining (NHEJ) and homologous recombination (HR)–mediated repair of telomeric DNA. The six-component core protein complex is shown schematically. Telomere extension by telomerase requires translocation of POT1 to an internal site, permitting telomerase to access the terminal G-residue. (B)Telomere dysfunction elicits a DNA damage response at telomeres and activation of ATM, resulting in p53-meduated cell cycle arrest/ replicative senescence or apoptosis to trigger the onset of premature aging symptoms

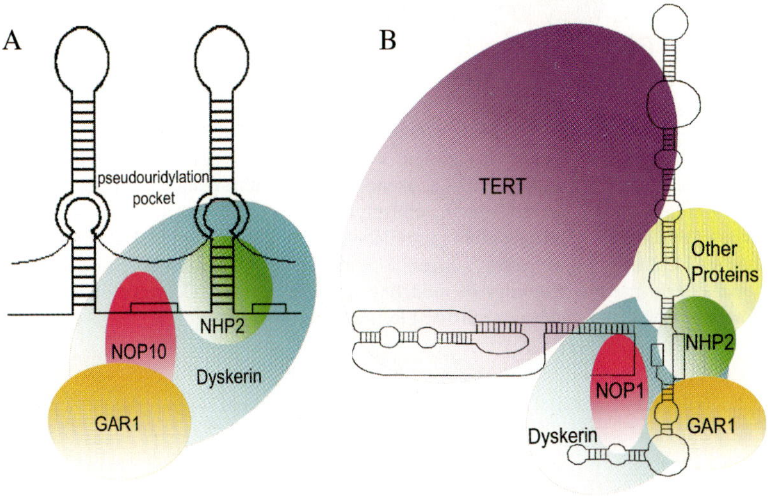

Fig. 5.1 Dyskerin is an important component of both the H/ACA small nucleolar (snoRNA) ribonucleoprotein (RNP) particles and telomerase complex. (A) Dyskerin, together with GAR1, NHP2, NOP10, and H/ACA snoRNA, comprises snoRNP particles. The H/ACA RNA-protein complex catalyzes pseudouridylation and processing of ribosomal (r) RNA. (B) *TERC* (telomerase RNA component) and TERT (telomerase reverse transcriptase) are the core components of telomerase complex. As an H/ACA snoRNA, *TERC* associates with dyskerin, GAR1, NHP2, and NOP10, which are also present in the telomerase complex

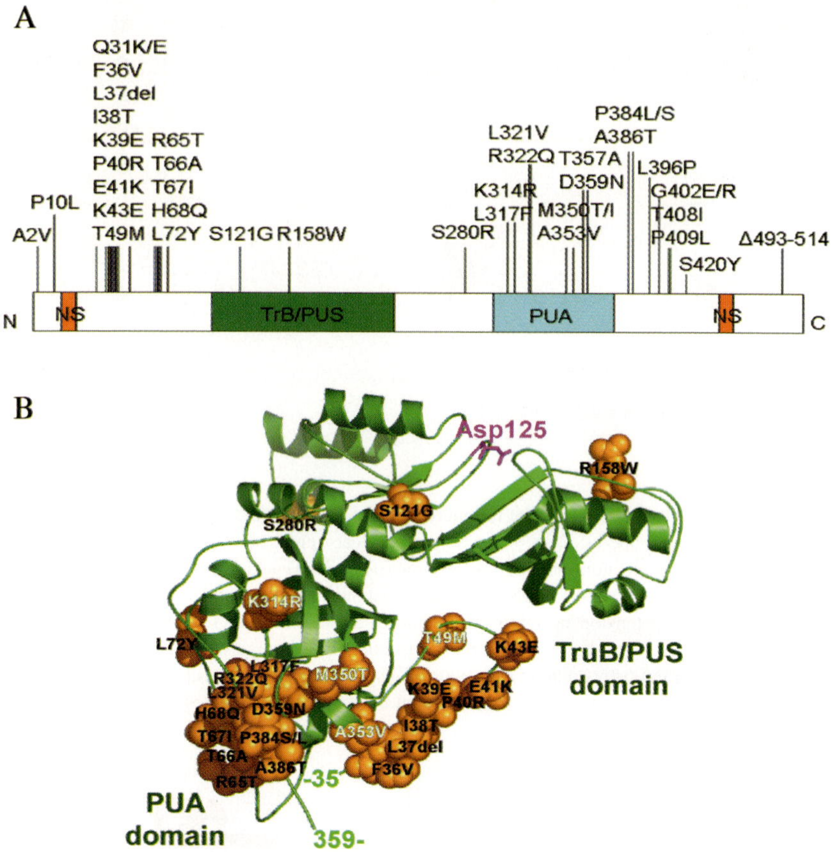

Fig. 5.2 Nonsynonymous amino acid substitutions in dyskerin affected in dyskeratosis congenita. (A) Linear representation of dyskerin protein that contains 514 amino acids. The functional domains: nuclear localization signals (NL), pseudouridine synthase domain (TruB/PUS), pseudouridine synthase, and archaeosine-spicific transglycosylase domain (PUA) are shown. Locations of nonsynonymous amino acid substitutions are labeled. (B) 3D structure of dyskerin with amino acid changes identified in dyskeratosis congenita. Mutations in the regions 1–35 and 359–513 are not shown due to the unavailability of modeling in these regions (Courtesy of Dr. Hong Li, Department of Chemistry and Biochemistry, Florida State University)

Fig. 5.3 (continued) labeled. The two polymorphisms G58A and G228A are marked in blue, and other nucleotide alterations are in red. The majority of the nucleotide alterations are located in the pseudoknot domain. While most alterations lead to a reduction of telomerase activity, G450A and Δ28–34 showed normal telomerase activity in both in vitro and in vivo assay

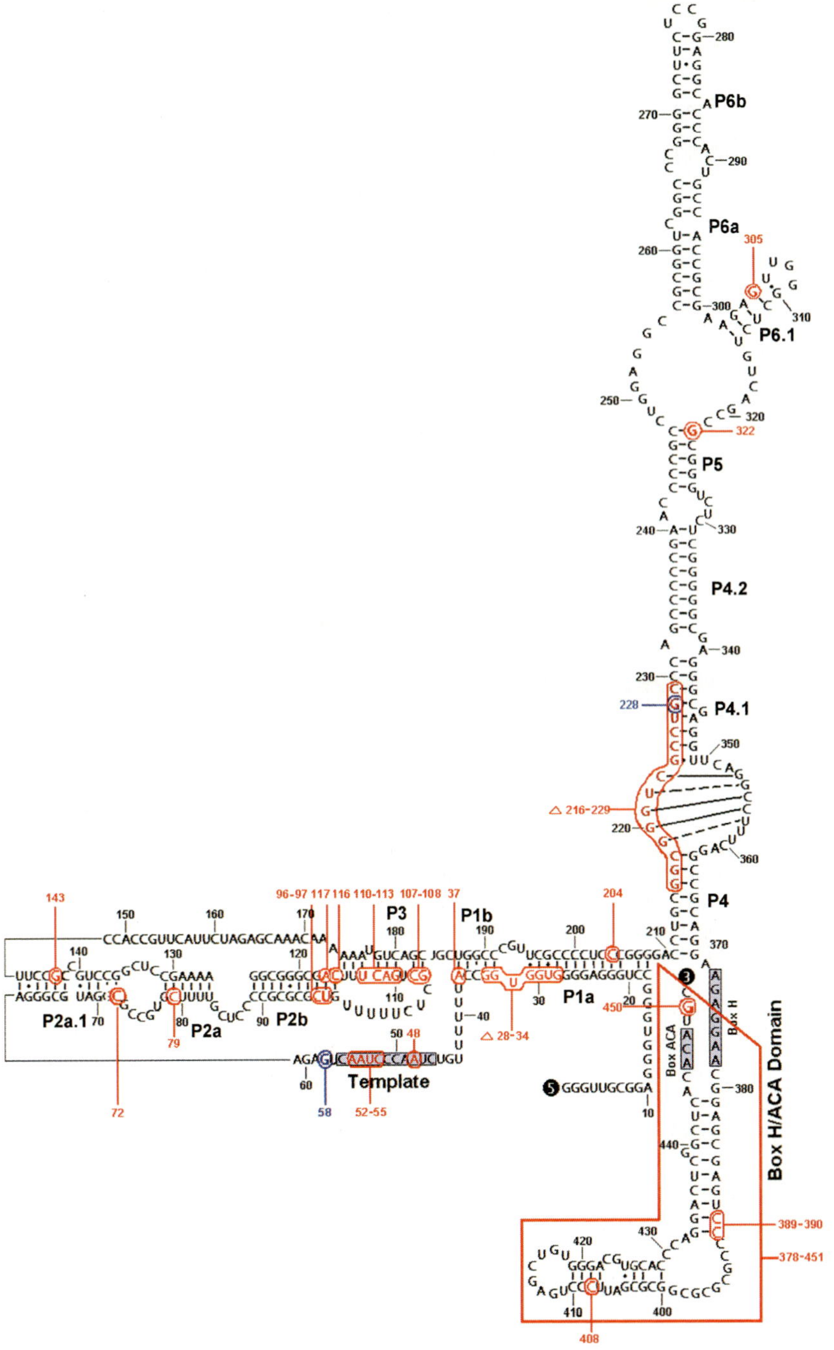

Fig. 5.3 Secondary structure of human *TERC* and nucleotide alterations. The mature human *TERC* consists of 451 nucleotides. The highly conserved secondary structure that is composed of four domains is shown: the pseudoknot domain, the CR4-CR5 domain, the hypervariable region, and the CR7 and H/ACA domain. Twenty-two nucleotide alterations within this molecule are

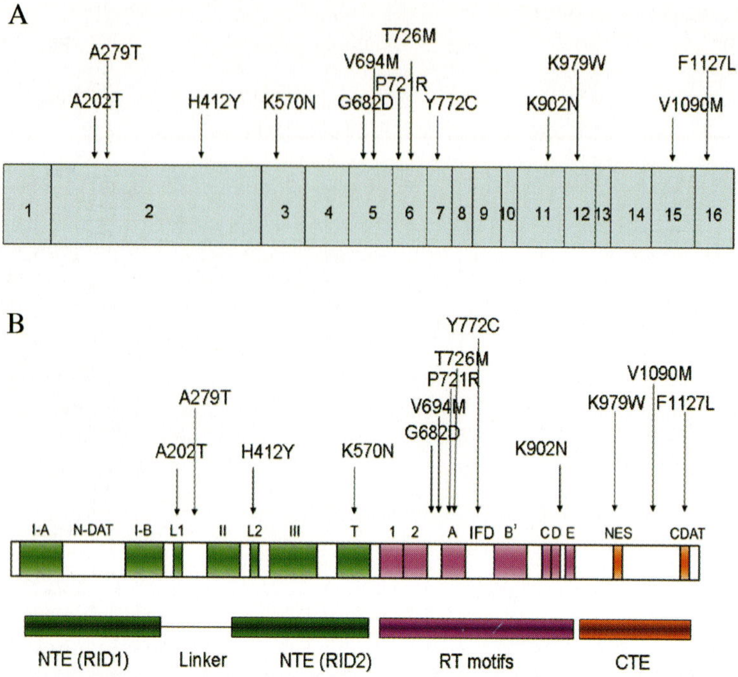

Fig. 5.4 Nonsynonymous amino acid substitutions identified in TERT. (A) Linear representation of the TERT transcript composed of 16 exons. Each box represents an exon. The location of nonsynonymous amino acid substitutions are indicated. (B) Structure of the TERT Protein represented with functional domains. The TERT protein includes the RT motif (1,2, A, B', C,D,E); the telomerase-specific motif (T); and N- and C- terminal extension (NTE, CTE). A linker domain separates the NTE; and IFD (insertion in fingers domain) separates A and B'. The functionally defined domains comprise I-A, DAT (disassociated activities of telomerase), I-B, II, III, RID (RNA interaction domain), NRS (nuclear export signal)

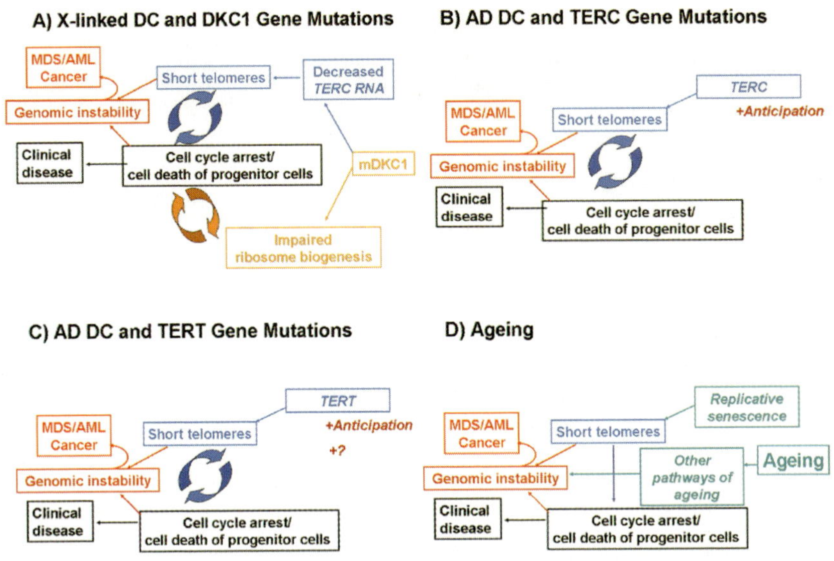

Fig. 5.5 Model of the pathogenesis of dykeratosis congenita and aging. (A) X-linked DC and *DKC1* gene mutation. (B) AD-DC and *TERC* gene mutation. (C) AD-DC and *TERT* gene mutation. (D) Ageing. Mutations in multiple components of the telomerase complex are identified in patients with dyskertatosis congenita, suggesting that excessive telomere shortening is the underlying cause of this disease. (A-C) The critically shortened telomeres lead to cell cycle arrest and/or cell death in rapid dividing cells including stem cells and early progenitor cells. The cell senescence/death results in the recruitment of more stem cells into cell cycle. Consequently, the increased proliferation of a decreased number of stem cells aggravates the rate of telomere shortening, which ends up with the depletion of the stem cells and the development of disease phenotype. Mutations in *TERC* or *TERT* directly reduce the enzymatic activity of telomerase and disrupt its function in telomere maintenance (B and C). Anticipation refers to early onset and more severity in successive generations in AD DC, which suggests that several generations may be required for haploinsufficiency for *TERC* or *TERT* to produce clinically detectable phenotype. Mutations in *DKC1* may either destabilize the structure of *TERC* that causes deregulation the telomerase activity or/and reduce pseudouridylation and retard the rate of ribosome biogenesis (which were observed in mice carrying *DKC1* mutation (A) that leads to cell cycle arrest/cell death of stem cells and their progeny. On the other hand, the few cells that survive the crisis accumulate genomic instability due to excessively short telomeres and become potentially malignant cells, leading to MDS (myelodysplastic syndrome), AML (acute myelogenous leukemia), or other types of cancer. In the ageing process (D), telomeres shorten in multiple organs, suggesting that normal somatic cells devoid of telomerase activity undergo telomere shortening during cell division, leading to cell senescence and organism aging. Critically short telomeres in aging populations, similar to those in dyskeratosis congenita, result in chromosomal instability and increased susceptibility to age-related disease and cancer. Nevertheless, telomere shortening is not the only cause for the pathogenesis of ageing. Other pathways such as accumulation of genetic mutations, unbalanced metabolic rate, and oxidative stress also contribute to the disease phenotype in ageing

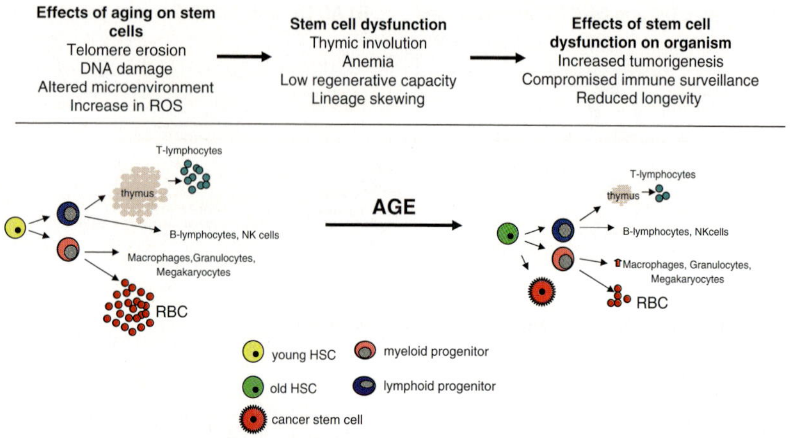

Fig. 6.1 This schematic represents the stem cell theory of ageing as it pertains to the hematopoietic system. The accumulation of damage and the series of changes that accompany ageing influence the HSC compartment in a way that results in compromised function of this system and reduced health of the entire organism. This illustrates the possible connection between adult stem cells and organismal longevity

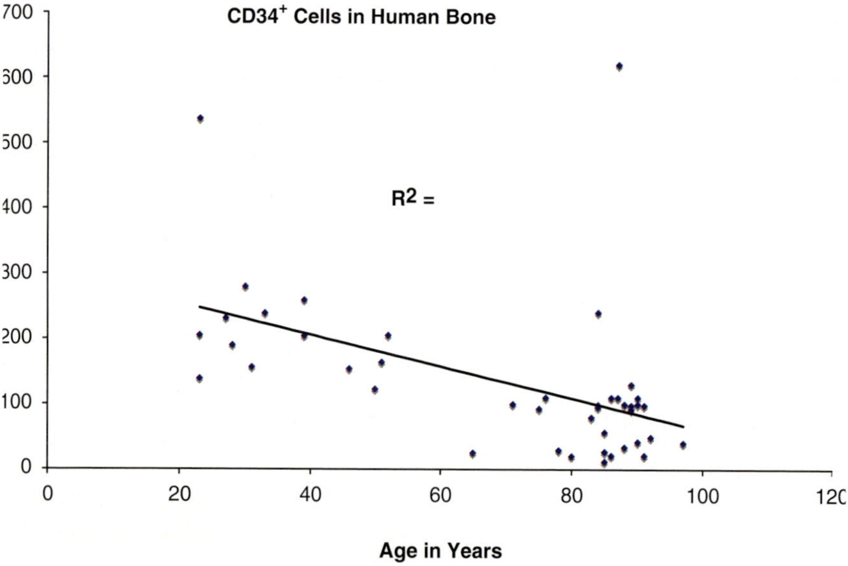

Fig. 6.2 Whole bone marrow aspirates were obtained from 43 adult humans. Mononucleated cells were collected by Ficoll-gradient centrifugation and stained for the presence of CD34. Flow cytometric analysis was used to identify the percentage of CD34+ cells in each sample. Linear regression analysis revealed that approximately 24% of the variation in the number of CD34+ cells in human bone marrow can be attributed to age

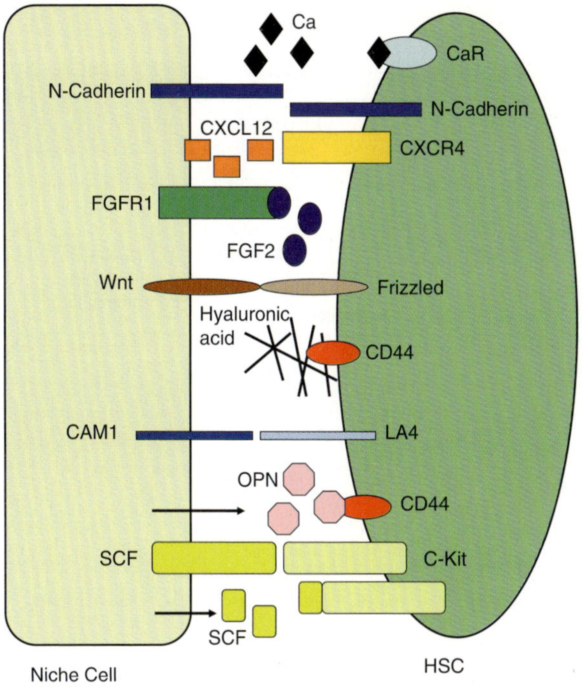

Fig. 6.3 HSCs engage in a number of cell-cell and cell-matrix interactions in the niche. Various cells of the niche (such as osteoblasts and stromal cells) have been shown to express integrins and signaling receptors on the cell surface, and secrete factors such as OPN, SCF, and CXCL12 that have binding partners on HSCs. HSCs also interact with extracellular matrix molecules such as hyaluronic acid via the CD44 molecule on the cell surface. Each of these molecular interactions has been shown to influence HSC functions such as adhesion, proliferation and quiescence

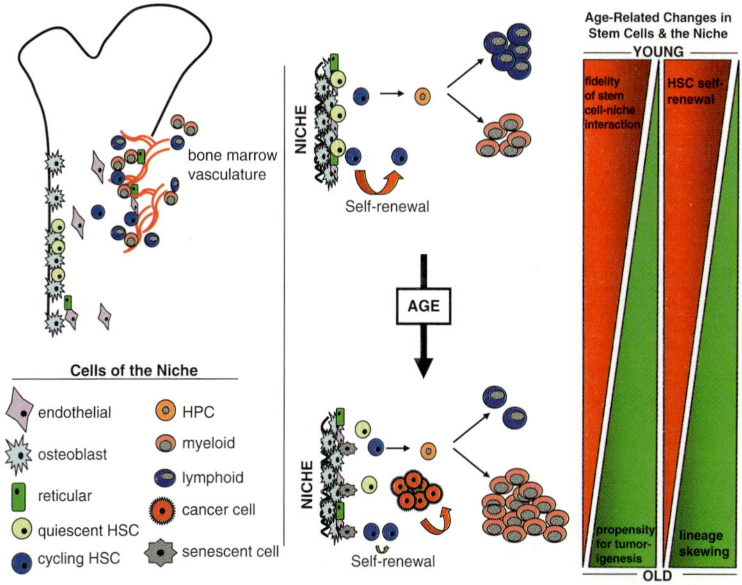

Fig. 6.4 HSCs reside in at least two known niches in the bone marrow, the vascular and endosteal niches. Within these specialized microenvironments, stem cells interact with a variety of cell types such as endothelial, osteoblast, and reticular cells. The fidelity of these interactions influences the

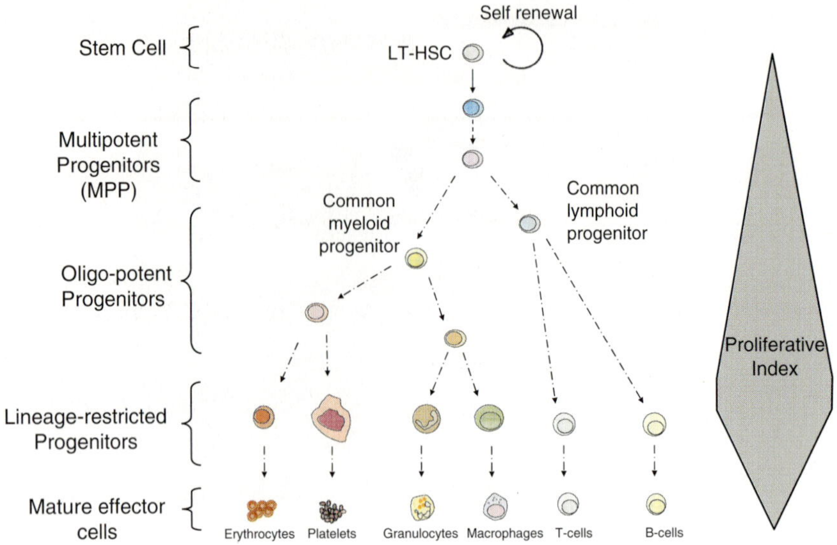

Fig. 9.1 Hematopoietic developmental hierarchy. Self-renewing HSC reside at the top of the hierarchy, giving rise to a number of multipotent progenitors. Multipotent progenitors give rise to oligopotent progenitors, including the common lymphoid progenitor (CLP), which gives rise to mature B lymphocytes, T lymphocytes, and natural killer (NK) cells. The common myeloid progenitor (CMP) gives rise to granulocyte/macrophage progenitors (GMP) that differentiate into monocytes/macrophages and granulocytes, and megakaryocyte/erythrocyte progenitors (MEP) that differentiate into megakaryocytes/platelets, and erythrocytes. Development from the oligopotent progenitors through to mature blood cells proceeds through a number of intermediate progenitors (not shown). The developmental passage of HSC through multipotent progenitors, oligopotent progenitors, and lineage-specific progenitors is generally associated with increases in proliferative index, although this trend is not absolute and has not been resolved for all stages of development

balance of quiescence verses cell division and cell fate decisions in more lineage committed progenitors. The niche also plays an important role in the repression or facilitation of tumorigenesis. While a healthy microenvironment can help limit the potential for tumorigenesis, the cells of a tumor microenvironement and premetastatic niche support cancer cell proliferation and metastasis. The fidelity of stem cell – niche interactions and the capacity for HSC self-renewal diminish with age. Propensity for tumorigenesis increases with age and an imbalance in lineage distribution develops, favoring myelopoiesis at the expense of lymphopoiesis

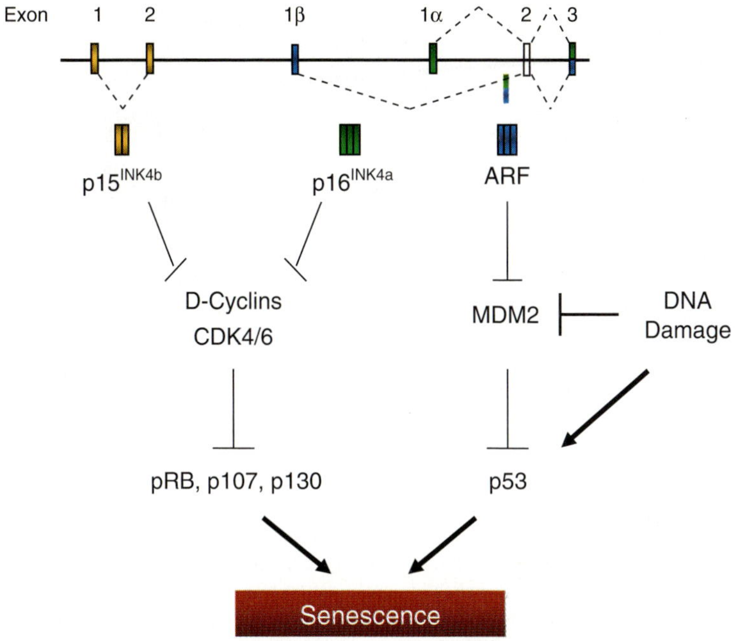

Fig. 9.2 The *INK4a/ARF/INK4b* locus and downstream targets. Three tumor suppressor genes are encoded in a common 40kb stretch of human chromosome 9p21: two highly related CDK inhibitor proteins (p15^{INK4b} and p16^{INK4a}) and ARF, a regulator of p53 stability. Expression of p16^{INK4a} and p15^{INK4b} inhibits CDK4/6 activity, leading to hypophosphorylation of Rb-family proteins (Rb, p107, po130) and growth arrest. Expression of ARF inhibits the ubiquitin ligase activity of MDM2, thereby stabilizing p53. ARF and p16^{INK4a} share exons 2/3, but are encoded in alternate reading frames and therefore have no amino acid homology. DNA damage inhibits p53 degradation by MDM2 in an ARF-independent manner, and can promote senescence independent of *INK4a/ARF* activation. DNA damage, however, likely also activates p16^{INK4a}; although this relationship is complex (see section 5.2)

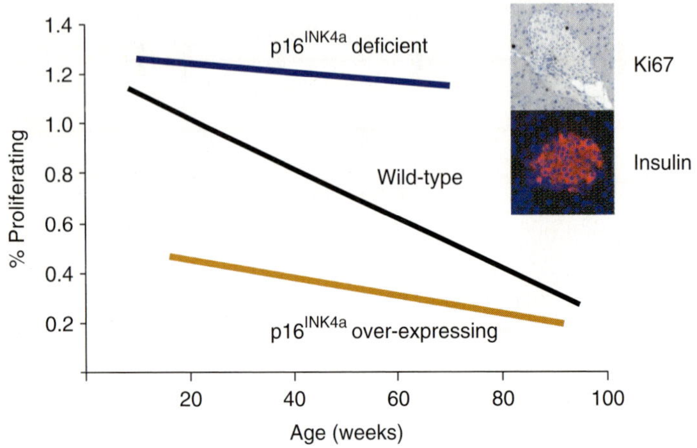

Fig. 9.3 β-cell proliferation with age. A decline in β-cell proliferation is modulated by the expression of the p16^{INK4a} tumor suppressor. Excess p16^{INK4a} expression correlates with reduced proliferation, while decreased p16^{INK4a} affords a resistance to β-cell ageing. The top panel in the inset shows an H&E image of an adult murine islet with immunohistochemical detection of Ki67, a marker of proliferation (brown), indicating proliferation in a rare subset of β-cells. The bottom panel shows an islet immunofluorescently labeled with insulin (red) and a nuclear stain (blue). Original magnification = 20X of both images

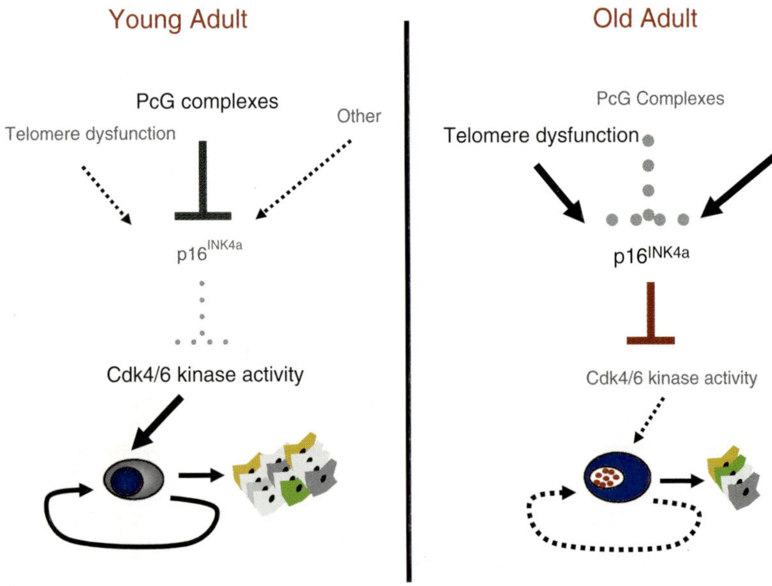

Fig. 9.4 Senescence and ageing. Diverse cellular stresses, including but not limited to telomere dysfunction and other forms of DNA damage, increase with age, inducing the senescence-promoting effects of p16^{INK4a} and p53. The PcG complexes appear to repress *Ink4a/Arf* activation and thereby modulate stem cell function, although to date, no direct proof has established a reduction in PcG activity with ageing. With ageing, activation of p16^{INK4a} is associated with the appearance of hypofunctional or senescent, formerly self-renewing cells (illustrated as blue cells with red SAHF)

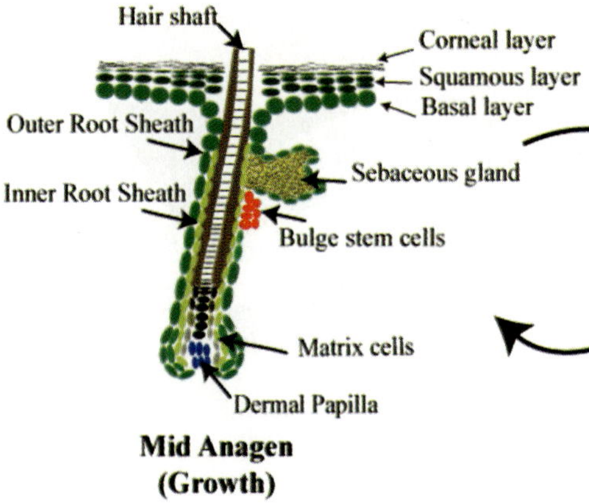

Fig. 10.1 Multipotent epidermal stem cells in the hair follicle bulge region. Hair follicles cycle between two states: telogen, the resting phase, and anagen, the active phase. Quiescent stem cells reside in the bulge region just below the sebaceous gland. A small number of these stem cells become activated, enter cell cycle, and give rise to the lower portion of the new anagen follicle. In anagen, proliferating progenitor cells (matrix cells) surround the dermal papilla and differentiate to produce the growing hair, which is pushed upward through the root sheaths. With injury to the skin, bulge stem cells can also repair interfollicular skin, although the epidermis of the skin is likely maintained by progenitors in the basal layer. Conditional expression of TERT in skin is sufficient to initiate a new anagen cycle and causes quiescent bulge stem cells to proliferate

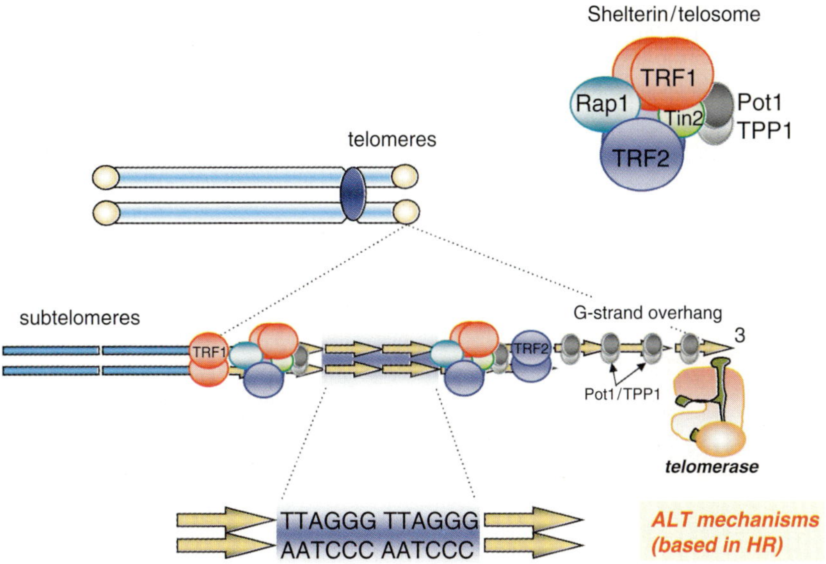

Fig. 12.1 Structure of mammalian telomeres. A) Double-stranded telomeric repeats are bound by a multiprotein complex known as *shelterin* or the *telosome*, which comprises TRF1, TRF2, Tin2, Rap1, Pot1, and TPP1. The G-strand overhang is also bound by the Pot1/TPP1 heterodimer. Telomerase is able to recognize the 3′end of the G-strand overhang and to elongate telomeres. In telomerase-negative cells, telomeres can be maintained by mechanisms involving homologous recombination between telomeric repeats, the so-called alternative lengthening of telomeres or ALT

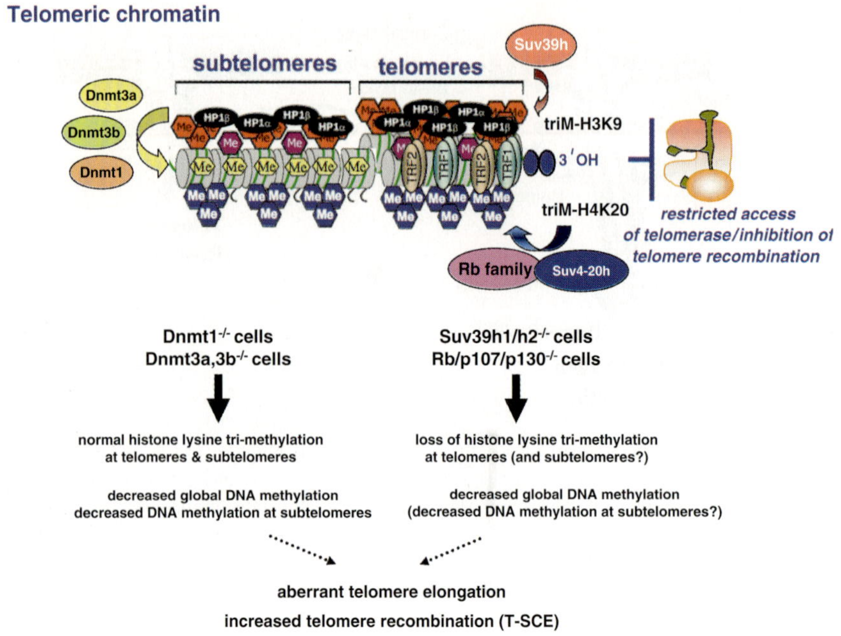

Fig. 12.2 Epigenetic modifications at mammalian pericentromeric and telomeric chromatin domains. Both telomeric and subtelomeric chromatin regions are enriched in trimethylated H3K9 and H4K20 and HP1 isoforms. In addition, subtelomeric DNA is heavily methylated by the DNMT1, DNMT3a,3b enzymes. Both histone trimethylation and DNA methylation have been shown to independently act as negative regulators of telomere length, concomitant with their ability to inhibit telomere recombination

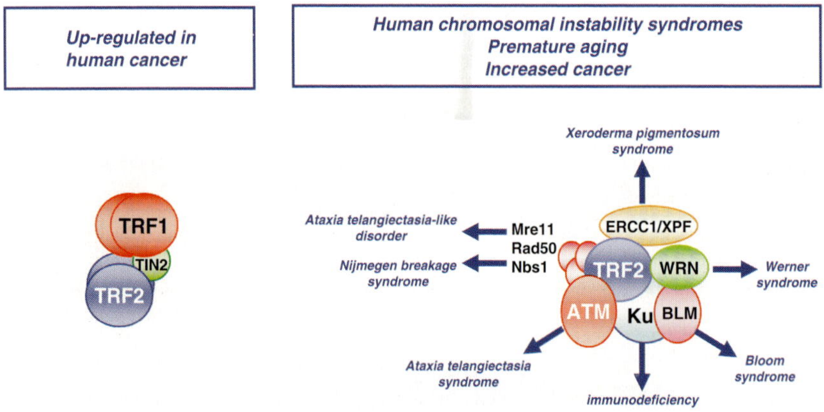

Fig. 12.3 Role of telomeric proteins in human disease. Components of the shelterin complex that are upregulated in human cancer are shown in the left panel. DNA repair and DNA damage signaling proteins that have been shown to interact with TRF2 are indicated in the right panel, together with the human chromosome instability syndromes where these proteins have been found to be mutated

Printing: Krips bv, Meppel, The Netherlands
Binding: Stürtz, Würzburg, Germany